Pollution Control for Agriculture

Second Edition

RAYMOND C. LOEHR

Department of Agricultural Engineering
Cornell University
Ithaca, New York

1984

ACADEMIC PRESS, INC.

(Harcourt Brace Jovanovich, Publishers)

ORLANDO   SAN DIEGO   NEW YORK   LONDON

TORONTO   MONTREAL   SYDNEY   TOKYO

ACADEMIC PRESS, INC.
Orlando, Florida 32887

United Kingdom Edition published by
ACADEMIC PRESS, INC. (LONDON) LTD.
24/28 Oval Road, London NW1 7DX

Library of Congress Cataloging in Publication Data

Loehr, Raymond C.
    Pollution control for agriculture.

    Includes bibliographical references and index.
    1. Agricultural wastes. I. Title.
TD930.L63   1984   628.5'4   84-6295
ISBN 0-12-455270-6

PRINTED IN THE UNITED STATES OF AMERICA

84 85 86 87   9 8 7 6 5 4 3 2 1

To the many individuals whose professional and personal interest and concern have made this book possible. Special recognition should be given to GEB, GAR, REM, NCB, WJJ, and with greatest warmth and appreciation to JML.

Contents

3   Environmental Impact

4   Waste Characteristics

5   Fundamentals of Biological Treatment

6   Ponds and Lagoons

Contents

7 Oxygen Transfer

8 Aerobic Treatment

9 Anaerobic Treatment

10 Utilization of Agricultural Residues

11   Land Treatment and Stabilization of Wastes

12   Nitrogen Control

13   Physical and Chemical Treatment

14   **Nonpoint Source Control**

15   **Management**

Appendix: Characteristics of Agricultural Wastes   438

Preface to the Second Edition

Through ignorance or indifference we can do massive and irreversible harm to the earthly environment on which our life and well-being depend. Conversely, through fuller knowledge and wiser action, we can achieve for ourselves and our posterity a better life in an environment more in keeping with human needs and hopes.

This statement is from the Declaration of the United Nations Conference on the Human Environment held in Stockholm in 1972 and very aptly summarizes the intent of this second edition. In updating all of the chapters, completely rewriting several chapters, and adding a new chapter, the goal was to identify (1) the environmental problems that can result from poorly managed agricultural operations and (2) the knowledge that can be used and the action that can be taken to avoid such problems and to sustain a viable agriculture.

The first edition of this book was published in 1977 and has been used as a text and reference throughout the world. The comments from colleagues and others who have used the book have been gratifying.

In the years since the first edition was written, there has been considerable progress and new knowledge in many areas, such as biogas production from manure and food processing residues, land treatment as a waste management alternative, and utilization of agricultural residues as a soil conditioner and supplemental animal feed. Nonpoint source management practices emerged as an important technical topic. In addition, many of the approaches and management methods that appeared feasible from laboratory and pilot-plant-scale studies a decade ago have been built and operated as full-scale technologies. Both new concepts and data from operating systems have resulted in the intervening years.

Another important change that has occurred in the past decade is the greater interest in energy conservation and natural resource utilization. Aeration systems have a high energy requirement, and consequently emphasis on that management approach has decreased and the emphasis on energy production, energy conservation, and residue utilization has increased.

The information in this second edition reflects the new knowledge and emphasis. As a result, this edition contains more material and better coverage of the topic. Every chapter has been critically reviewed and revised to be as up-to-date and as succinct and readable as possible. A large number of new tables and figures have been included to summarize the pertinent information and to make it more understandable. Wherever possible, examples and data from operating systems have been included to illustrate important fundamentals and to report operating data.

Three chapters have been completely rewritten because of the extensive amount of new material that has become available in the past 10 years. These are Chapter 9, Anaerobic Treatment; Chapter 10, Utilization of Agricultural Residues; and Chapter 11, Land Treatment and Stabilization of Wastes. These chapters provide much broader coverage and are a larger proportion of this volume than they were in the first edition.

A completely new chapter, Chapter 14, Nonpoint Source Control, has been added. This chapter provides information on sources of such problems, on approaches to identifying and implementing the best management practices (BMPs) for agricultural nonpoint source control, and on specific BMPs that can be used for such control.

In addition to the above changes in each chapter, every attempt has been made to have the index be more comprehensive and useful.

The intent of this edition remains the same as that stated in the first, i.e., to provide the reader with basic information needed to understand the concern with this subject and with potential solutions to specific problems. The book also is intended to meet the needs of an upper-level undergraduate or a graduate-level course. Therefore, the second edition is intended to serve both as a text and as a useful reference and introduction for individuals interested in this problem and in appropriate solutions.

Individuals throughout the world have provided the information that is in this book—comments and ideas that have hopefully improved this edition—and the stimulation and enthusiasm that have provided progress in this technical and environmental area. The contribution and cooperation of these individuals are gratefully acknowledged and appreciated.

Raymond C. Loehr

Preface to the First Edition

Enhancement of environmental quality is an accepted national goal. Historically, the major efforts to maintain and enhance environmental quality have focused on problems caused by urban centers. This emphasis has been due to pressing problems in controlling industrial pollution, in treating domestic liquid wastes, in disposing of municipal solid wastes, and perhaps to an instinctive feeling that agriculturally related environmental quality problems were uncontrollable and/or minor.

The quantity of wastes generated by agriculture and specific environmental problems that have resulted from agricultural operations have illustrated that greater emphasis must be given to the management of agricultural residues and by-products. Data on fish kills from feedlot runoff, nutrient problems due to runoff from cultivated lands, the quantities of animal and food-processing wastes produced nationally, the pollutional strength of these wastes, the possible contamination of groundwaters from crop production and land disposal of wastes, and the increasing size of agricultural production operations indicate that attention must be given to the development of a number of alternative methods to handle, treat, and dispose of agricultural wastes with minimum contamination of the environment.

The past decade has seen increasing funds and manpower devoted to finding solutions to the management of agricultural wastes. In this period, the first recognition of pollution caused by agriculture and the need to control such pollution occurred in federal water pollution control legislation (PL 92-500). Considerable knowledge of the magnitude of the national problem and of possible technical solutions to specific problems now exists, and detailed management methods to prevent contamination of the environment by agriculture are available.

Agriculture is being faced with a number of constraints as the nation attempts to improve the quality of the environment. Since the concept of a totally unimpaired or totally polluted environment is not meaningful, feasible compromises must be obtained between agricultural production and environmental quality control to assure adequate food for the nation, adequate profit for the producer, and an acceptable environment for the public.

Both long- and short-term solutions must be found. Certain solutions will be of a stopgap nature until more definitive and comprehensive agricultural waste management systems are developed. These systems will contain the fundamentals of all successful waste management systems which include being technically feasible and economically attractive, requiring minimum supervision and maintenance, being flexible enough to allow for the varied and seasonal nature of the waste and the agricultural operation, and producing consistent results.

A workable strategy to develop satisfactory agricultural waste management systems will include imaginative developments and concepts from sanitary engineering, soil science, agricultural engineering, agricultural economics, poultry and animal sciences, and food and crop sciences. The solution to specific problems and the development of feasible agricultural waste management systems will occur by combining the fundamentals of these disciplines with the practice of agricultural production.

This book presents a summary of the processes and approaches applicable to agricultural waste management problems. In the context of this book, agricultural wastes are defined as the excesses and residues from the growing and first processing of raw agricultural products, i.e., fruits, vegetables, meat, poultry, fish, and dairy products. Implications and possible management systems for crop production are also discussed.

Emphasis is placed on those processes that appear most adaptable to the treatment, disposal, and management of agricultural wastes. Fundamental concepts are followed by details describing the use of processes and management approaches. Examples in which the processes or approaches were used with agricultural wastes are included to illustrate the fundamentals as well as the design and operational facets.

This book is a substantial revision of "Agricultural Waste Management" published in 1974 by Academic Press, and is intended to meet the needs of an upper level undergraduate or a graduate level course. It aims to provide the reader with basic information needed to understand the concern with this subject and potential solutions to specific problems. Extensive references, nonessential details, and appendixes are kept to a minimum. The reader can refer to the earlier book ("Agricultural Waste Management," 1974) for greater detail, references, and tables of specific waste characteristics. This edition incorporates pertinent information that has become available in recent years.

The book attempts to place the agricultural waste problem in reasonable perspective, to illustrate engineering and scientific fundamentals that can be applied to the management of these wastes, to illustrate the role of the land in waste management, and to discuss guidelines for the development of feasible waste management systems. It is also intended to serve as an introduction for individuals in many organizations and agencies who are interested in knowing and applying feasible agricultural waste management concepts and approaches. It is hoped that this work will stimulate further inquiry and that the material presented will minimize environmental quality problems that may be associated with agricultural production.

Cooperation of my many colleagues at Cornell and in governmental and industrial organizations who took the time and interest to review portions of the book is gratefully acknowledged and appreciated.

Raymond C. Loehr
1977

1

Constraints

Introduction

Agriculturally caused pollution is but one part of the national environmental quality problem. All pollution sources, i.e., municipal, industrial, marine, agricultural, and mining sources, must be considered in an integrated manner to improve the quality of the environment. Recent changes in agricultural production methods have caused natural interest in agriculturally related pollution to escalate. Such pollution is no longer considered minor or uncontrollable.

Agricultural production is becoming more intensive. Had agricultural production practices remained static, food production and the standard of living of the American public would not have reached the high levels enjoyed today. However, remarkable changes in the efficiency of United States agriculture have occurred. This increased efficiency has generated or been associated with a variety of environmental problems.

Examples of adverse problems attributed to agriculture include excessive nutrients from lands used for crop production or waste disposal that unbalance natural ecological systems and increase eutrophication, microorganisms in waste discharges that may impair the use of surface waters for recreational use, impurities in groundwater from land disposal of wastes, contaminants that complicate water treatment, depletion of dissolved oxygen in surface waters causing fish kills and septic conditions, and odors from concentrated waste storage and land disposal.

The causes and concerns of agricultural waste treatment and disposal are analogous to the environmental problems caused by people. When people were fewer in number, when agricultural production was less concentrated, and when both were better distributed throughout the land, their wastes could be absorbed without adversely affecting the environment. Aggregations of people in cities and the development of large-scale industrial operations have caused the air and water pollution as well as health problems of which we are increasingly aware. While the problem of municipal and industrial wastes has been increasing for centuries, the agricultural waste problem has been more recent.

Some of the most dramatic changes in agricultural production have occurred in animal production, which is changing from small, individual farm operations into large-scale enterprises. Small animals, such as chickens and hogs, are confined within small areas and buildings in which the environmental conditions are controlled to produce the greatest weight gain in the shortest period of time. There is an increasing trend for cattle to be finished in similarly controlled areas, dry lot feedlots. Under such conditions it is no longer possible for these animals to drop their wastes on pastures where the wastes can be absorbed by nature without adversely affecting the environment.

Concentrated animal feeding operations have resulted in water pollution problems in some areas of the country. Storm runoff can transport organic and inorganic matter from unconfined feedlots to streams where the wastes are unable to be assimilated. Fish kills and lowered recreational values have resulted. Runoff from manured fields and effluents from animal waste disposal lagoons also can affect the quality of adjacent streams.

Odors can be another problem associated with animal production facilities, especially during land spreading of the wastes. Some odors are inevitable near such facilities. Odors can be reduced by proper sanitation in production facilities and by proper waste management.

There is an increasing concern with the environmental effect of the disposal of wastes on the land. Land application of wastes has been a traditional method of agricultural waste disposal and remains the best approach. The high concentrations of animal manure in a small area and the disposal of the manure on the soil have raised questions about surface runoff and groundwater quality problems. In areas of several European countries, the quantity of wastes from concentrated animal operations, generally pig, dairy, and veal, is at or close to the capacity of the available land to assimilate the wastes.

Food processing wastes are an agricultural waste problem. In contrast to animal wastes, which are very high strength and low-volume wastes, food processing wastes are lower-strength, higher-volume wastes. The discharge of untreated food processing wastes to streams has caused pollution problems and the improper use of land disposal by irrigation can cause resultant runoff, soil clogging, and odor problems.

In the last few years, nutrient budget studies have indicated other concerns. The quantity of nutrients in a number of stream and river basins has been inferred to be a result of agricultural practice, especially of crop fertilization and animal waste disposal. Undoubtedly, some of the excess nutrients from crop production and waste disposal are reaching surface waters and groundwaters and contributing to the eutrophication and nutrient concentrations in some areas. The magnitude of this contribution will vary between basins.

The primary focus of agricultural waste management continues to be on the obvious problems such as odor control, land runoff, and food processing wastes. There is, however, awareness of potential long-range problems associated with contaminant leaching to groundwaters and salt buildup where land is used for waste disposal or where water reuse is practiced.

Legal Constraints

The increase in environmental problems has required protection of public interest in the form of state and federal statutes and regulations. No longer are agricultural contributions considered as ''background'' or uncontrollable pollution. As a result, the available legal restraints are being applied increasingly to agricultural operations.

Federal

Federal pollution control legislation has been developed since the turn of the century but with increasing rapidity since 1948. The basic policy and philosophy of water pollution control in this country include the following: (a) Congress has the authority to exercise control of pollution in the waterways of the nation; (b) both health and welfare are benefited by the prevention and control of water pollution; and (c) a national policy for the prevention, control, and abatement of water pollution shall be established and implemented.

The Water Quality Act of 1965 greatly expanded the scope of federal activities. One of the far-reaching effects of the Act was the provision for establishing water quality standards. Each state was to develop water quality criteria and a plan for implementation and enforcement. The water quality criteria adopted by a state were to be the water quality standards applicable to the interstate waters or portions thereof within that state. In establishing the standards, consideration was to be given to their use and value for public water supplies; recreation and aesthetics; fish, other aquatic life, and wildlife; agricultural uses; and industrial water supplies.

A change in the water pollution control philosophy of the nation has occurred with the establishment of the standards. The emphasis is now on the amount of wastes that can be kept out of the water rather than on the amount of wastes that can be accepted by the waters without causing serious pollution problems. This philosophy will guide acceptable waste treatment and disposal methods as well as legal actions in the future.

The standards were to protect the public health and welfare and enhance the quality of water. Enhancement of water quality was taken to mean that all existing water uses would be protected and all wastes amenable to control would be controlled. All wastes amenable to treatment were to receive secondary treatment, or its equivalent as a minimum requirement. The criteria may vary within a state depending upon water use, i.e., public water supply, industrial water supply, recreation, agricultural purposes, or receipt of treated waters.

Many states now are assuming that wastes from agricultural operations are controllable. Agricultural producers will have to consider the impact of existing legal restraints on both site selection and waste treatment and disposal in all expansion plans and the establishment of new facilities. The facilities should be located where treatment and disposal can be obtained at minimum cost and without adversely affecting the environment.

With the increased focus on environmental matters, a need existed to centralize the fragmented federal pollution control activities into a single agency. In 1970, the United States Environmental Protection Agency (EPA) was established. The EPA contains the federal activities dealing with solid waste management, air and water pollution control, and water hygiene among others and has the responsibility to integrate such environmental quality activities to minimize pollution transferral. The federal government has the legislation to control agricultural pollution and to encourage states to take a more aggressive role in their abatement activities.

In 1972 the Federal Water Pollution Control Act (PL 92-500), as amended, was enacted. The Act contained the first specific federal requirements for the abatement of agriculturally related sources of pollution. Among many requirements of this highly complex statute, the Act stipulated that each state must develop a waste treatment management plan that was to incorporate identification of agricultural nonpoint pollution sources, including runoff from manure disposal areas, and to set forth feasible procedures to control such sources. Similarly, the "source control" philosophy indicated earlier was advanced by requiring technology-based effluent limitations for control of point sources of waste discharges from all industrial operations, including many that are agriculturally based (Table 1.1).

The burden was placed upon industry to implement water pollution controls in a manner that represents the maximum use of technology within its economic capability. In determining the effluent reduction possible by the application of

TABLE 1.1

Parameters Controlled by Effluent Limitations in Agricultural Industries

Industry category	Parameters[a,b]
Dairy products	BOD, TSS, pH
Grain mills	BOD, TSS, pH, no discharge
Canned and preserved fruits and vegetables	BOD, TSS, pH
Canned and preserved seafood	TSS, O&G, pH, BOD
Sugar refining	BOD, TSS, pH, temp, fecal coliform
Feedlots	No discharge, BOD, fecal coliform
Fertilizer manufacturing	TP, F, TSS, pH, NH$_3$, Org N, NO$_3$, no discharge
Timber products processing	No discharge, pH, BOD, phenol, TSS, O&G, Cu, Cr, As, COD
Meat products and rendering processing	BOD, TSS, O&G, pH, fecal coliform, NH$_3$

[a]Parameters controlled in one or more of the subcategories in the major industrial category for BPCTCA, BATEA or BCPCT.

[b]The abbreviations used are BOD, biochemical oxygen demand; TSS, total suspended solids; O&G, oil and grease; temp, temperature; TP, total phosphorus; NH$_3$, ammonia; Org N, organic nitrogen; NO$_3$, nitrate; TKN, total Kjeldahl nitrogen; F, fluoride; Cu, copper; Cr, chromium; As, arsenic; COD, chemical oxygen demand.

existing technology, a "no-discharge" limitation is possible in cases where no discharge is found to be the best technology.

With the passage of the 1972 amendments, national pollution control emphasis shifted from water quality standards to effluent limitations. This approach was taken because of difficulties in linking waste discharge quality with stream quality and in enforcing previous legislation. The basic assumption, constrained only by the availability of economical control technology, is that the nation will strive toward elimination of water pollution. For many industries, this philosophy will result in a shift from wastewater treatment plants to closed-loop systems that recycle wastewater.

The Act declared, "It is the national goal that the discharge of pollutants into the navigable waters of the United States be eliminated by 1985." To achieve this goal, effluent limitations on point waste sources were to be achieved first by the application of best practicable control technology currently available (BPCTCA) and then by the application of best available technology economically achievable (BATEA). The difference between these two levels of technology is that BATEA is to represent an increased level of technology or combination of treatment and in-plant controls achieving (or likely to achieve) the best effluent quality current technology will permit. The intent was to increase the utilization of existing and new technology rather than rely on status quo for pollution control.

Existing sources are required to achieve effluent limitations based on the BPCTCA and BATEA technologies. New sources are required to meet new source performance standards (NSPS). NSPS represents a level of treatment which can be accomplished by the best available demonstrated control technology processes, operating methods, or other alternatives, including, where practicable, a standard permitting no discharge of pollutants.

In 1977, the 1972 Act was revised with the passage of the Clean Water Act (PL 95-217). Three distinct classes of pollutants—conventional, nonconventional, and toxic—are now regulated by the EPA. Conventional pollutants include biochemical oxygen demand (BOD), suspended solids, pH, oil and grease, and fecal coliforms. Toxic pollutants are defined by the legislation to be chemicals and compounds contained on an initial list of 65 pollutants and classes of pollutants. This list now contains 126 chemicals and compounds which frequently are referred to as "priority pollutants." Nonconventional pollutants include all pollutants not classified by the EPA as either conventional or toxic.

Different types of technology are used to control these three classes of pollutants. Conventional pollutants are controlled by limitations based on best conventional pollutant control technology (BCPCT or BCT). BCT replaces BATEA for the control of conventional pollutants. Both toxic and nonconventional pollutants are controlled by limitations based on BATEA.

Industry has three options for the discharge process wastewater: (a) no discharge, (b) direct point source discharge, and (c) discharge to a municipal wastewater treatment plant (publicly owned treatment works—POTW) (Fig. 1.1). No-discharge technologies are those that do not result in a direct discharge to surface waters. These include controlled land treatment and recycle and reuse of wastewaters. Land treatment of wastewaters and sludges can be a cost-effective and environmentally sound technology for the wastes from many agricultural production processes.

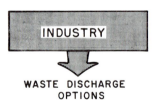

Fig. 1.1. Waste discharge options for industrial wastes.

Direct point source discharges are controlled by limitations based on BPCTCA, BATEA, BCT, and NSPS. Because there are few toxic pollutants in agricultural wastes, there are few limitations based on BATEA.

Discharge of wastes to a POTW can be a feasible approach for many agricultural industries such as food processing plants. Such discharges, however, are controlled by pretreatment standards. These standards require the removal of pollutants that (a) create problems such as clogging of sewers, (b) interfere with the sewage treatment processes at the POTW, (c) pass untreated through the POTW in concentrations that have an adverse effect on a stream, or (d) contaminate the treatment plant sludge, rendering it unfit for disposal or beneficial use. In addition to general pretreatment standards, there can be categorical pretreatment standards. Categorical standards are to control toxic pollutants that may be in the wastewater from a specific industry category or subcategory.

The legislative history of PL 92-500 makes it clear that Congress (a) intended that POTWs be used for the treatment of industrial pollutants that can be treated by treatment processes at a POTW, and (b) did not intend that there be duplication of treatment, i.e., that industry pretreatment requirements should not duplicate the treatment that can be achieved at a POTW. Discharge of agricultural industry wastes to a POTW can be a technically and economically feasible approach provided that the appropriate general pretreatment standards are met.

Dischargers are required to treat their discharges to the degree indicated by the respective technologies, but they are not required to use a particular method of pollution control. Any method is satisfactory as long as the treated effluent meets the limitations on specific parameters.

The specific effluent limitations for industry categories and subcategories were determined from a detailed evaluation of the industry, factors affecting the wastes produced, in-plant process changes to reduce wastes, pollutant parameters that should be controlled, technologies that can be used for pollution control, non-water-quality environmental impact such as solid wastes, air pollution, or energy usage that may result, and the economic effect of imposing the limitations. The pollutant parameters that are controlled by the effluent limitations in the agricultural industries vary (Table 1.1) but are the critical parameters for a category. Requirements of no discharge are among the effluent limitations identified as technologically feasible for one or more agricultural industry categories.

The effluent limitations for industry are identified in terms of mass units, quantity of pollutant per unit of production. As an example, one of the effluent limitations for the food processing industry is in terms of pounds of BOD_5 per ton of food product processed. This approach eliminates the role of dilution in meeting limitations based on concentration units.

The technologies identified as available to meet the effluent limitations are common available methods used to treat municipal and industrial wastes. The

technology that can be applied to achieve the effluent limitations for the agricultural industries (Table 1.2) has been demonstrated in many pollution control
systems. There are many technical approaches that can be utilized and those
seeking solutions in this area should become familiar with all of them. A specific
facility can use any technology that would permit it to best meet the limitations
for its location, production processes, raw materials, management, and economic position.

Permits to discharge wastes are required of all industrial point sources. The
conditions in the permits are based on the effluent limitations and categories
outlined above or on water-quality-based limitations for the stream to which the
treated wastes are discharged.

State

Legal controls on water quality have been recognized as traditional state
responsibilities. Considerable variation has existed among states in their regulation of agricultural activities. While most state legislation does not mention
agricultural pollution specifically, such pollution is not excluded. Where live-

TABLE 1.2

**Technology Identified as Available to Meet Effluent Limitations in Agricultural
Industries**

Industry category	Technology[a]
Dairy products	Biological treatment, in-plant control
Grain mills	Biological treatment, in-plant control, biological solids removal by filtration
Canned and preserved fruits and vegetables	Biological treatment, solids removal from effluents, disinfection, improved in-plant control
Canned and preserved sea-food	In-plant control, screens, grease traps, biological treatment
Sugar refining	Biological treatment, separation of barometric condenser cooling water, solids removal from effluents, in-plant control, land application
Feedlots	Runoff control, land application
Fertilizer manufacturing	Chemical treatment, runoff control, ammonia steam stripping, urea hydrolysis, oil separators, ion exchange, denitrification
Timber products processing	Screening, settling, biological treatment, reuse, in-plant changes and control, oil separation
Meat products and render-ing processing	In-plant control, biological treatment, grease recovery, nitrifica-tion, ammonia stripping, effluent solids removal

[a]Technology as used in one or more subcategories of the industrial point source category for
development of effluent limitations.

stock production is an important factor, specific legislation or guidelines for animal waste management have been developed. Moreover, many states and local jurisdictions have air pollution statutes or guidelines that deal with air quality problems associated with agricultural production, such as odors and particulate matter arising from feedlot dust and animal production activities.

Certain states have developed requirements for control of runoff from animal feedlots. While the specific requirements vary, they usually include provisions for a feedlot operator to (a) obtain a permit from the appropriate state agency, (b) correct any pollution hazard that exists, and (c) assure that the operation conforms to all applicable federal, state, and local laws. Specific minimum pollution control methods or criteria such as runoff retention ponds, dikes, and distance from dwellings may be incorporated into the requirements. Approved facilities normally include (a) diversion of runoff from nonfeedlot areas, (b) retention ponds for all wastewater and runoff contacting animal wastes, (c) application of liquid waste to agricultural land, and (d) application of solid wastes to agricultural land.

To assist the control of agricultural wastes while the most practical and economical waste management approaches are being developed, many states have suggested or prepared "Codes of Practice" or "Good Practice Guidelines" for the disposal of agricultural wastes and the expansion of existing or expansion of new livestock facilities. Key items in the codes or guidelines include suggestions of land area to dispose of wastes, waste storage capacity, distance to human dwellings, criteria for satisfactory waste handling and treatment facilities, odor control, solid waste disposal, and runoff control (Chapter 15).

Where guidelines are available, an agricultural producer would be in a vulnerable position if he did not adhere to them. Such guidelines do not ensure that pollution or nuisance problems will not result. The advantage of codes or guidelines is that they represent best available practice yet can be altered more simply than regulations when better procedures become known and proved. The following of existing guidelines may help eliminate punitive damages in a lawsuit since the operator was following good practice and using the best available knowledge rather than intentionally committing a wrongful act.

Violation of air pollution codes can be of concern for concentrated agricultural operations. The definition of air pollution in these cases is not always clear; however, one used in many states is

"Air pollution" means the presence in the outdoor atmosphere of one or more air contaminants in quantities, of characteristics and of a duration which are injurious to human, plant or animal life or to property or which unreasonably interfere with the comfortable enjoyment of life and property throughout the state or throughout such areas of the state as shall be affected thereby.

The terms "unreasonable" and "comfortable enjoyment" require individual interpretation; however, the definition provides justification for complaints due to odors and dusts.

The odors generated by the confined agricultural production facilities have been a source of complaints and an important consideration in legal action against these operations. Enforcement of air pollution codes offers another avenue to assure maintenance of environmental quality but represents an additional constraint for agriculture.

Local

Nuisances caused by agricultural operations can be controlled by city or county governments or by individual action where the injured party seeks compensation for damages. Experience has shown that local legislation rarely is effective.

Zoning can keep complaints to a minimum by requiring that nonfarm residents be located some distance from agricultural operations. Zoning laws do not, however, provide absolute protection against the application of nuisance laws. Automatic protection also is not provided because the agricultural operation existed prior to residential or recreational uses of adjacent land. An important aspect is whether changes in the operation have caused nuisances to arise.

Individual action can be an effective legal restraint on agricultural operations. The basis for such action includes trespass, nuisance, negligence, and strict liability. Private regulation of environmental quality problems can occur by civil lawsuits based on nuisance laws. All persons have the basic right to enjoy their property. Unreasonable interference with such enjoyment is legally a nuisance. A nuisance may invove air, water, solid waste, and/or noise pollution. Cases have been brought against agricultural operations for odors, water pollution, air pollution, and aesthetic reasons and in a number of cases, injunctions were issued and/or damages awarded the plaintiff.

Social Aspects

Suburban development and the increase in farm technology have increased the interest in agricultural wastes. The number of urban-oriented rural residents and open country recreational activities is increasing and the agricultural population of the United States continues to be a lesser proportion of the total population. Individuals can live and die without growing their own food or knowing what is involved in producing a gallon of milk, a dozen eggs, or a pound of meat. They are not accustomed to and may not tolerate the liquid and gaseous effluents associated with agricultural production. Concentrated agricultural production has intensified these effluents in the same period that the public has joined the quest for a pollution-free environment.

When public concern is voiced about the manner in which agricultural operators dispose of their wastes, apply fertilizers to crops, or manage their waste systems, what is implied is that the social costs of producing these products may have equaled or exceeded the benefits. The national search for an improved quality of the environment will require agriculture to give a high priority to wastes in decisions concerning production, i.e., to continually "think waste management." The heightened public interest in the environment and the increase in the urban and suburban population will be an effective reminder to those operations that do not place a high priority on pollution control activities.

2

Changing Practices

Introduction

The quantity and quality of the world's food supply will be among man's most critical future problems. Efforts to increase the quantity of available food have led to the substitution of mechanical energy for human and animal energy, increased fertilizer use to boost crop yield per acre, genetic upgrading of animals and crop varieties, and advances in production technology. As a result, there have been significant increases in agricultural production. Farm size and productivity per farm worker have increased. Crop and livestock production per acre and total crop production also are increasing with less land used for production. It is unlikely that the ultimate in agricultural productivity has been reached, and further increases in agricultural production and productivity will be achieved.

One of the benefits of these efficiencies has been reasonable food prices. U.S. consumers spend a smaller portion of their disposable income on food than do consumers in any other country (Table 2.1). In 1978, U.S. consumers spent about 13% of their total expenditures on food compared with the 17–23% spent by Western Europeans and the more than 50% spent in many developing countries.

Agricultural production continues to be one of the more successful components of the U.S. economy. Currently, American farmers produce about 13% of the world's wheat, 29% of the coarse grains (mainly corn), 17% of the meat, and 62% of the soybeans (2). Because the United States has less than 5% of the

12

TABLE 2.1

Consumer Expenditures for Food[a]

Country	Food expenditures as a percentage of total expenditures
Philippines	57
Korea	44
South Africa	23
Finland	23
West Germany	20
Sweden	20
France	19
United Kingdom	18
Netherlands	17
Canada	15
United States	13

[a]Data for 1978; U.S. Department of Agriculture (1).

world's population, there continues to be a large surplus for export. In 1981, the U.S. agricultural export volume was 163 million metric tons.

Only 3.1% of the American labor force produces the food and fiber for domestic consumption and for export. This production has been accomplished by a number of technological advances: mechanization, improved plant and animal species, increased use of fertilizers, improved pest and disease control, and better management. These factors also have the potential for an adverse impact on the environment.

Many potential environmental quality problems are associated with the residues from agricultural production operations. These residues can be in the form of animal waste at animal production facilities, runoff and leachate from fertilized and manured fields, and liquid and solid wastes generated at food processing operations. Residues of this nature always have been associated with agriculture but have become more noticeable because the natural cycles associated with agriculture have been altered and in some situations broken.

The basic needs of man include food, fiber, fuel, fertilizer, and shelter. In underdeveloped regions and with early forms of agriculture, small farming operations could and did cycle available resources (Fig. 2.1). The basic cycle was one of the land furnishing fuel, fiber, shelter, and food in the form of crops and animals to the consumer. A portion of the residues of the harvesting and processing were used directly by the consumer and the remainder were returned to the land for further utilization in additional crops. This approach had the benefit of optimum utilization of residues or wastes but resulted in a low level of total agricultural productivity.

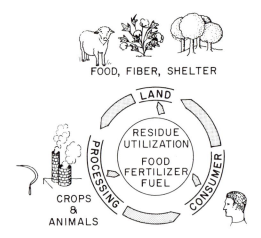

Fig. 2.1. Resource utilization cycle of nonintensive agriculture.

Inexpensive fossil fuel energy, the input of inorganic fertilizers, and greater mechanization were utilized to increase agricultural productivity and raise the standard of living in many countries. However, the consumer–land link was weakened and was broken with certain types of agricultural activity. Residues were no longer as well utilized and the agricultural waste problems became more apparent.

Animal production offers an opportunity to examine the alteration of a natural cycle and the resultant environmental effects. The basic cycle consists of the land providing for the animals, whose waste is returned to the land to help produce more feed (Fig. 2.2). The wastes also can be processed and reutilized as part of the ration of other animals and for other purposes. In modern agriculture certain amounts of fertilizer, pesticides, and external energy are necessary to

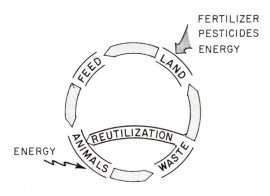

Fig. 2.2. Schematic animal production pattern.

obtain needed crop and animal production levels. As long as the resultant animal manures are utilized in the cycle as noted, the cycle remains essentially closed, external inputs are low, and environmental concerns related to waste management are minimal. However, when there is a lag in the cycle or improper management of one of the components, environmental problems increase.

As an example, when the wastes are allowed to accumulate prior to application to the land, undesirable odors can result and natural precipitation can transport a portion of the wastes to streams. Improper management of the wastes when applied to the land can result in nutrient losses and surface water and groundwater pollution. The large, concentrated animal feedlots have increased the efficiency of animal protein production but can accumulate large quantities of waste, can weaken the waste–land link, and can be a cause of environmental problems.

In the United States, adequate land is available so that animal wastes generally can be returned to land and the cycle remain essentially intact. The press to increase animal production to satisfy the demand for animal protein requires that decisions on feedlot site selection carefully reflect the availability of land for waste application and utilization. If the waste–land link is broken or is inadequate for all of the animal wastes at a feedlot, then environmental costs in the form of increased fertilizer use, waste treatment, resource loss, and alteration of other natural cycles occur (Fig. 2.3). If adequate land is not available to apply the wastes and grow feed for the animals, then feed must be imported. This in turn creates a further dislocation in the basic cycle since nutrients in the feed grown in one area are transported to another and result in waste management concerns in the second geographical area.

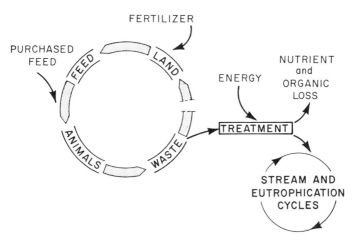

Fig. 2.3. Schematic intensive animal production pattern.

The further from the basic cycle the situation is permitted to go, the more unbalanced the situation becomes, and the greater the waste management problems and costs to maintain desired environmental quality conditions. The better solution to animal waste management lies in keeping to the basic cycle (Fig. 2.2) as much as possible, and in the best utilization of the resources in the residues rather than in wasting or losing them.

The basic cycle is easier to maintain with animal production than it is with food and fiber processing. Such processing is further removed from the land and a closed cycle in which the wastes are returned to the land is less possible. The food and fiber processing pattern is a sequence involving many natural cycles (Fig. 2.4). Even where closed cycles are not possible, the residues or by-products should be considered as resources to be utilized rather than merely disposed of.

The basic nature of waste management is how to utilize the by-products that

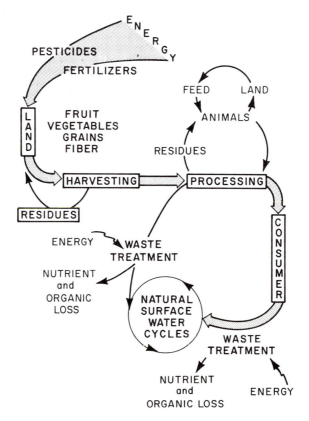

Fig. 2.4. Schematic food and fiber processing pattern.

result from processing, how to reduce the quantity of wastes that are generated, and how to economically handle, stabilize, treat, or dispose of the wastes that ultimately result.

Agricultural Productivity

The productivity of American agriculture in recent decades has been re-markable, especially since about 1940 (Fig. 2.5). In 1940 one farm worker supplied farm products for about 11 individuals including himself. In 1980 that number had increased to over 60 individuals. There does not appear to be any decrease in the rate of change (Fig. 2.5) in the past two decades.

The increases in agricultural production have been accompanied by changes in the size of the average farm and the number of people living on farms. Between 1950 and 1980, the number of people living on farms decreased from 23 million to about 7 million. In the same time period, the average farm size increased from 215 to 431 acres. By the year 2000, it has been estimated that the 50,000 largest farms in the United States will account for 63% of all agricultural sales.

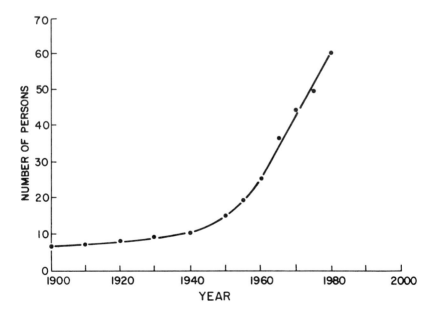

Fig. 2.5. Number of persons supplied farm products by one farm worker in the United States. Data from (3) and (4).

Labor and Energy

To obtain increases in agricultural production, mechanical energy has been substituted for human labor (Fig. 2.6). In the early 1800s, before the widespread mechanization of agriculture, the production of 100 bushels of wheat took more than 250 hr of labor. Now it takes less than 10 hr of labor.

The decreases in the number of labor hours spent per unit of agricultural production have been dramatic in all areas of food production (Figs. 2.7 and 2.8). It has been especially dramatic in the production of broilers, soybeans, and corn. It now takes only 0.1 hr of labor to produce 100 lb (liveweight) of broilers.

Livestock and Crop Production

Population growth and consumer desire are important factors in determining future demand for agricultural products and trends in the growth or decline of components of the industry and in providing an estimate of future concerns. This section outlines changes that are occurring in agricultural production as well as the comparative size of specific segments of the industry. A pattern of the waste

Fig. 2.6. Wheat harvesting equipment illustrates the mechanization and labor-saving equipment in American agriculture. (Courtesy of the U.S. Department of Agriculture.)

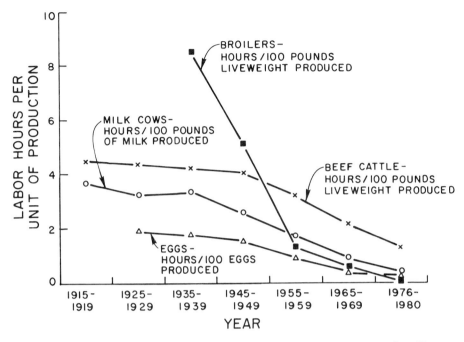

Fig. 2.7. Number of labor hours spent per unit of agricultural production. Data from (4).

management challenges will emerge as the segments are described and as they are combined with information in Chapter 4 on the quantity and quality of the resultant wastes.

National crop and livestock requirements will continue to increase due both to population growth and to increase in per capita consumption of certain products. Changes in consumer income, age distribution of the population, and developments of new uses for farm products will influence the actual future requirements.

Food Consumption

In general, there has been a greater per capita demand for meat protein, especially poultry, and for sugar and sweeteners and a decreased per capita demand for cereal products in the United States (1). There has been a variable demand for fresh and processed fruits and vegetables. In the 1978–1981 period, there has been a decreased per capita demand for processed vegetables and an increased per capita demand for fresh fruits and vegetables. Prior to 1978, there was an increased per capita demand for processed fruits and vegetables.

Throughout the world, as the standard of living and income levels of people

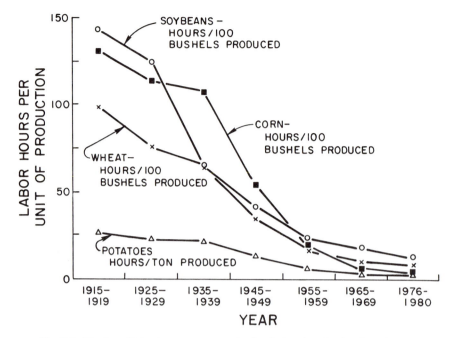

Fig. 2.8. Number of labor hours spent per unit of agricultural production. Data from (4).

increase, they spend more of their income on protein foods, especially proteins derived from animal origin. In the United States, about 92% of the daily protein consumption is animal protein. This compares to 55% in Western Europe, 39% in Eastern Europe, 41% in Japan, and 20% in Africa. The next decades will see the per capita demand for animal products increase, and countries will attempt to bring large-scale animal protein consumption within the reach of their population. This, in turn, will increase the problem of animal waste management in other countries as concentrated, large-scale animal production units are utilized for this purpose.

Livestock Production

Total animal production in the United States has shown a continued increase because of demands for meat noted earlier. The most dramatic increase has occurred in the broiler industry where the numbers of broilers raised in the United States increased from 1.8 billion in 1960 to almost 4 billion in 1980.

To be economically competitive, livestock production has become concentrated in larger operation, has become specialized in certain geographical areas, and has increased confinement feeding of livestock and increased numbers of animals per livestock operation. Mechanization, improved production methods, and better nutrition and disease control have made it possible for the livestock producers to handle more animals with a minimum increase in help. The scarcity of inexpensive farm help has influenced the trend.

These changes have had an impact on milk production (Fig. 2.9). The total milk production has stayed reasonably constant between 115 and 125 billion pounds per year over the past 40 years. However, there has been a dramatic change in the milk production per cow and the number of milk cows. The number of dairy cows has decreased from 21 million head in 1955 to 10.8 million head in 1980. In the same period, average milk production per cow increased from 5840 to 11,800 lb per year to keep total production reasonably constant. Similar relationships are likely to exist in the future, e.g., decreased numbers of dairy animals with little change in total milk production. Fewer but larger dairy farms will exist in the future as dairymen improve techniques to produce livestock products more efficiently and profitably.

The changes in both the numbers and sizes of beef cattle feedlot operations

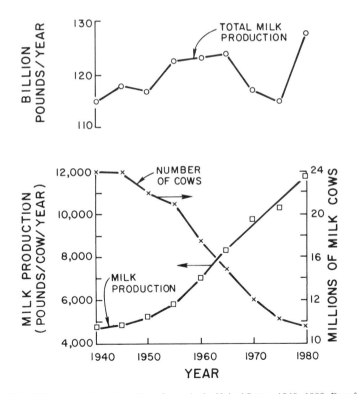

Fig. 2.9. Milk production and number of cows in the United States, 1940–1980. Data from (4).

have been dramatic. Specialization has removed cattle from pastureland and grassland and has resulted in confinement of large numbers in small areas with an average density of 1 animal per 50–150 ft². The number of cattle on feed for slaughter was over 11 million head in 1981.

The increase in commercial confined feedlot operations occurred as a result of the availability of relatively inexpensive feed grains, proximity to an adequate supply of feeder cattle, and the strong demand for beef. The advantages of confinement include less space per animal, less labor, and economies of scale (Fig. 2.10).

Two types of cattle feeding operations exist, the farmer–feeder who has a feedlot with less than 1000-head capacity and the commercial feedlot, which generally has greater than 1000 head and whose business is primarily cattle production. Neither the number of cattle fed nor the capacity of the feedlot is the most important criterion to indicate the pollution potential of these two operations. From the pollution standpoint, the important factor is whether land is available for waste disposal. Generally the smaller cattle-feeders are farmers and have adequate land available to integrate waste disposal with crop production. The farmer–feeder operations are abundant in the midwestern cornbelt and the large commercial feedlots are abundant in the Texas, Oklahoma, Kansas, and Colorado high plains area.

The large beef feedlots will continue to market a greater proportion of the

Fig. 2.10. A typical large-scale commercial beef feedlot. (Courtesy of the United States Department of Agriculture.)

beef. Feedlots with 1000-head capacity or more account for slightly more than 1% of all cattle feedlots in the United States although they market over 60% of all the beef. The total number of feedlots is decreasing as the smaller, inefficient lots go out of business. In the 10-year period from 1962 to 1972 the number of feedlots with a capacity of 8000 head or greater almost tripled, and those with a capacity greater than 32,000 increased by a factor of almost 12. Fig. 2.11 identifies the pattern of feedlot capacity over the past 15 years. The number of feedlots with a capacity of 16,000 head or more continues to increase.

The poultry industry is another example of confined, intensive livestock production. In the major poultry-producing regions, most laying hens and broilers are raised in confinement. Large poultry operations are highly mechanized and are able to handle over 100,000 birds per operation. Commercial egg production is almost 100% from confinement poultry houses. Present poultry management permits the concentration of egg-laying hens in buildings housing several hundred thousand birds on a site consisting of small acreage (Fig. 2.12).

About 70 billion eggs were produced in 1980 by about 287 million birds. Egg production is distributed throughout the United States. The top five egg-producing states in 1980 were California, Georgia, Pennsylvania, Arkansas, and Indiana. Over 90% of U.S. egg production is disposed of as shell eggs. Between 8 and 10% of the eggs are broken for commercial use as frozen eggs.

The marketing of poultry and eggs has undergone substantial changes which have been facilitated by advanced production and marketing technology. Per unit profits on these items have remained stable or declined even though wage rates and other prices have increased. Improvements in plant operating efficiency, larger production units, and higher density of production have been responsible for reduction of operational costs.

The swine industry represents some 10% of the total cash farm receipts, ranking only behind beef and dairy cattle among all animal commodities. Swine

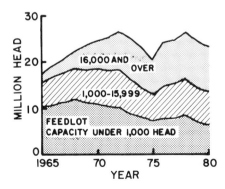

Fig. 2.11. The relationship of cattle marketed as a function of feedlot capacity (4).

Fig. 2.12. Commercial egg production; manure held beneath cages until disposal. (Courtesy of the United States Department of Agriculture.)

production is moving toward confined feeding although less the 20% of the hogs were produced in total confinement in 1980. Separate farrowing and finishing herds are common in operations having 1000–10,000 animals. A few 10,000- to 50,000-hog capacity operations exist. More large, specialized operations will develop, incorporating complete feeding, waste treatment, and disposal in a single operation.

About 80% of the U.S. hogs are produced in the north central states, 10% in the south central states, 8% in the south Atlantic states, and the rest throughout the country. Over 96 million hogs were slaughtered commercially in 1980.

The increased production per animal and the trend to confinement animal production operations produce situations that are analogous to modern industries. Inputs (the feed) and frequently raw material (the young stock) may be purchased elsewhere and brought to the production operation where under controlled conditions optimum production of meat, eggs, or milk occurs. Although the producer may be more interested in the quality and quantity of the product, the waste material that results must be disposed of without adversely affecting the environment.

Summary

Animal and dairy product production illustrate the type of changes that have occurred in agriculture in recent years. Meat, milk, dairy products, and eggs increasingly will be produced in confined large industrial-type facilities. The trend toward controlled and enclosed facilities, which is virtually complete for smaller animals such as hens and broilers, is increasing even for the larger animals such as beef and dairy cattle. Decreased numbers and increased size of production operations are expected to continue because of the enlargement of existing units and the closing of uneconomic operations, of operations less specialized, and of those unable to adjust their size.

Livestock Processing

While livestock production is becoming more centralized, marketing and slaughtering of animals, especially beef and hogs, has become more decentralized. Beef and hog processing and packaging facilities are being established near the sources of supply. The livestock slaughtering industry has changed in a functional manner. More slaughtering plants are specializing in only one kind of livestock. Processing plants show a trend to specialize in one activity, e.g., slaughter or processing only, rather than both.

The meat-packing industry has had three major phases. Initially, it was a small industry with widely scattered local plants throughout the western United States. In the early part of this century, it became centralized in major midwestern cities which served as major terminal markets. The third phase is the decentralization now underway. As a part of the third phase, new regionally based concerns are capturing an important share of the market from the established packers.

Decentralization has been accelerated by improvements in truck transportation, rail freight rate adjustments, and refrigerated trucks. Weight loss in transportation of live animals lends a comparative advantage to the transportation of carcass meat. Another factor has been the age of the facilities of major packing concerns and the efficiency and automation obtainable in newer facilities.

An implication of this decentralization is that livestock processing wastes are being concentrated near the source of livestock supply. Wastes from livestock production and from slaughtering and processing operations are increasing around the smaller urban areas and towns of the nations. These industries are an economic advantage to the communities. However, their wastes represent a challenge for satisfactory management since they frequently are many times the magnitude of the wastes from the community.

Fruit and Vegetable Production

As with other agricultural products, Americans have changed their desires for fruits and vegetables. Processed fruit and vegetables have shown the greatest increase in total production and per capita consumption. A further increase in processed food consumption, especially frozen foods, will occur in the future as Americans increase their preference for convenience foods.

Fresh vegetable production in the United States has increased in proportion to increases in population and per capita consumption during the past 15 years. Processed vegetable production also has increased, with processed tomatoes now accounting for over half of the processed vegetable tonnage. Per capita consumption of fresh vegetables has risen since the 1970s, with lettuce, tomatoes, broccoli, cauliflower, celery, and cucumbers becoming increasingly popular.

During the past decades, there has been a constant consolidation of smaller operations into larger, more centralized ones. The larger plants offer a greater opportunity for better control of waste discharges and for in-plant changes to reduce wastes. The processing plant wastes frequently are generated during a relatively small harvest period. Waste management systems must be geared to fluctuations in waste quantity and quality.

Seafood Production

Americans are tending to eat more fish and fish products. In 1980 the total quantity of fish caught by commercial fisheries was about 6.5 billion pounds. Of this amount, about 2.6 billion pounds was frozen or sold fresh, 1.2 billion pounds was canned, and the rest was used as by-products or bait.

Of the U.S. catch about 35% is rendered, 30% is marketed fresh, 20% is frozen, 1% is dried, and 4% is handled by miscellaneous methods (5). Frozen fish products are increasing in popularity and a 150% increase has been predicted in the next 15 years.

The catfish industry has the potential to develop along lines analogous to that of livestock production. Farm-pond-grown catfish are grown in southern and south central states. Intensive catfish farming began in a significant manner in the 1960s. Commercial catfish pond acreage increased from about 400 acres in 1960 to 45,000 in 1970 (6). Yield per acre increased from about 800 to 1200 lb during that period and was about 3000 lb in 1980 in Mississippi. Scientifically managed pond systems can have an annual yield of over 2 tons of fish per acre. In pond culture of catfish, either fry or fingerling fish may be stocked. The fish are fed a commercial feed ration until maturity, at which time they are processed into fillets or steaks which are marketed fresh or frozen.

Fertilizer Production

The sizable increases in crop and food production would not have been possible without the greater use of fertilizers. The increased inorganic fertilizer use in the United States has accounted for over half of the increase in crop production per acre since 1940. Fertilizer has been substituted for cropland to increase total agricultural production. The fertilizer nutrients used per acre (Fig. 2.13) have increased rapidly in the past 25 years.

Although the United States has increased its use of fertilizers, other countries had much greater rates of application to agricultural land (Table 2.2). The Netherlands has had the most intensive use of fertilizers in the world.

The use of fertilizers varies among nations and regions (Table 2.2). Europe, excluding the U.S.S.R., was the heaviest user of fertilizer, applying an average of 158 kg of nitrogen, phosphate, and potash per hectare of arable land under permanent crops. The world fertilizer consumption increased from 7.5 to 17 kg per capita over the period of 1955 to 1969. The most developed nations were the largest users of fertilizers. The largest consumers, in terms of quantity per person (kilograms per capita) in 1969, were North America, 66; Western Europe, 45; and Eastern Europe, 40 (8). The consumption per hectare varies widely depending upon crop, farming intensity, and crop management. The changes in fertilizer use reflect the substitution of fertilizer for land, producing higher yields on fewer acres.

In the United States, fertilizer use varies among states, regions, crops, and farming areas. About 40% of the plant nutrients used in the United States were applied to corn. Hay and pasture accounted for about 15% of the total nutrients used because of the large acreage involved. Cotton was the third highest user, accounting for about 8% of the plant nutrients consumed. Crops having the highest value per unit land utilize the greatest amount of fertilizer per unit of land area. Potatoes, fruits and nuts, vegetables, and sugar beets are among the crops on which the fertilizer use has been the highest.

Fertilizers have been profitable for farmers and have played an important

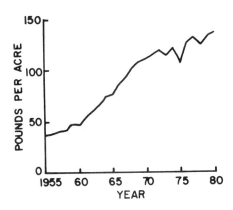

Fig. 2.13. Fertilizer nutrients used per acre of cropland in the United States (4).

TABLE 2.2

National and Regional Fertilizer Use (kg/ha)

Region or nation	Use[a]	Region or nation	Use[a]
Europe	158	Worldwide	42
North and Central America	66	Netherlands	616
U.S.S.R.	35	Japan	400
Oceania	33	Latin America	22
Mainland China	30	India	10
Asia	22	Western Europe	161
South America	14	Eastern Europe	49
Africa	7		

[a]1969–1970; source, FAO (7).

part in providing consumers with an abundant and low-cost food supply. On soils of low natural fertility, initial response in yields is large for a relatively small amount of fertilizer. Additional amounts of fertilizer produce successively smaller increases in yields. Ultimately, there is some point at which additional fertilizer usage provides no additional production. Amounts applied beyond this point can result in decreased yield. In the United States increased fertilizer utilization will occur from increased fertilizer application on crops and in areas now using low amounts of fertilizer.

Previous low costs of fertilizers may have caused farmers to be less concerned with nutrients that do not become part of the growing crop. However, the higher cost and tighter supplies of fertilizers in recent years have caused farmers to be more concerned with better utilization of inorganic fertilizers as well as available animal manures. This concern should have the beneficial effect of reducing the quantity of agricultural nutrients that can cause an adverse environmental impact.

Pesticides

The use of pesticides in agriculture has permitted the agricultural producer to produce food at a lower cost. As with fertilizers, pesticides have been substituted for labor and equipment. A marked increase in the use of pesticides started in the 1940s and lasted until the late 1960s. United States production peaked at about 1.2 billion pounds in 1968.

Pesticides have been categorized according to their use as insecticides, herbicides, fungicides, rodenticides, and fumigants. Pesticide use patterns have changed over the years and may be expected to change further as environmental constraints are imposed on their application. In 1970, although production of fungicides and insecticides decreased, that of herbicides increased. Herbicides are now applied to more than 90% of the acreage planted in corn. Ten years ago they were applied to only 57% of the corn acreage.

Pesticides are primarily applied to corn and cotton (over 60%), with other major crops on which they are used being vegetables, field crops, and apples (about 20% combined). Future pesticide use will decrease and be selective. Concern of the environmental effect of pesticides has increased and caused greater scrutiny of need, use, and dosages of pesticides. The regulatory powers of the federal government over the manufacture, sale, and use of pesticide products will become greater to provide more effective protection for the public and environment while permitting the continued use of selected pesticides for beneficial uses.

References

1. U.S. Department of Agriculture, "1982 Handbook of Agricultural Charts," Agriculture Handbook No. 609. Washington, D.C., 1982.
2. Batie, S. S., and Healy, R. G., The future of American agriculture. *Sci. Am.* **248** 45–53 (1983).
3. U.S. Department of Agriculture, "Agricultural Statistics—1970." U.S. Govt. Printing Office, Washington, D.C., 1970.
4. U.S. Department of Agriculture, "Agricultural Statistics—1981." U.S. Govt. Printing Office, Washington, D.C., 1982.
5. Soderquist, M. R., Williamson, J. J., Blanton, G. I., Phillips, D. C., Law, D. K., and Crawford, D. L., "Current Practice in Seafoods Processing Waste Treatment," Rep. Proj. No. 12060 ECF. Water Quality Office, U.S. Environmental Protection Agency, Washington, D.C., 1970.
6. Anonymous, "Producing and Marketing Catfish in the Tennessee Valley." Tennessee Valley Authority, Oak Ridge, Tennessee, 1971.
7. Food and Agriculture Organization, "Production Yearbook," Vol. 28.1. FAO United Nations, New York, 1974.
8. Harre, E. A., Kennedy, F. M., Hignett, T. P., and McCune, D. L., "Estimated World Fertilizer Production Capacity as Related to Future Needs—1970 to 1975." Tennessee Valley Authority, Oak Ridge, Tennessee, 1970.

3

Environmental Impact

Introduction

Real and potential environmental quality problems have accompanied the changes in agricultural productivity in recent decades. Difficulties in obtaining precise relationships between agricultural practices and environmental problems occur because of the diffuse source of residuals from agriculture and the many factors that are involved, some of which are not controllable by man. The factors include topography, precipitation, cover crop, timing and location of chemical or fertilizer application, and cultivation practices. The variability of waste discharges complicates assessment of the environmental impact of agricultural production. Agricultural wastes frequently are not discharged on a regular basis. Food processing wastes tend to be seasonal, with production a function of crop maturity; feedlot runoff is a function of rainfall frequency and intensity; fertilizer applications are timed for ease of distribution and to maximize crop production; and land disposal of wastes are related to need for disposal and ability to travel on the land.

Available information suggests that potential environmental quality problems due to agricultural operations may be more dependent upon the production practices and waste management techniques utilized by farmers and processors than the size of the operation, the number of animals fed, or the amount of waste involved.

All of the environmental impacts described in this chapter that result from agriculture can be minimized by the use of suitable production and waste management methods. Common sense and the elevation of the concept of environmental impact to a higher level of consciousness in agricultural operations are important first steps in the control of real or potential environmental quality problems caused by agriculture.

This chapter will attempt to summarize the environmental problems that can or have been associated with agricultural production, to put these problems in perspective with those caused by other segments of society, and to indicate procedures to minimize these problems. Detailed waste management methods are described in subsequent chapters.

Water Quality

Nutrients

One of the most challenging problems is that of excessive nutrients and the conditions they can cause. Of particular concern are nitrogen and phosphorus compounds. Although these elements are needed in small amounts for all living matter, excessive amounts in surface waters can result in overfertilization and can accelerate the process of eutrophication. Other concerns include excessive amounts of nitrates in groundwaters and surface waters, ammonia toxicity to fish, altered effectiveness of chlorination by ammonia, and the nitrogenous oxygen demand of reduced nitrogen compounds in surface waters.

Until recently the major reason for investigating the nutrient content of the soil water below the crop root depth was to evaluate the loss of fertilizer nutrients and to estimate the loss of production or the need for more fertilizer to adjust for the losses. The increased concern about environmental quality has changed this emphasis to the study of nutrients and other contaminants in soil water because of potential water pollution. The quantity of rainfall or drainage water and the permeability of the soil are key factors in the leaching of contaminants from the soil.

The processes governing the transformation and movement of nitrogen and phosphorus must be understood as waste management alternatives are considered. A schematic summary of nitrogen in the soil–plant system (Fig. 3.1) illustrates some of the important factors and transformations.

The nitrogen in sludges, food processing wastes, manure, and other biological residues is primarily in the organic and ammonium form. Fertilizer nitrogen may be in the ammonia, ammonium, nitrate, or urea form. The largest quantity of nitrogen in soil occurs as organic nitrogen. Microbial decomposition of the organic matter results in the production of ammonia nitrogen, which, with good

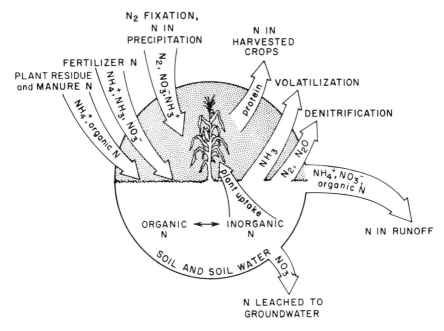

Fig. 3.1. Nitrogen inputs, outputs, and transformations (1).

aeration and suitable temperatures, can be oxidized to nitrites and then to nitrates. Unless inhibition of the nitrate-forming bacteria occurs, practically no nitrite accumulation will occur. Nitrates can be reduced to gaseous nitrogen compounds and lost to the atmosphere if the nitrates are held under conditions of poor aeration and if suitable carbon sources are available for denitrification. Ammonium ions are held to the cation exchange sites in soils, and the ammonium ion concentration in soil solution and leaching to the groundwater is low. Nitrate ions are soluble in the soil water. The form of nitrogen of greatest environmental quality concern in the soil is nitrate nitrogen since it is subject to leaching and will move with the soil water. Losses of organic nitrogen are related to surface runoff.

A generalized summary of phosphorus interactions in soils and plants (Fig. 3.2) indicates the important inputs, transformations, and outputs. Fertilizers and organic wastes add to the organic and inorganic forms of phosphorus in the soil. The inorganic forms mainly are iron and aluminum phosphates in acid soils and calcium phosphates in alkaline soils. Any phosphorus added as fertilizer or released in the decomposition of organic matter is converted rapidly to one of these compounds. All inorganic forms of phosphate in soils are extremely insoluble, and concentrations of phosphorus in soil water solutions are low, generally

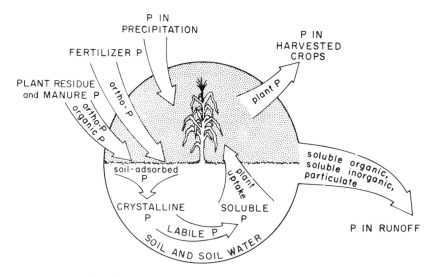

Fig. 3.2. Phosphorus inputs, outputs, and transformations (1).

less than 0.2 mg per liter. The phosphorus of environmental quality concern is associated with the soil itself and with the interchange of phosphorus from bottom deposits in bodies of water with the upper waters.

Soil management practices can reduce the losses of phosphorus caused by erosion. Losses of soil nitrogen are more difficult to control since soluble nitrate nitrogen can be transported through the soil. Soil conservation can reduce the quantity of organic nitrogen reaching surface waters by erosion.

Nutrient budget estimates of agricultural operations have indicated that in the humid East and parts of the West, nutrients added to cropland are generally taken off in crops. Nitrogen applied in excess of the crop requirements can end up in surface waters or groundwaters or be lost to the atmosphere by denitrification. In the prairie and plains area, residual fertility remains high and more nutrients are removed by the crops than are added as fertilizer. These high-fertility soils can contribute nitrates to surface waters and groundwaters even in the absence of added fertilizers. In western states where irrigation is practiced, salts, including nitrates, tend to build up in the groundwater. Finally in the semiarid and western states where intensive agriculture is not practiced, nutrients added to land may be about equal to the amount in the crop or brush (2).

Fish Kills

Actual pollution caused by agriculture is difficult to document. The effect of agricultural contaminants can be observed in the number of fish kills that have

TABLE 3.1

Source of Fish Kills in the United States (1961–1975)[a]

Source	Fish kill reports attributable to noted source	Fish killed (%)
Agricultural activities		
Insecticides	813	4.4
Fertilizers	85	0.1
Manure–silage drainage	424	1.5
Food products industry	362	9.2
All industries	2338	21
Municipal	1407	42

[a]Data from (4).

occurred in the United States (Table 3.1). Comparable data from other countries are less available. For the period 1962–1967, a total of 1218 fish kills was reported in fishing waters of Switzerland (3). Manure liquids caused the greatest number of fish kills (30.3%). The readily identified agricultural sources, manure liquids, silage juice, insecticides, and herbicides, caused 39.4% of the fish kills.

The data in Table 3.1 (4) represent an assessment of the source of pollution by the investigating organizations and as such are based on oral investigations and studies conducted after fish kills were noted. In the official reports, the term "agriculture" was used to describe the cause of fish kills that resulted from farming operations, i.e., use of insecticides and herbicides, fertilizers, and manure–silage drainage. Problems due to food production were included in the industrial waste category. In keeping with the terminology used in this book agricultural sources noted in Table 3.1 are more broadly defined to include both farming and food processing sources.

Although agriculture can be a significant contributor to surface water pollution problems, municipal and industrial sources continue to be the major sources of such pollution (Table 3.1). In the 1961–1975 period, low dissolved oxygen was the leading known specific cause of fish kills with pesticides, petroleum products, inorganic chemicals, acids, organic chemicals, and fertilizers, in that order of importance, as the other known causes. The majority of the agriculturally related fish kills have been in the midwestern and southern states.

Groundwater

Agricultural practices have the capacity to produce wastes that can enter the groundwater and alter its quality. The magnitude of agricultural contamination of groundwater is difficult to ascertain. There is limited information on the amount of agricultural contaminants that enters subsurface waters under given cropping and disposal practices and on the changes in groundwater quality caused by various agricultural contaminants. In addition, there is a scarcity of adequate groundwater monitoring programs to investigate these factors.

Comparative studies in Colorado and Wisconsin have indicated that land use affected the nitrogen content of soils (Table 3.2). The highest soil nitrate nitrogen occurred under feedlots. The nitrogen content in virgin soil and under native grassland represents natural decomposition of organic matter as well as natural deposits in the soil rather than man's activities. The lower quantities below alfalfa are because of the ability of the crop to remove nitrates deep in the soil and the fact that this crop rarely is fertilized with nitrogen. Data of this type need not reflect groundwater nitrate concentrations since that concentration is a function of the water moving through the soil and the hydraulic characteristics of the aquifer. The nitrate nitrogen concentrations in the groundwaters under the lands studied in Colorado (Table 3.2) ranged from 0 to more than 10 mg per liter.

Water from shallow wells is more likely to contain contaminants than water from deeper wells, particularly shallow wells near barnyards, feeding lots, and manure piles. Occasionally, deep drilled wells contain considerable nitrate. The nitrate enters the well either by surface leakage through poor well seals or from nitrogen-rich deposits within lower soil zones.

At present, groundwater contamination from agricultural sources is, for the

TABLE 3.2

Soil Nitrogen Concentration in Agricultural Land

Land use	Average nitrate nitrogen (lb/acre)[a]
Irrigated land, alfalfa	80
Native grassland	90
Nonirrigated cropland	260
Irrigated cropland not alfalfa	506
Feedlots	1436

Land use	Nitrate nitrogen (lb/acre)[b,c]	Ammonium nitrogen (lb/acre)[b,c]
Virgin soil–Jack Pine	56	21
Cultivated	142	31
Barnyard	407	2200

[a]From (5).
[b]20-ft cores.
[c]From (6).

most part, below levels that have been demonstrated to cause disease, concern, or aesthetic nuisance. In establishing enlarged agricultural operations and in choosing sites for the land disposal of wastes, the possibility of groundwater contamination should receive thorough investigation.

Sediment

Nationally, over 6.4 billion tons of soil per year are lost in water and wind erosion. The majority of this amount (83%) is from agricultural and forest land, and more than half of the amount is from cropland (7).

The rates of erosion and resultant sediment have been accelerated by human use and management of lands, vegetation, and streams. The rates of erosion relate to how the land is being used and the characteristics of the soil. Land covered with permanent natural vegetation has lower erosion rates than does land that is intensively used for agriculture. On cropland, the erosion can range from less than 3 to more than 50 tons/acre/year. Over half of the cropland erosion occurs in the midwestern and northern plains states. Areas that have easily erodible soil and steep slopes can have very high erosion rates.

Sediment can reduce the recreational value of water; is a carrier of plant nutrients, crop chemicals, and plant and animal bacteria; and increases water treatment costs. Proper erosion control is a positive solution to sediment problems.

Soil erosion and sediment yield are a function of rainfall, soil properties, land slope length, steepness, and cropping practices. While little can be done to change the amount, distribution, and intensity of rainfall, measures can be taken to reduce the ability of rainfall to cause erosion. Examples include decreasing the amount and velocity of overland flow and decreasing the impact of the rain on the soil. Both flow and impact can be reduced by maintaining vegetative cover on the land through proper crops or mulch. The incorporation of crop residue and animal wastes in the soil increases soil porosity and aggregation and decreases runoff. Slowing runoff by vegetative cover and land modification (terraces, ponds, waterways) offers opportunities to provide greater infiltration time, conserve moisture, and decrease erosion. Wise land use planning and careful use and management of land can reduce erosion and control certain water quality problems arising from agricultural activities. Practices and structures to conserve soil and water are identical to those producing a reduction of pollutants found in agricultural runoff. Approaches to control erosion and nonpoint source pollutants are discussed in Chapter 14.

Land Use

Each watershed will have a unique combination of land uses, which will contribute to the flow and constituents in a stream (Fig. 3.3). Precipitation, as well as being an important factor in the quantity of runoff that occurs, will also contribute to the constituents. The contribution of each land use will vary.

In the following sections, the characteristics of several point and nonpoint sources are reported in concentration units (milligrams per liter) and unit area load values (kilograms per hectare per year). The characteristics represent the reported data of many investigators and should not be utilized or extended beyond the conditions of the original study. Each situation will have its own runoff or leachate characteristics as well as potential control approaches.

The unit area load information should be interpreted as potential yield data. It would be incorrect to multiply these yield data by the area of a particular land use category and infer that the result is the actual amount of material entering a specific surface water body. The actual amount that enters a specific water body is a function of interrelated hydrologic, geochemical, agricultural, and human activities on the land.

The concept of unit loads is very useful in attempting to estimate pollutant or nutrient contributions from various sources and to organize and compare data in a watershed or basin. Unit area loads express the pollutant or nutrient generation per unit area and per unit time for each type of input. For land-based inputs, such as nonpoint sources, unit load values usually are expressed as kilograms per hectare per year. For point source inputs such as from a city, unit load values will be related to the source of the parameter, such as kilograms per capita per year.

Although unit load values have considerable utility in estimating and comparing nutrient inputs, such information should be used with an understanding of the factors that affect the values and the variation that can occur. Unit load values

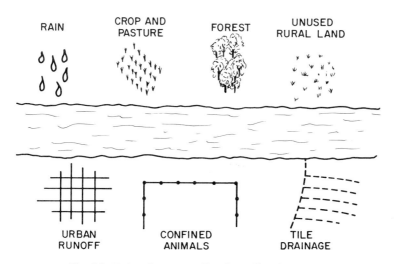

Fig. 3.3. Various inputs can affect the quality of a stream.

are influenced by factors that contribute to the movement of the pollutants or nutrients from their source to surface waters. For nonpoint sources, these factors include climate, precipitation patterns and intensity, land use, soil characteristics, land cover, and transport in tributaries.

The unit load concept implies that information exists on land use and land areas devoted to specific processes and that the type of conditions can be extrapolated from one location to another. Unit load values are most applicable to long-term estimates of nutrient inputs, such as average annual loadings. They do not adequately estimate event or other short-term inputs.

As noted in subsequent sections, unit area loads can have a broad range, sometimes several orders of magnitude. However, in spite of the factors that affect the values and broad range of the values, the unit load concept can be of value for estimating the relative contribution from various sources and for comparing data obtained from a specific watershed. When more precise estimates are needed, it is best to obtain data on actual nutrient inputs for specific subareas in the watershed.

Forest Runoff. Forested areas represent areas that have not been grossly contaminated by man's activities. The runoff from these areas can indicate characteristics that result from natural conditions. However, the increased world demand for wood fiber has accelerated cultural production practices in managed forested areas. Practices such as forest fertilization, block cutting of mature trees, and other forest management practices will alter the characteristics of runoff from forested areas.

Nutrient problems caused by forest fertilization do not appear to be a problem under current conditions and practices. Illustrative nitrogen and phosphorus unit loads are presented in Table 3.3.

Rangeland. Where rainfall is sparse or the land not very fertile, intensive crop production is not practiced and few animals can be supported per acre of land. Some domestic animals are permitted to range over the land in search of available food. Rangeland, with its low level of animals and fertilized acreage,

TABLE 3.3

Nitrogen and Phosphorus Unit Loads from Forest Land (kg/ha/year)[a]

Location	Total nitrogen	Total phosphorus	Reference
Canada	—	0.007–0.183	(8)
Sweden	1.3	0.06	(9)
United States	1.4–6.3	0.04–0.86	(10)

[a] 1.0 lb per acre = 1.12 kg per ha.

represents a natural situation. Runoff from these lands can represent background or "natural" contamination.

Runoff from rangeland that had low intensity agriculture and no evidence of chemical fertilizers contained 0.65 kg of NO_3-N, 0.76 kg of total P, and 0.024 kg of soluble P per hectare per year (11). The characteristics of the runoff were due to the erosion and leaching of the soil in the area.

Rural Land and Cropland. Rural land refers to land that is or could be under agricultural production. Examples include grassland and idle farmland. Cropland refers to land under cultivation.

Constituents of the runoff from rural land and cropland originate in rainfall, wastes from wildlife, leaf and plant residue decay, applied nutrients, nutrients and organic matter initially in the soil, and wastes from animals on pasture. It is difficult to separate the natural from other contaminant sources. To do so requires that water volumes be accurately measured, storm characteristics described, and representative sampling initiated. Constituents are released at varying rates from all soils and exposed geological formations. The natural weathering of rocks and minerals and the oxidation and leaching of organic matter will contribute to runoff and leachate characteristics, even in the absence of man's activities.

The time distribution of nitrates in a number of rivers suggests that the nitrate concentration and load are the greatest when the flow is high, i.e., when rainfall has the greatest opportunity to leach the soluble nitrates from the soil. A seasonal variation in nitrate concentrations was found, with the highest levels in surface waters in spring and late summer. At many sampling stations, the nitrate concentrations closely paralleled stream flow (12). The relatively direct correlation between the two parameters indicated that the nitrate reached the stream seasonally by way of surface runoff, drainage, or interflow and originated from diffuse sources.

Data from a small agricultural watershed in Kansas indicated a relationship between total phosphate and nitrate concentrations and stream flow (13). Total phosphate concentrations were from 0.2 to 0.4 mg per liter during low flows but increased to maximums of 2.4–4.0 mg per liter during storm runoff. The phosphate load followed the turbidity pattern. The nitrate concentrations ranged from 0.2 to 4.2 mg per liter as nitrate in low flows but increased to maximums of 40–45 mg per liter during storm runoff.

Many factors affect the release of nitrate from the soil organic matter and the existing nitrogen pool in the soil. Conditions that favor decomposition of organic matter and oxidation of nitrogen are neutral-to-slightly-basic conditions, warm temperatures, and aeration. When drainage losses increase so does the rate of loss of soluble constituents, such as nitrate, from the soil. A study of nitrate concentrations in English rivers (14) was unable to correlate nitrogen fertilizer usage in adjacent areas with the nitrate concentrations. A high proportion of the total nitrogen was carried by the rivers in a few periods of high flow.

Phosphorus unit area loads can be related to the type of surface soil (Table 3.4)(15). Due to the fine particle size of clay soils, higher soil loss and therefore loss of total phosphorus will occur. The data in Table 3.4 resulted from large-scale studies. Unit loads from small plots (<1 ha) are usually higher than those from large-scale studies.

Rural noncropland such as pastureland, grassland, and idle land have a more stable ground cover and generally contribute less nutrients than do croplands. Ranges of unit loads of nitrogen and phosphorus from cropland and rural noncropland are identified in Table 3.5.

Fertilized Lands.   One of the requirements for satisfactory crop production is the availability of nutrients when the crop is growing. Intensive farming operations are designed to satisfy this requirement either by the addition of fertilizers or by the incorporation of readily decomposable organic material such as manures in the soil.

Crop fertilizers have been blamed as major contributors of nutrients in surface waters. Generally only 10–30% of the fertilizer phosphorus added to a soil is taken up by the following crop. The remaining applied phosphorus is converted to insoluble forms and may become a source of available phosphorus for crops in subsequent years. Rarely does phosphorus from commercial fertilizers or spread manures occur in groundwater except with very porous soils in areas with heavy rainfall or irrigation. Phosphorus added to soils as a fertilizer or released in the decomposition of organic matter will be rapidly converted to iron and aluminum phosphates in acid soils and to calcium phosphates in alkaline

TABLE 3.4

Total Phosphorus Unit Loads from Rural Land (kg/ha/year)[a]

Land use and intensity		Surface soil texture			
	Sand	Coarse loam	Medium loam	Fine loam	Clay
Rural noncropland					
Pasture/range—dairy	0.05	0.05	0.10	0.40	0.60
Grassland/idle land	0.05	0.05	0.10	0.15	0.25
Rural cropland					
Cultivated fields—row crop (low animal density)	0.25	0.65	0.85	1.05	1.25[b]
Cultivated fields—mixed farming (medium animal density)	0.10	0.20	0.30	0.55	0.85

[a]Reprinted with permission from *Environmental Science and Technology* **14,** 148–153. Copyright 1980 American Chemical Society.

[b]Unit area loads can be higher for soils with very high clay content.

TABLE 3.5

Nitrogen and Phosphorus Unit Loads from Rural Lands (kg/ha/year)

Location	Total nitrogen	Total phosphorus	Reference
Canada	4.3–31	0.15–1.7	(16)
Sweden	16	0.2	(9)
United States	0.1–13	0.06–2.9	(10)

soils. With proper erosion control measures, little of the applied phosphorus should reach surface waters.

Efforts have been made to find a direct relationship between fertilizer nitrogen applied and the concentration of nitrogen in stream and drainage waters. A direct relationship is not likely, because of a large preexisting reservoir of nitrogen in the soil and the rate of biological transformation the reservoir can undergo. In the upper Rio Grande River, the nitrate concentration in the river had not increased, although fertilizer use increased considerably over a 30-year period (17). Crop usage and denitrification were probable explanations for the disappearance of the applied nitrogen. In other areas where organic and inorganic nitrogen has accumulated for centuries, the nitrate concentrations of the drainage waters are more a result of natural conditions than of fertilizer applications.

A study was made of the extent to which the increase in fertilizer usage in the Netherlands had affected the quality of groundwater over the years. It was not possible to determine whether observed increases in chlorides and nitrates were caused by the increased use of fertilizers or by the increase in well water demand. In the period from 1920 to 1966, the average chloride and nitrate concentration of unpurified groundwater in fertilized sandy soils increased by about 3 and 2.5 mg per liter, respectively. Over two-thirds of wells had no nitrates in their water during this period (18). During the same period, nitrogen fertilizer usage increased from 28 to 178 kg/ha/year. Denitrification was suggested as a principal reason for the low nitrate content of the groundwater. The change in nitrate and chloride content of the deeper groundwater in the Netherlands during a period of increased fertilizer usage was not large enough to reduce the quality of potable water prepared from the groundwater.

Conditions required to leach nitrates from soil depend upon the water percolating through the soil and the presence of excessive nitrate in the soil when percolation occurs. Situations conducive to nitrate leaching include conditions in which the following occur: infiltration exceeds water lost by evaporation and transpiration, soils have low water-holding capacities, irrigation is practiced, and soils have high infiltration rates.

It is evident that factors such as natural nitrogen content of the soil, soil type and topography, crop type and rate of fertilizer application, and climatic conditions complicate the relationship between fertilizer use and water quality changes. It is inevitable that there is a risk of nitrate loss during runoff and leaching. It is difficult to conceive of any approach that will reduce that risk to zero. It is possible to decrease the risk by proper application of inorganic fertilizers in periods when the crops are growing, coupled with the use of slow-release organic fertilizers.

Cropland Subsurface Drainage.   The soils of many farms have poor drainage characteristics. Subsurface tile drains can be installed to permit better water movement and crop yields. The drainage from the lands will contain soluble constituents from the soil and materials added to the soils. The drainage can enter surface streams at many places and essentially is a nonpoint source of contaminants for the streams.

Pastureland and Manured Lands.   These lands either contain animals for production or are used for disposal of animal wastes. In either case, the runoff of these lands can contain contaminants from animal wastes. Where animals have direct access to streams, animal urine and feces can be discharged directly to these waters.

Studies on the relative pollutional effects of manure disposal and pastured animals (19) indicated that the extent of water pollution from farm animal production units is more dependent on waste management methods than on the volume of the waste involved. To minimize water pollution from pastured animals, points of animal concentration should be located away from streams and hillsides leading directly to streams, and vegetation should be provided between areas of animal concentration and drainage paths or surface waters to intercept contaminants.

The access of animals and animal wastes to surface waters represents a possible controllable nonpoint pollution source. Animal manures should not be disposed of where rainfall or snow melt will result in their direct discharge to watercourses. The manure should be integrated with cropland or disposal land shortly after disposal.

When dairy cattle manure was surface-applied to coastal bermudagrass on a sandy loam at a rate of 45 metric tons/ha/year for 4 years, the BOD and nitrate nitrogen of the surface runoff did increase (20). However, the mean BOD and nitrate nitrogen concentrations in the runoff were less than 10 mg per liter and 8 mg per liter, respectively. Based upon other data, the unit loads from land that had received manure were estimated to be 4–13 kg of total nitrogen per hectare per year and 0.8–2.9 kg of total phosphorus per hectare per year (10).

Manure Storage.   When manure cannot be disposed on land routinely, such as in the winter, it may be stacked or stored until conditions permit land disposal and integration with crop production. During storage, seepage can result, which can contaminate surface waters. Table 3.6 (21) indicates the magni-

TABLE 3.6

Characteristics of Seepage from Stacked Dairy Cattle Manure and Bedding[a]

	Winter		Summer	
Parameter	Average	Range	Average	Range
Total solids (%)	2.8	1.8–4.3	2.3	1.7–2.9
Volatile solids (% TS)	55	52–59	53	50–58
Suspended solids (%)	0.35	0.2–0.8	0.24	0.2–0.3
BOD (mg/liter)	13,800	4200–31,000	10,300	4400–21,700
COD (mg/liter)	31,500	21,000–41,000	25,900	16,400–33,300
Total N (mg/liter as N)	2350	1500–2900	1800	1200–2770
NH_4-N (mg/liter)	1600	980–1980	1330	780–2200
Total P (mg/liter as P)	280	64–560	190	90–340
Potassium (mg/liter as K)	4700	3000–7200	3900	3000–4900
Total precipitation (in.)	15.0		9.4	
Seepage volume (gal/cow/day)	3.0		1.2	

[a]From (21).

tude of contaminants from stacked dairy manure. Although the volume of seepage is small, the quantity of contaminants is not insignificant.

The release of stored manure seepage to surface waters should not be permitted. The seepage can be controlled by retention ponds and distribution on cropland in a nonpolluting manner. Even though manure seepage can occur in a large number of locations throughout the country, and as such approximates a nonpoint source of pollution, it can be considered as a controllable source of pollution.

Animal Feedlots.   In many areas of the United States, animals remain an integral part of cropping systems, and their wastes are returned to the fields that produce feed for the animals. In the Midwest and West, large feedlots now produce animals for slaughter, and the manure tends to accumulate in the feeding areas. Until these wastes enter groundwaters or surface waters, they do not represent a serious water pollution problem. The characteristics of the livestock wastes are such that they will stay within a confinement area until the area is cleaned or until runoff washes them away. The water pollution problem associated with livestock wastes is a drainage problem. Runoff from uncovered confinement areas and from land used for disposal of waste occurs during and following rainfall.

Surface drainage from cattle feedlots and other areas where confined animals are housed in the open has caused considerable concern. In the early development of open confined animal operations, feedlots were situated in locations where the natural drainage facilitated the removal of the wastes. The water pollution potential was given little if any consideration. This situation has changed.

The quantity and quality of feedlot runoff will depend upon previous weather conditions, the number of cattle per feedlot area, the method of feedlot operation, soil characteristics, topography of the area, and intensity of rainfall. Studies that have included natural and simulated rainfall events have provided data on the characteristics of feedlot runoff under a variety of environmental conditions.

An early investigation illustrated the water quality changes that can occur when feedlot runoff reaches surface waters (22). Considerable lengths of streams were devoid of oxygen due to the runoff. BOD and ammonia concentrations were as high as 90 and 12 mg per liter, respectively (Table 3.7). Ammonia concentrations from the runoff were detectable before other parameters. Such runoff provides little warning to downstream users and can trap game fish in the polluted waters.

Detailed studies on the environmental factors affecting the quantity and quality of feedlot runoff have been conducted. The greatest pollutant concentrations occurred during warm weather, during periods of low rainfall intensity, and when the manure had been wet.

Part of the organic matter in animal manure is stabilized by bacterial action after it is deposited in the feedlot. Some nitrogen contained in the manure and urine can be lost to the atmosphere. About 50% of the organic matter can be decomposed into carbon dioxide and water. The amount of decomposition that takes place is dependent upon temperature and moisture conditions. The greater the bacterial action and moisture content of the wastes, the greater the degree of solubilization and the greater the amount of soluble constituents in subsequent

TABLE 3.7

Water Quality Changes Caused by Feedlot Runoff, Fox Creek, Kansas (mg/liter)[a]

Time after rainfall (hr)	DO[b]	BOD$_5$	COD	Cl	NH$_3$
13	7.2	8	37	19	12.0
20	0.8	90	283	50	5.3
26	5.9	22	63	35	—
46	6.8	5	40	31	0.44
69	4.2	7	43	26	0.02
117	6.2	3	22	25	0.08
Dry weather average	8.4	2	29	11	0.06

[a]From (22).
[b]DO, dissolved oxygen.

runoff. Because of a decrease in bacterial action in the winter, there will be a greater accumulation of wastes in an open feedlot during this period than during the rest of the year.

The loss of carbon and nitrogen from the waste by bacterial action increases the mineral content, reduces the organic content, and results in an accumulation of undegradable material. This residual material is not readily amenable to treatment by biological treatment methods and will result in a high COD/BOD ratio. While the manure is decomposing on the lot, it is constantly being mixed with fresh wastes and soil if the feedlot is unsurfaced. The characteristics of the actual waste in a feedlot will be a mixture of these physical and biochemical changes.

The highest concentration of pollutants in the feedlot runoff can occur in the initial runoff and decreases to "equilibrium" conditions as the runoff continues. Figure 3.4 (23) illustrates the pattern for a number of rainfall rates and antecedent feedlot conditions. The "equilibrium" conditions were caused by the runoff continually dissolving the surface layer of manure on the lot. Once the ground surface was covered with manure, the depth of manure was not an important parameter of runoff water quality.

The characteristics of runoff from other beef cattle feedlots (Fig. 3.5) (24) showed even larger contaminant concentrations. Data from additional studies corroborate the strength of feedlot runoff. The ranges of the reported data are summarized in Tables 3.8 and 3.9. Intensive livestock operations, such as beef cattle and dairy feedlots, contribute soluble forms of pollutants. The impact of these sources is site dependent. On a unit area basis, nutrient losses may be high and therefore have a large impact on a local stream. However, the total land area used for these sources on a national basis is small, and the overall pollutant load, compared to other sources, also is small.

In addition to the beef cattle feedlots, there are a large number of cattle herds of less than 100 head of beef cattle. These herds are confined in barn lots in the winter and pastured in the summer. The barn lots have intermixed mud and manure as a surface in the wet, cool seasons and a hard-packed surface in the warm, dry periods. The runoff from barn lots also can be a source of surface water contamination. The runoff from a small barn lot housing 60 beef steers had the following range of characteristics: total solids, 0.1–1.2% on a wet basis; volatile solids, 17–38% of total solids; COD, 350–5650 mg per liter; and BOD, 9–560 mg per liter (25).

Runoff from feedlots can have a detrimental effect on surface waters. Runoff collection ponds can reduce the sediment load in such runoff and can produce an effluent with less variable characteristics. The quality of the effluent from runoff retention ponds is inadequate for release to streams or reservoirs and the collected runoff should be irrigated on adjacent cropland.

The concentrations of pollutants in feedlot runoff are relatively independent of the type of ration fed the animals. In the semiarid regions of the southwestern

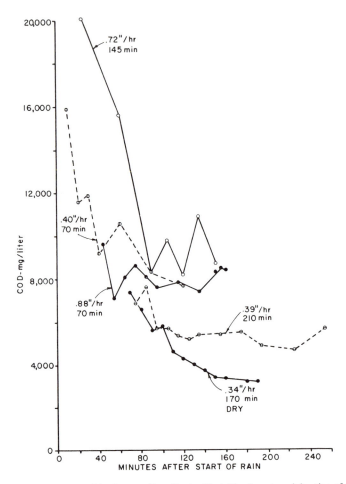

Fig. 3.4. The variation of feedlot runoff quality is affected by the rate and duration of precipitation. Except as noted, the waste on the feedlot was slightly moist before the rainfall event. The first sample was collected as runoff began. [Prepared from data from (23).]

United States, cattle wastes are dehydrated in a short time and remain that way until wetted by precipitation. The wetting may reconstitute the wastes to almost original composition. In more humid climates, the wastes may remain moist for longer periods of time before natural drying occurs. The longer the wastes remain moist, the greater the opportunity for bacterial action and solubilization of the solid matter.

Pollution from an uncovered livestock area, such as a feedlot, is related to

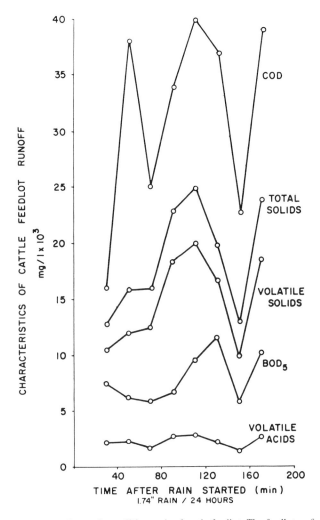

Fig. 3.5. Characteristics of natural runoff from a beef cattle feedlot. The feedlot surface was moist before the rain (24).

the fraction of precipitation that becomes runoff and reaches surface streams. Only after a portion of the rainfall soaks into the manure does runoff occur. This fraction will depend upon previous weather and precipitation conditions. The relationship between precipitation and runoff at feedlots can be used to estimate the quantity of feedlot runoff that will occur for a given storm and therefore to design control systems for the runoff.

TABLE 3.8

Range of Reported Feedlot Runoff Characteristics

Parameter	Concentration (mg/liter)	Unit load (kg/ha/year)
Total solids	3100–28,800	640–60,000
COD	1300–77,000	720–16,000
BOD	300–12,700	130–3800
TKN	9–1070	45–3110
NH_4-N	2–2020	—
Total N	11–8600	100–1600
Total P	4–5200	10–620

The general relationship relating precipitation (inches) and runoff (inches) from cattle feedlot surfaces is

$$\text{Runoff} = A \times \text{rainfall} - B \qquad (3.1)$$

Results obtained from small surfaced and unsurfaced cattle feedlots are shown in Fig. 3.6. The data represent both natural and simulated rainfall events occurring over separate 2- to 3-hr periods from June to November in Kansas.

Other precipitation–runoff data at cattle feedlots have been analyzed in separate studies. In general, runoff was shown to occur only after precipitation

TABLE 3.9

Major Ions in Cattle Feedlot Runoff

Location	Ion	Mean (mg/liter)	Range (mg/liter)
Nebraska[a]	Na	840	40–2750
	K	2520	50–8250
	Ca	790	75–3460
	Mg	490	30–2350
	Zn	110	1–415
	Cu	7.6	0.6–28
	Fe	765	24–4170
	Mn	27	0.5–146
Texas[b]	Na	408	130–655
	K	760	226–1350
	Ca	700	194–1620
	Mg	69	28–89

[a]From (26).
[b]From (27).

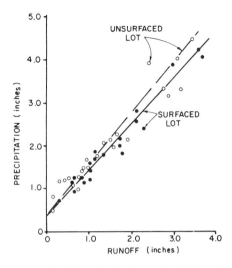

Fig. 3.6. Precipitation–runoff relationships for beef cattle feedlots. [Developed from data from (23).]

exceeded 0.2–0.5 in. The proportion of the precipitation that ends up as runoff will vary at different locations but generally is greater than 50% and has ranged to 90%.

The rainfall–runoff relationships at beef feedlots indicate that, after a minimum amount of rainfall, much of the rainfall ends up running off the feedlot. The surface of the feedlot acts much the same as a paved surface in this regard. A dry, hard crust frequently is observed below the manure surface. The silt, clay, and manure mixtures of feedlots result in the clogging of the porous soil voids and in the development of a less permeable layer.

These conditions suggest that infiltration to groundwaters can be neglected in determining the total runoff from operating feedlots. When feedlots are abandoned, the accumulated wastes may remain, the hard crust breaks up, and greater downward motion of pollutants may occur.

Runoff and percolation are not the only ways that feedlot contaminants can affect water quality. Beef cattle feedlots can contribute ammonia, amines, and odorous sulfur compounds to the atmosphere (28, 29). Such compounds are potential pollutants to waters in the area and can have an effect on nearby plants and animals. Aliphatic amines were a significant fraction of the volatilized nitrogen. Similar results can be expected from other accumulation of organic wastes exposed to the atmosphere.

The water pollution potential of livestock feedlots is related to the waste production per animal, the number of animals in the confinement unit, days confined, frequency of cleaning, climate, waste characteristics, and waste degradation in the lots. The contribution of feedlot runoff to surface water pollution will be a function of the temperature, magnitude of rainfall, slope of the confinement area, surface area of the feedlot, type of lot surface, and management practices. Range-fattened cattle represent a smaller runoff pollution problem than feedlot cattle since they are more widely distributed on the land. As the density of animals per acre decreases, the wastes are less concentrated and the nature can absorb more of the wastes. The trend, however, is toward confinement livestock feeding.

Irrigation Return Flows.   Irrigation return flow is any water that finds its way back to a water supply source after it has been diverted for irrigation purposes. The term includes (a) tailwater runoff, the portion of the applied water that runs off the land surface; (b) seepage, water seeping from the distribution system; (c) deep percolation, applied water that contributes to groundwater recharge or to a subsurface drainage system; and (d) bypass water, water that is diverted but returned directly to the source of supply without being used for irrigation. The soluble salt concentration in the irrigation return flows is the major contaminant of concern. The environmental irrigation return flow problems primarily are located in the arid areas. However, both the salt and nutrient content of irrigation return flows can contribute to the contaminant load of water sources wherever irrigation is practiced.

An increase in soluble salts is an unavoidable result of the use of water for irrigation. As water is transpired by the crop and evaporated from the soil, the salts in the applied water are concentrated. Most of the salts in the applied irrigation water remain soluble in the soil solution. The amount of mineral uptake by crops rarely is significant in the overall salt balance of irrigated agriculture. Almost the entire amount of salts present in the applied irrigation water is contained in the soil solution. A favorable salt balance is obtained when the output of salts in the drainage water equals or exceeds the input to the soil. To maintain a favorable salt balance in the root zone of the irrigated crops, i.e., to prevent an accumulation of soluble salts in the soil, a more saline water must leave the land by percolation below the root zone to the groundwater or to a natural or artificial drainage system. The quantity of water applied for irrigation includes the amount required by the crop and a sufficient amount to leach the soluble salts from the soil.

The characteristics of surface and subsurface irrigation return flows will be different. Surface runoff from irrigated land generally will have the following characteristics: total dissolved salts only slightly greater than in the applied water; presence of precipitated or adsorbed contaminants will be a function of the amount of erosion; colloidal and sediment load will be greater than in subsurface flow; and characteristics will be variable for any given area.

The general characteristics of subsurface return flows will include a greater concentration of dissolved solids than in the applied water; little colloidal or particulate matter; a different distribution of cations and anions than those in the applied water; a nitrate content that will depend upon the concentration in the applied water, the nitrogen content of the soil, and the nitrogen uptake by the crop; and low phosphorus concentrations.

Fruit and Vegetable Processing.   The food processing industry creates considerable quantities of liquid and solid wastes for disposal. Food processing wastes not disposed of as solid waste are disposed of as part of the plant wastewater.

Food processing wastes are organic and can be treated by processes generally used at municipal treatment plants. If, however, these plants are not designed to treat the variable and high-strength food processing wastes, the municipal plants may be overloaded and unable to attain the efficiency for which they were designed. Excessive food processing wastes in municipal plants can tax the solids handling and disposal facilities of a municipal waste treatment plant as well as the oxygen capacity of aerobic units.

The seasonal nature of the food processing industry compounds the problem of designing treatment facilities for the wastes or joint treatment of the wastes in municipal treatment plants. The preponderance of processing activity takes place during the summer and fall months (Fig. 3.7)(30).

Dairy Product Processing.   The dairy food industry represents an important part of the food industry and contributes significant liquid wastes for treatment and disposal. There has been a decrease in the number of plants processing dairy foods, with increased waste loads per plant.

Wastes from dairy plants consist primarily of varying quantities of milk solids, sanitizers, detergents, and human wastes. The wastes produced in urban plants are sent to municipal waste treatment facilities with a minimum of prior treatment. Dairy plant wastes produced in nonurban areas are treated in aerated treatment facilities, by irrigation on the land, or by uncontrolled disposal to land and surface waters.

Approximately 32 billion gallons of wash water, 3.5 billion gallons of cheese whey, and 66 billion gallons of condenser water are discharged from the nation's dairy plants each year. Whey wastes contain about 4% solids and have a high oxygen demand (30,000–45,000 mg per liter of BOD), which makes disposal to streams undesirable. Only about one-third of the whey produced each year is used to any extent. The remainder is discharged from the plants. The nutritive content of whey offers interesting possibilities for by-product recovery and utilization if suitable equipment and markets are available.

Nonagricultural Sources

Agriculture is not the only source of contaminants in a watershed. Effluent from sewage treatment systems, septic tank drainage, urban land runoff, pre-

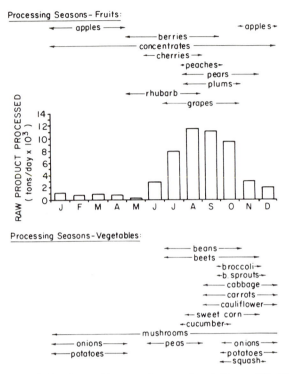

Fig. 3.7. Seasonal fruit and vegetable processing operations in Oregon (30). [Reproduced from *Journal of Environmental Quality* **1**(1), 82 (1972), by permission of the American Society of Agronomy, Crop Science Society of America, Soil Science Society of America.]

cipitation, and the effluent from industries also contribute to the contaminants in the watershed. This section will attempt to present the characteristics of other nonpoint sources so that agricultural nonpoint sources can be placed in perspective.

Precipitation. The constituents of precipitation are influenced by both man-made and natural events. Fuel burning, automobiles, manufacturing operations, forest and other fires, volcanic eruptions, and wind erosion are examples of activities that contribute to the constituents in dry, dust fallout and in rain or snow that are deposited on the sample collecting surfaces. The distribution of contaminants between dry fallout and rainfall is rarely indicated by various investigations.

The actual sources of the nutrients found in atmospheric deposition are difficult to trace. Large atmospheric inputs appear to be associated with urban areas, possibly as a result of combustion emissions. Wind erosion of soils is likely a major source of phosphorus in agricultural areas. Ammonia volatilization from feedlots and manure storage areas adds to the nitrogen content of the atmosphere. Atmospheric deposition from forested regions contains fewer nutrients than from agricultural or urban regions. However, atmospheric nutrients can originate well outside an individual watershed. A summary of the total nitrogen and total phosphorus unit loads that result from atmospheric deposition is presented in Table 3.10.

A comparison of the range of data indicates that the level of human activity in an area affects the atmospheric unit loads. This also relates to agricultural activities. Table 3.11 indicates that the atmospheric unit loads were higher in areas adjacent to barnyards than in areas removed from the barnyards.

Precipitation is a variable and intermittent source of constituents in surface waters. Once contaminants are in precipitation, they are uncontrollable. Man, however, can exert some control over the contaminants that are released to the atmosphere through management practices that minimize particulates and gases from combustion processes, particulates resulting from disturbing the land, and volatile gases from industrial operations.

Urban Land Drainage. Street litter, gas combustion products, ice control chemicals, rubber and metals lost from vehicles, decaying vegetation, domestic pet wastes, fallout of industrial and residential combustion products, and chemicals applied to lawns and parks can be sources of contaminants in urban

TABLE 3.10

Atmospheric Deposition of Total Nitrogen and Total Phosphorus as a Function of Land Use (kg/ha/year)

Land use	Total nitrogen	Total phosphorus
Estimated inputs to lake surfaces[a]		
Forested regions	0.99–11.3	0.07–0.54
Rural–agricultural	10.5–38.0	0.12–0.97
Urban–industrial	4.7–24.8	0.26–3.67
Dominant land use in area of sampling stations[b]		
Coastal	5.8	0.31
Urban	7.2	0.48
Rural (nonagricultural)	6.2	0.27
Rural (agricultural)	8.8	0.66

[a]Adapted from (31).
[b]Florida—adapted from (32).

TABLE 3.11

Atmospheric Nitrogen Unit Loads near Barnyards (kg/ha/year)[a]

Nitrogen compound	Area adjacent to barnyard	Area away from barnyard
Organic N	14.4	7.5
Ammonia N	12.2	2.9
Nitrate N	3.5	2.7
Total N	30.2	13.1

[a]Adapted from (33) *Journal of Environmental Quality* **1,** 203–208 (1972). [By permission of the American Society of Agronomy, Crop Science Society of America, Soil Science Society of America.]

runoff. A portion of the urban runoff can drain to sewerage systems while the remainder may reach surface waters by natural drainage channels without receiving treatment. The contaminant load to streams or sewage treatment plants results in a slug load effect that the receptors may be unable to handle. Besides the conventional water pollution parameters, a number of other contaminants such as fecal coliform bacteria, chlorinated hydrocarbon and organic phosphate compounds, a number of heavy metals, and polychlorinated biphenyls have been found in urban runoff (34).

Unit area loads from urban lands generally are of the same magnitude or greater than the loads from rural croplands. Table 3.12 identifies total phos-

TABLE 3.12

Total Phosphorus in Storm Water from Different Urban Land Uses (kg/ha/year)[a]

Land use	Total phosphorus
Single family—residential	0.77–1.44
Multifamily—residential	1.29–1.61
High rise—residential	1.96–2.18
Commercial	0.1–7.6
Industrial	0.9–4.1
General urban	0.3–2.1

[a]Adapted from (35) Whipple, Grigg, Grizzard, Randall, Shubinski, Tucker, "Stormwater Management in Urbanizing Areas," © 1983, p. 85. [Adapted by permission of Prentice-Hall, Inc., Englewood Cliffs, New Jersey.]

phorus loads for different urban land uses. Data on nitrogen unit loads from urban areas are minimal. Total nitrogen unit loads from residential, commercial, and industrial areas have been estimated to range from 5 to 7.3, 1.9 to 11, and 1.9 to 14 kg/ha/year, respectively (16). The phosphorus and nitrogen unit loads from developing urban land were found to be several times higher than losses from other urban sources.

Comparison of Sources

It is difficult to compare data from various studies because of the variations in sampling methods and analytical methods and the fact that all studies did not measure comparable parameters. The characteristics of nonpoint sources are the result of complex interactions in and on the soil, making definitive comparisons of these nonpoint sources difficult. For a comparison of these sources, the order of magnitude of the characteristics and the differences between sources are more significant than the values themselves. The range of unit loads reported for many parameters and for different sources is summarized in Table 3.13.

The National Eutrophication Survey (36) collected stream samples from 928 watersheds in the United States once a month for 1 year. The samples were analyzed for total nitrogen, inorganic nitrogen, total phosphorus, and orthophosphate. Good correlations were found between the general land use that contrib-

TABLE 3.13

Summary of Nonpoint Source Characteristics[a]

Source	Unit area loads (kg/ha/year)				
	COD	BOD	NO_3-N	Total N	Total P
Precipitation	124	—	1.5–4.1	5.6–10	0.05–0.06
Forested land	—	—	0.7–8.8	3–13	0.03–0.9
Rangeland	—	—	0.7	—	0.08
Agricultural cropland	—	—	—	0.1–13	0.06–2.9
Land receiving manure	—	—	—	4–13	0.8–2.9
Irrigation tile drainage, western United States					
Surface flow	—	—	—	3–27	1.0–4.4
Subsurface drainage	—	—	83	42–186	3–10
Cropland tile drainage	—	—	—	0.3–13	0.01–0.3
Urban land drainage	220–310	30–50	—	7–9	1.1–5.6
Seepage from stacked manure	—	—	—	—	—
Feedlot runoff	7200	1560	—	100–1600	10–620

[a]Data do not reflect the extreme ranges caused by improper waste management or extreme storm conditions.

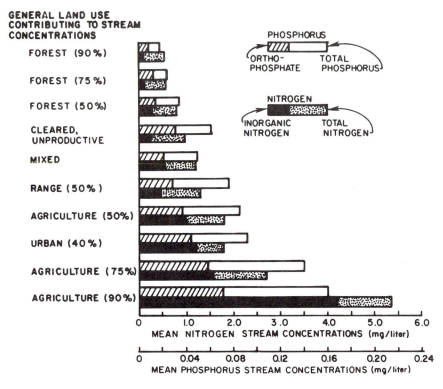

Fig. 3.8. Relationship between general land use and nutrient concentrations in streams [Adapted from (36).]

uted surface runoff to the stream and the mean annual nutrient concentrations in the streams. The general relationship is shown in Fig. 3.8. In the figure, the predominant land use was equal to or greater than the percentage indicated. Nutrient concentrations were lower in streams draining forested watersheds than in those draining agricultural or urban watersheds.

The nitrogen and phosphorus unit loads from various sources have been compared graphically (Fig. 3.9). Although there is considerable variation in the unit loads, the ranges can be compared to provide some perspective on the relative contributions from various sources. The total nitrogen unit loads from precipitation, forestland, cropland, land receiving manure, surface irrigation return flows, and urban land drainage span a comparable range. The total phosphorus unit loads of precipitation and range land are comparable, as are the unit loads of land receiving manure, surface irrigation return flows, and urban land drainage. The total phosphorus unit loads from forest land and cropland span a wide but comparable range, again undoubtedly due to the effect of erosion. The

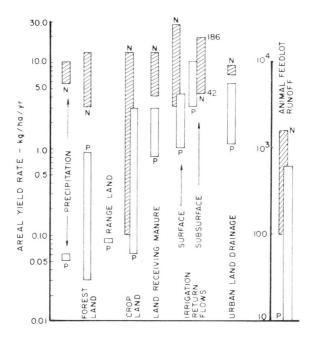

Fig. 3.9. Range of potential unit loads of total nitrogen and total phosphorus from various sources (kilograms per hectare per year) (10).

nitrogen and phosphorus unit load of animal feedlot runoff also is many orders of magnitude greater than that of the other nonpoint sources.

Although these comparisons cover wide ranges and may be considered gross, they do permit an assessment of the nonpoint sources that may require control. Actual decisions on the control of nonpoint sources should be made on the basis of the relative importance of the respective sources in specific locations and on what is technically and logically controllable.

Summary

The review of available data illustrates the difficulty of quantifying the contribution of agricultural operations to characteristics of surface waters. The differences between the types of agricultural situations are apparent. Forest land and rangeland runoff represent baseline, natural conditions that do not need to be controlled. The concentrations and yield of constituents from these sources are comparable to that of precipitation. Precipitation, forest runoff, and rangeland runoff can be considered as background, nonpoint contaminant sources.

Nonpoint agricultural sources that may require control include cropland runoff, land receiving manure, cropland tile drainage, and irrigation return flow. Routine control of cropland runoff and runoff from land receiving manure technically is difficult and may not be needed in all cases. Each situation requires individual evaluation before a decision is made about the need for control and the type of control, if it is needed. In most cases, agricultural nonpoint sources can be controlled through proper management of the land rather than through the application of treatment techniques. When erosion of cropland is controlled, the contaminants in cropland runoff should be comparable to that of background, nonpoint sources. If manure is disposed of on land under good crop and land management conditions, contaminants from these lands should be comparable to that of cropland.

Cropland tile drainage and irrigation return flows are difficult to intercept and control. The nitrogen content of this drainage is the major item of concern.

Nonpoint agricultural sources that should be controlled include manure seepage, feedlot runoff, and the direct discharge of animal wastes to surface waters. Nonsophisticated control procedures are available for the control of these sources and should be included in water pollution control policies (Chapter 14).

Urban land runoff and the effluent from municipal and industrial waste treatment plants can be major sources of contaminants in urban and suburban areas. In rural areas, septic tank leakage, manure runoff, and cropland runoff can be important contributors to contaminants in surface waters. The sources in each watershed need to be evaluated, relative contributions identified, and relative costs estimated, before pollution control policies are implemented. Uniform policies in each watershed may not be cost-effective nor produce the desired pollution abatement.

Bacteria

The list of infectious disease organisms common to man and livestock is lengthy, including a number that can be waterborne. When drainage or runoff from animal production units reaches a watercourse, a potential chain for the spread of disease has been initiated. *Salmonella* organisms have been isolated from animal fecal specimens, runoff from animal confinement operations, carcasses of dead animals, and waterholes from which the animals drank (37–39). Two organisms, *Salmonella dublin* and *S. typhimurium* were the salmonella organisms most commonly found in the cattle and contaminated water investigated. *Salmonella dublin* is essentially a pathogen of cattle but can cause meningitis and septicemia in humans. Children are more susceptible than adults. *Salmonella typhimurium* can infect practically all species of birds, animals, and man. Animal wastes contain other organisms, such as *Staphylococcus aureus,* which also can be human pathogens. Evidence indicates, however, that the animal waste environment or the soil environment is not very hospitable to the survival of these pathogens. The disease potential inherent in the disposal of animal wastes on land is not considered to be critical. Organism survival in soil is determined by the viability of organisms in the soil environment, their interaction with the soil, and the nature of their transport.

A study of the effectiveness of soil to remove bacteria in animal wastes indicated that coliform and enterococcus bacteria were removed by adsorption during soil percolation and by die-off because of the inability to compete against the established soil microflora (40). Greater than 98% removal occurred in 14 in. of soil. It was concluded that there was little concern that the bacteria would move any great distance from the point of application of the manures or manure water on undisturbed soil.

The ratio of fecal coliform to fecal streptococci has been used to delineate between human and animal pollution. The ratio in human and domestic wastes was shown to be over 4, while for fecal material from animals, the ratio was less than 0.6. A ratio of 0.2 has been noted in cattle feedlot runoff (23).

Agricultural operations such as livestock and crop production are natural contributors to bacteria in surface waters. Runoff from cropland can increase the bacteria in surface waters, which must be removed in water treatment plants to produce potable water. The possible presence of pathogens in runoff and effluent from livestock waste treatment systems suggests that caution should be exercised in reusing such water without bacterial control.

Air Quality

In the broad sense, air pollution can be defined as the presence of air contaminants under conditions that are injurious to human, plant, or animal life or property or that unreasonably interfere with the enjoyment of life and property. Agriculture can be a recipient of the air contaminants of municipalities and industries and sustain the associated injuries. It is rare that agriculture is the producer of air contaminants that are injurious to plants, animals, humans, or property. Agriculture can, however, produce odors and dusts that interfere with the enjoyment of life and property and thus can be a source of localized air pollution. An agricultural industry that offends its neighbors with the production of air pollutants can be forced to cease if the nuisances are not corrected.

Odors

The subjective nature of odors is a difficulty in evaluating their offensiveness and effects. Individuals have varying personal response to odors, depending upon age, sex, smoking habits, occupation, mental attitudes, and respiratory differences. Each can have varying thresholds of detection. Regular and prolonged exposure to an odor can raise the detection threshold. However, for any group of people, as the concentration of odor increases, the odor eventually will be detected and become obnoxious or discomforting.

In spite of the sophisticated technical equipment that has been developed, the human odor-sensing system remains the most suitable detector available. Most odors are a complex mixture of chemical compounds. In some odors a specific compound such as ammonia or hydrogen sulfide can be predominant and detected instrumentally to obtain a measure of its strength. Because most odors are caused by small amounts of a number of compounds, a human odor evaluation panel remains one of the best odor detection and evaluation systems. Methods to develop odor evaluation panels have been standardized, and such panels are helpful for comparable studies.

Odors from animal wastes are a persistent problem arising from the confinement of large numbers of animals. The problem is especially prevalent near feedlots, in and around confined and enclosed animal production operations, and where field-spreading of unstabilized waste is practiced.

The conditions under which the odorous compounds are produced offer suggestions as to possible control measures. With current practice, most agricultural wastes are held in piles, storage tanks, or pits prior to disposal. The storage period varies from weeks to years, depending on the flexibility desired by the operator. The environmental conditions during such storage are controlled, and anaerobic products such as ammonia, sulfides, mercaptans, amines, organic acids, and methane will result. Many of the unoxidized compounds have their own unique and, even in low concentrations, objectionable odors. Some of the odorous compounds are tenacious, can cling to clothing and other articles, persist for long periods, and carry great distances.

The fact that the odor-causing materials are reduced organic compounds suggests the possibility of an oxidative microbial process for odor control. If the stored wastes are aerated, adequate oxygen can be added so that anaerobic end products are not produced. Under these conditions the primary odorous compound will be ammonia since during the aeration of animal wastes, ammonia can be released from the mixture. If sufficient oxygen is added to the aerated mixture the ammonia can be microbially oxidized to nitrites and nitrates, eliminating ammonia as an air pollution problem in these circumstances. The key in reducing anaerobically caused odors is to be sure that the oxygen supplied is equal to or greater than the oxygen demand. If the oxygen supply is less than that of the demand, objectionable odors may still result.

The greater part of the odor associated with confined animal housing originates from the manure. Liquid manure handling and disposal have generated many of the odor nuisance conditions that have resulted in complaints. Anaerobic liquid systems, in which the waste is diluted with water for ease of handling, provide a better environment for the production of the odorous compounds than does a "dry" or an adequately aerated system. Odors from diluted, anaerobic "liquid" manure are more offensive than odors from the undiluted, fresh manure.

Compounds identified as part of the odors from anaerobic poultry manure included ammonia, two- to five-carbon organic acids, indole, skatole, diketones, mercaptans, and sulfides. An odor panel indicated that the organic acids, mercaptans, and sulfides were the major malodorous components of the odor.

Because many of the odor-causing compounds contain sulfur, the volatile sulfur compounds resulting from the anaerobic and aerobic decomposition of beef, dairy, sheep, and swine manure were evaluated (41). Hydrogen sulfide, methyl mercaptan, and dimethyl sulfide were released under anaerobic conditions. Dimethyl disulfide, carbonyl sulfide, and carbon disulfide also were generated from some of the manures. Only trace amounts of one sulfur compound (dimethyl disulfide) could be detected in the gaseous products of aerobic decomposition. No evidence was obtained that indicated that sulfur gases contributed to the odors of dried manures.

Gases

The gases generated by microbial degradation of the wastes are of concern to the health of animals and humans in confined livestock environments. The gases of greatest concern are carbon dioxide, ammonia, hydrogen sulfide, and methane. Under conditions with adequate ventilation, there has been no evidence that these gases reached harmful concentrations for animals or humans. Under abnormal conditions, gas concentrations have exceeded threshold values and have had adverse effects on animals.

The conditions leading to accidents have been animals in close proximity to anaerobic wastes, inadequate ventilation, and the release of anaerobic gases as the anaerobic wastes were agitated for movement or removal. Toxic gas concentrations can exist in a waste storage unit and the unit should be adequately ventilated prior to human entry after mixing anaerobic wastes.

References

1. Porter, K., ed., "Nitrogen and Phosphorus—Food Production, Waste, and the Environment," Chapter 4. Ann Arbor Sci. Publ., Ann Arbor, Michigan, 1975.

2. Frink, C. R., Plant nutrients and water quality. *Agric. Sci. Rev.* pp. 11–25, Second Quarter (1971).

3. Anonymous, "The Pollution of Waters by Agriculture and Forestry," Proc. Semin. (organized by the Committee on Water Problems). United National Economic Commission for Europe, Vienna, Austria, 1973.

4. U.S. Environmental Protection Agency, "Fish Kills Caused by Pollution—Fifteen Year Summary—1961–1975," EPA-440/4-78-011. Office of Water Planning and Standards, Washington, D.C., 1978.

5. Stewart, B. A., Viets, F. G., Jr., and Hutchinson, G. L., Agriculture's effect on nitrate pollution of ground water. *J. Soil Water Conserv.* **23**, 13–15 (1968).

6. Beatty, M. T., Kerrigan, J. E., and Porter, W. K., What and where are the critical situations with farm animal wastes and by-products in Wisconsin? *In* "Proceedings of the Farm Animal Wastes and By-Product Management Conference," pp. 36–57. University of Wisconsin, Madison, 1969.

7. Conservation Foundation, "State of the Environment—1982," Chapter 6. Washington, D.C., 1982.

8. Dillon, P. J., and Kirchner, W. B., The effects of geology and land use on the export of phosphorus from watersheds. *Water Res.* **9**, 135–148 (1975).

9. Ryding, S. O., and Forsberg, C., Nitrogen, phosphorus and organic matter in running waters—studies from six drainage basins, *In* "Research on Recovery of Polluted Lakes, Loading, Water Quality and Responses to Nutrient Reduction." Inst. of Limnology, University of Uppsala, Sweden, 1978.

10. Loehr, R. C., Characteristics and comparative magnitude of non-point sources. *J. Water Pollut. Control Fed.* **46**, 1844–1871 (1974).

11. Campbell, F. R., and Webber, L. R., Contribution of range land runoff to lake eutrophication. *Proc. Int. Conf. Water Pollut. Res., 5th, 1970.*

12. Harmeson, R. H., Sollo, F. W., and Larson, T. E., The nitrate situation in Illinois. *J. Am. Water Works Assoc.* **63**, 303–310 (1971).

13. Stoltenburg, G. A., Water quality in an agricultural watershed. *In* "Transactions of the 20th Annual Sanitary Engineers Conference," pp. 15–22. University of Kansas, Lawrence, 1970.

14. Tomlinson, T. E., Trends in nitrate concentrations in English rivers in relation to fertilizer use. *Water Treat. Exam.* **19**, 277–293 (1970).

15. Sonzogni, W. C., Jeffs, D. N., Konrad, J. C., Robinson, J. B., Chesters, G., Coote, D. R., and Ostry, R. C. Pollution from land runoff. *Environ. Sci. Technol.* **14**, 148–153 (1980).

16. Pollution from Land Use Activities Reference Group, "Environmental Management Strategy for the Great Lakes." Great Lakes Regional Office, International Joint Commission, Windsor, Ontario, Canada, 1978.

17. Bower, C. A., and Wilcox, L. V., Nitrate content of the Upper Rio Grande as influenced by nitrogen fertilization of adjacent irrigated lands. *Soil Sci. Soc. Am. Proc.* **33**, 971–973 (1969).

18. Kolenbrander, G. J., Does leaching of fertilizers affect the quality of groundwater at the waterworks? *Stikstof (Engl. Ed.)* **15**, 8–15 (1972).

19. Howells, D. H., Kriz, G. J., and Robbins, J. W. D., "Role of Animal Wastes in Agricultural Land Runoff," Final Rep. Proj. 13020 DGX. U.S. Environmental Protection Agency, Washington, D.C., 1971.

20. Long, L. F., Runoff water quality as affected by surface applied dairy cattle manure. *J. Environ. Qual.* **8**, 215–218 (1979).

21. Cramer, C. O., Converse, J. C., Tenpas, G. H., and Schlough, D. A., The design of solid

manure storages for dairy herds. *Winter Meet., Am. Soc. Agric. Eng., 1971* Paper 71–910 (1971).

22. Smith, S. M., and Miner, J. R., Stream pollution from feedlot runoff. *In* "Transactions of the 14th Annual Sanitary Engineers Conference," pp. 18–25. University of Kansas, Lawrence, 1964.

23. Miner, J. R., Water pollution potential of cattle feedlot runoff. Ph.D. thesis, Kansas State University, Manhattan, Kansas, 1967.

24. Loehr, R. C., Drainage and pollution from beef cattle feedlots. *J. Sanit. Eng. Div. Am. Soc. Civ. Eng.* **96**, SA6, 1295–1309 (1970).

25. White, R. K., and Edwards, W. M., Beef barnlot runoff and stream water quality. *In* "Proceedings of the Agricultural Waste Management Conference" (R. C. Loehr, ed.), pp. 225–235. Cornell University, Ithaca, New York, 1972.

26. McCalla, T. M., Ellis, J. R., Gilbertson, C. B., and Woods, W. R., Chemical studies of solids, runoff, soil profile and groundwater from beef cattle feedlots at Mead, Nebraska. *In* "Proceedings of the Agricultural Waste Management Conference" (R. C. Loehr, ed.), pp. 211–223. Cornell University, Ithaca, New York, 1972.

27. Kreis, R. D., Scalf, M. R., and McNabb, J. F., "Characteristics of Rainfall Runoff from a Beef Cattle Feedlot," Final Rep., Proj. 13040 FHP, Environ. Protect. Tech. Serv. EPA-R2-72-061. U.S. Environmental Protection Agency, Washington, D.C., 1972.

28. Hutchinson, G. L.,and Viets, F. G., Nitrogen enrichment of surface water by absorption of ammonia volatilized from cattle feedlots. *Science* **166**, 514–515 (1969).

29. Agricultural Research Service, "Pollution Abatement from Cattle Feedlots in Northeastern Colorado and Nebraska," EPA-660/2-75-015. U.S. Environmental Protection Agency, Corvallis, Oregon, 1971.

30. Soderquist, M. R., Waste management in the food processing industry. *J. Environ. Qual.* **1**, 81–86 (1972).

31. Reckhow, K. H., Beaulac, M. N., and Simpson, J. T., "Modeling Phosphorus Loading and Lake Response under Uncertainty: A Manual and Compilation of Export Coefficients," EPA 440/5-80-011. Clean Lakes Section, U.S. Environmental Protection Agency, Washington, D.C., 1980.

32. Hendry, C. D., Brezonik, P. L., and Edgerton, E. S., Atmospheric deposition of nitrogen and phosphorus in Florida. *In* "Atmospheric Pollutants in Natural Waters," S. J. Eisenreich, ed., pp. 199–215. Ann Arbor Sci. Publ., Ann Arbor, Michigan, 1981.

33. Hoeft, R. G., Keeney, D. R., and Walsh, L. M., Nitrogen and sulfur in precipitation and sulfur dioxide in the atmosphere in Wisconsin. *J. Environ. Qual.* **1**, 203–208 (1972).

34. Sartor, J. D., Boyd, G. B., and Agardy, F. J., Water pollution aspects of street surface contaminants. *Water Pollut. Contr. Fed. Conf.,* **48**, 458–468 (1974).

35. Randall, C. W., and Grizzard, T. J., Runoff pollution. *In* "Stormwater Management in Urbanizing Areas" (W. Whipple, N. S. Grigg, T. Grizzard, C. W. Randall, R. P. Shubinski, and L. S. Tucker, eds.). Prentice-Hall, Englewood Cliffs, New Jersey, 1983.

36. McDowell, T. R., and Omernik, J. M., "Nonpoint Source–Stream Nutrient Level Relationships: A Nationwide Study—Supplement 1," EPA-600/3-79-103. Corvallis Environmental Research Laboratory, U.S. Environmental Protection Agency, Corvallis, Oregon, 1979.

37. Oglesby, W. C., Bovine salmonellosis in a feedlot operation. *Vet. Med. Small Anim. Clin.* **59**, 172–174 (1964).

38. Gibson, E. A., Disposal of farm effluent—animal health. *Agriculture (London)* **74**, 183–192 (1967).

39. Hibbs, C. M., and Foltz, V. D., Bovine salmonellosis associated with contaminated creek water and human infection. *Vet. Med. Small Anim. Clin.* **59,** 1153–1155 (1964).
40. McCoy, E., Removal of pollution bacteria from animal wastes by soil percolation. *Annu. Meet. Am. Soc. Agric. Eng., 1969* Paper 69-430 (1969).
41. Banwart, W. L., and Bremner, J. M., Identification of sulfur gases evolved from animal wastes, *J. Environ. Qual.* **4,** 363–366 (1975).

4

Waste Characteristics

Introduction

Knowledge of agricultural waste characteristics is fundamental to the development of feasible waste management systems. Treatment and disposal methods that have been successful with other industrial wastes may be less successful with agricultural wastes unless the methods are modified to accommodate the characteristics of specific agricultural wastes. The wastes produced by agriculture vary in quantity and quality. Wastes from food processing are low-strength, high-volume liquid wastes, while those from livestock operations tend to be high-strength, low-volume wastes. Both liquids and solids result from food processing and livestock production, requiring that both liquid and solid waste management possibilities be considered.

An understanding of the characteristics of a waste permits judgments on the type of treatment and/or disposal methods that may be effective. With a liquid waste containing dissolved organic solids, biological treatment is appropriate. Solid wastes with a high organic content are amenable to incineration or composting. Other alternatives are noted in Table 4.1. In general the bulk of the oxygen-demanding material is in the dissolved state for food processing waste waters, whereas with livestock wastes most of the oxygen-demanding material is in the form of particular matter.

Information on the frequency and quality of waste discharges allows design of facilities to handle constant as well as intermittent discharges such as those

TABLE 4.1

Feasible Treatment and Disposal Methods with Wastes of Different Characteristics

Wastes	Treatment and disposal methods
Liquid	
Dissolved organic wastes	Biological treatment, land treatment
Dissolved inorganic matter	Land disposal, physical or chemical treatment
Suspended organic wastes	Sedimentation, biological treatment, chemical precipitation, land treatment
Suspended inorganic matter	Sedimentation, land treatment, chemical treatment
Solid	
Organic wastes	Incineration, composting, land treatment, dehydration, animal feed
Inorganic wastes	Land disposal

due to the seasonal nature of fruit and vegetable processing and those due to the variable nature of livestock waste runoff. Identification of the waste sources within a processing plant provides information for in-plant separation of the waste streams in the plant, for reuse of the less-contaminated waters, and for changes in processes that produce large quantities of wastes and/or concentrated wastes. Knowledge of the waste characteristics permits consideration of reuse of waste liquids for the transport of raw products, recovery of specific waste components, irrigation, by-product development, and fertilizer value. Similar reuse possibilities exist with solid wastes, such as for animal feed and as a soil conditioner. The prevention of waste is of economic importance in all processing operations. The quantity and quality of the waste affects the complexity and size of waste treatment systems as well as recovery opportunities.

Agricultural wastes consist of food processing wastes, liquid and solid animal wastes, waste packaging materials, agricultural chemical losses, crop and field residues, greenhouse and nursery wastes, dead livestock and obsolescent vehicles, equipment, and buildings. The characteristics of food processing wastes and animal wastes receive major emphasis in this chapter. Data on the characteristics of specific wastes are presented in the Appendix.

Food Processing

General

The function of the food processing industry is to serve the farm operations which produce perishable products on a seasonal basis and consumers who

desire a variety of nutritional foods throughout the year. Changes in food marketing have resulted in alterations of food processing methods to accommodate
consumer desires. Requirements of supermarkets have encouraged producers to
wash, prepare, and prepack fruits, vegetables, and meats at the source. Greater
water volumes may be used, and more wastes are left at the processing plant and
in the fields.

Each food processing plant has wastes of different quantity and quality. No
two plants are the same, and predictions of loadings expected from proposed
plants must be regarded as approximate. Few plants have adequate knowledge of
the volumes, characteristics, and fluctuations of their wastes. At each processing
plant, balances of materials and water usage in the plant can provide reasonable
estimates of the total waste flow and contaminant load. These balances, coupled
with in-plant surveys of individual waste flows, can pinpoint opportunities for in-
plant changes to reduce wastes and help identify unnoticed waste discharges.

Food processing wastes generally are low in nitrogen, have high BOD and
suspended solids, and undergo rapid decomposition. Some wastes, such as those
from beet processing, are highly colored. Fresh wastes have a pH close to
neutral. During storage the pH decreases. In addition to the organic content, fruit
and vegetable processing wastes can contain other pollutants, such as soil, lye,
heat, and insecticides. Food processing wastes result from the washing, trimming, blanching, pasteurizing, and juicing of raw materials; the cleaning of
processing equipment; and the cooling of the finished product. In most plants,
cooling waters have low contamination and may be reused for washing and
transport of the raw product.

The wastes from food processing plants are largely organic and can be
treated in municipal waste treatment plants. The flow and organic load fluctuations must be adequately evaluated when these wastes are permitted to enter
municipal treatment facilities. Fruit and vegetable processing can cause overloading in these facilities because of the variable characteristics and short processing season. Meat, poultry, and milk processing are less seasonal.

Fruit and Vegetable Processing

The fruit and vegetable processing industry provides a market for a large
part of the nation's fruits and vegetables. Approximately 90% of the beets; 80%
of the tomatoes; 75% of the asparagus, lima beans, and leafy vegetables; 70% of
the apricots, cranberries, and pears; 60% of the green or snap beans, peas, and
sweet corn; and 50% of the peaches and cherries are preserved by the industry.
Over 30 million tons of raw fruits and vegetables are processed annually.

Wide ranges of wastewater volume and organic strength are generated per
ton of raw product among plants processing the same product. Different waste
volumes and strengths are generated from different styles of the same product,

such as peeled versus pulped or sliced versus whole style. Wastewater volume and organic strength vary throughout the operating season and operating day. Thus, facilities to treat these wastewaters must be designed to handle large volumes intermittently rather than at constant flow rates and constant organic concentrations.

Ranges of characteristics for fruit and vegetable processing wastes are presented in the Appendix, Tables A-1 to A-4. Waste flows and pollutional characteristics can be reduced by in-plant changes and close control over processing operations. Increased attention to the following can help reduce the quantity and pollutional content of the wastewaters requiring treatment and disposal:

1. Reduction of freshwater requirements through use of recycle systems
2. Segregation of strong wastes for separate treatment
3. Modification of processes to minimize waste generation
4. Education of plant personnel regarding pollution control and water conservation

The high pH values of many fruit and vegetable processing wastes result from the use of caustics such as lye in peeling. These caustic solutions can have a pH of about 12–13 and are discharged intermittently as they lose their strengh, resulting in slug loads to the waste treatment facilities. Pickle and sauerkraut wastes are acidic and contain large chloride concentrations as well as organic matter.

There are many in-plant process modifications that can be used in fruit and vegetable processing operations to reduce the waste load that must be treated. These include dry versus wet peeling, hot gas or individual quick blanching (IQB) methods, reduction of cleaning water pressure, and dry collection and transport of the product and waste solids.

Data on dry caustic peeling methods indicate the reduction in waste load that can occur. When the process was used for peaches, the wastewater volume was reduced by 93%, the COD by 70%, and the suspended solids by 70% (1). With the dry caustic peeling method, the caustic, softened peel is removed by the mechanical abrasion of rotating rubber disks rather than by high-pressure water sprays as in the conventional wet method. The dry peeling method has been used successfully for white and sweet potatoes, beets, carrots, peaches, apricots, pears, and tomatoes.

Meat and Poultry Processing

The slaughtering and meat-packing industry consists of a number of small plants. Meat-packing plants range in size from plants with annual kills of from less than 1 million pounds to those with greater than 800 million pounds. The main wastes originate from killing, hide removal or dehairing, paunch handling, rendering, trimming, processing, and cleanup operations. The wastes contain blood, grease, inorganic and organic solids, and salts and chemicals added during processing operations. The BOD and solids concentrations in the plant effluent will depend on in-plant control of water use, by-product recovery, waste separation at the source, and plant management.

Meat-packing plants carry out the slaughtering and processing of cattle, calves, hogs, and sheep for the preparation of meat products and by-products. These plants range from those that carry out only one operation, such as slaughtering (slaughterhouses), to plants that not only slaughter but also process the meat to products such as sausages, cured hams, and smoked products (packinghouses). The amount of processing at a plant varies considerably, because some plants process only a portion of their kill, while others process not only their kill, but also that from other plants. Most packinghouses and many slaughterhouses also render by-products. Edible and inedible by-products are rendered from edible fats and trimmings and from inedible materials, respectively.

The nonmeat portion of an animal results in significant waste loads. Wherever possible, such material should be separated from the general waste stream at the source. Numerous opportunities exist for recovery of this material for animal feed, rendering, and other by-product use.

At a beef cattle slaughtering plant, as high as 50 lb of blood, 50 lb of paunch manure, and 40 lb of animal manure per animal can result. Blood from beef cattle had a BOD_5 of 156,500 mg per liter, a COD of 218,300 mg per liter, a moisture content of 82%, and a pH of 7.3 (2). The average weight of wet blood produced per 1000 lb of beef animal was 32.5 lb. Recovery of the blood is an important aspect of pollution control and should be a part of all animal processing plants.

Rendering recovers many of the inedible portions of an animal. This process converts animal by-products into fat, oils, and proteinaceous solids. Heat is used to melt the fat from the tissue, to coagulate cell proteins, and to evaporate the moisture of the raw material. Rendering can produce animal feed constituents from animal blood, feathers, bones, fat tissue, meat scraps, inedible animal carcasses, and animal offal.

Dry rendering, i.e., cooking under vacuum and low temperature with no water added, results in a minimum of pollutants added to the plant waste stream. Wet rendering results in tank water containing dissolved organic matter and having a high BOD, about 30,000 mg per liter.

The poultry processing industry slaughters, dresses (eviscerates), and ice- or freeze-packs broilers, mature chickens, turkeys, ducks, and other poultry. Some plants only slaughter and dress; others slaughter, dress, and further process; and still others further process only. Further processing converts dressed poultry into a variety of cooked, canned, and processed poultry meat items such as precooked roasts, rolls, patties, meat slices in gravy, and canned boned chicken. The amount of further processing performed in these plants varies considerably.

At poultry processing plants, wastes originate from killing, scalding, defeathering, evisceration, washing, chilling, and cleanup operations. Waste quantity and quality depend on the manner in which the blood, feathers, and offal are handled, the type of processing equipment used, and the attitude of the plant management concerning pollution control. In most modern plants wastes discharged from evisceration and the feathers, dirt, and blood from the defeathering machines are carried in the wastewater streams. These streams normally pass through screens which remove the larger solids. Usually the chilling and packing waters and cleanup waste also pass through the screens. About 75% of the daily waste volume, BOD, and suspended solids are discharged during processing, with the remainder discharged during cleanup periods.

In the processing of broilers, about 70% of the original weight of the bird represents the finished product. The remaining 30% includes feathers, intestines, feet, head, and blood, which require liquid and solid disposal at the processing plant. The waste of greatest pollutional significance is the blood from the killing operation. About 8% of the body weight of chickens is blood, and about 70% is drainable and collectable.

Reasonable values of the characteristics of meat processing, poultry processing, and rendering wastewaters are summarized in Table 4.2. Additional characteristics of these wastewaters are presented in the Appendix, Tables A-5 to A-9.

Milk Processing

Milk processing wastes result from manufacturing and transfer operations after the milk from a dairy farm reaches a central receiving station. The wastes consist of whole and processed milk, whey from cheese production, and wash water. Fresh-milk processing wastes are high in dissolved organic matter and very low in suspended matter. The BOD of whole milk is about 100,000 mg per liter and can exert a significant oxygen demand, even in small quantities. The BOD and COD of milk plant wastewater will be a function of the type of product manufactured.

A survey of over 50 plants revealed the range of unit flow and BOD values reported in the Appendix, Table A-10. These plants used either advanced technology or a mixture of advanced and typical processing technology. Over half of the plants produced more than one type of dairy product. The controlling factor in waste volume and BOD production appeared to be management. Under extremely good management, about 0.5 lb of waste flow per pound of milk processed and 0.5 lb of BOD per 1000 lb of milk processed were obtainable. A realistic average of 1.5 lb of waste flow per pound of milk processed and 2.0 lb of BOD per 1000 lb of milk processed appeared to be achievable under good

TABLE 4.2

Illustrative Characteristics of Meat and Poultry Processing and Rendering Wastewaters[a]

Parameter	Red meat slaughtering	Red meat processing	Poultry slaughtering	Poultry processing	Rendering
Water use					
m³/tonne FP[b]	3–27	10–16	18–30	15–100	1–28
BOD					
mg/liter	200–6000	200–1200	400–600	100–2400	100–30,000
kg/tonne FP	1.5–25	0.2–24	10–20	4–32	1–22
Suspended solids					
mg/liter	750–5000	100–1500	200–400	75–1500	300–4000
kg/tonne FP	0.6–22	0.1–12	5–25	1–25	0.03–8
Grease					
mg/liter	800–2200	10–550	150–250	100–400	200–7000
kg/tonne FP	0.2–20	0.1–8	4–18	1–16	0.01–15
Kjeldahl nitrogen					
mg/liter	30–300	ND[c]–10	5–300	50–100	60–100
kg/tonne FP	0.2–2.2	0.6–9.0	—	1–15	0.1–13
Phosphate					
mg/liter	—	ND–100	—	—	—
kg/tonne FP	ND–1.3	—	0.1–13	0.1	ND–0.4

[a]From (3). [Reproduced with the permission of the Minister, Supply and Services Canada, from "Biological Treatment of Food Processing Wastewater—Design and Operations Manual," Report EPS 3-WP-79-7, Environment Canada, Ottawa, 1979.]

[b]Tonne of finished product.

[c]Nondetectable.

management. Factors above 3.0 in both categories were considered excessive and an indication of poor waste management. Plants producing whey will have BOD factors considerably in excess of the aforementioned values. Other data on the characteristics of milk plant processing wastes are noted in the Apendix, Table A-11.

Milk processing plants will have numerous short-term fluctuations in characteristics and flow during the day. There are seasonal peaks due to milk production variability throughout the year. The wastes are amenable to biological treatment and can be treated in municipal waste treatment facilities provided that the plant has the capacity for the fluctuation and increased oxygen demand.

The largest pollutants in dairy food plant wastewaters are whey from cheese production operations followed by wash water and pasteurization water. The manufacture of cheese from either whole or skim milk produces cheese and a greenish yellow fluid known as whey. Whole milk is used to produce natural and processed cheeses, such as cheddar, and the resultant fluid is called sweet whey, with a pH in the range of 5 to 7. Skim milk is used to produce cottage cheese and the by-product fluid is acid whey with a pH in the 4–5 range. The lower pH is the result of the acid developed during, or employed for, coagulation. Each pound of cheese produced results in 5–10 lb of fluid whey.

The BOD of whey ranges from 32,000 to 60,000 mg per liter, depending on the specific cheese-making process used. Whey contains about 5% lactose, 1% protein, 0.3% fat, and 0.6% ash. Because the protein has been precipitated as cheese, evaluation of the nitrogen content of whey should be made to assure adequate nitrogen, if biological treatment is contemplated.

Seafood Processing

The degree of waste in seafood processing varies widely. Fish that are rendered whole, such as menhaden, for the production of fish meal result in no solid waste. Crab processing results in up to 85% solid waste. Each fish processing operation will produce significant liquid flows from the cutting, washing, and processing of the product. These flows contain blood and small pieces of fish and skins, viscera, condensate from cooking operations, and cooling water from condensers. The latter two flows have large volumes but contain small organic loads.

Variations in a freshwater fish processing plant are caused by the type of fish processing techniques, plant size, water usage, and the time the solid wastes were in contact with the wastewater. The pollutional strength of perch and smelt wastewater increase considerably with increased contact time between the solid waste material and the liquid waste (4).

Few comprehensive, detailed studies have been made to determine fish processing waste strengths. General ranges of reported values for various types of fish processing operations are presented in Table 4.3. Additional information on the characteristics of various seafood processing wastewaters are presented in the Appendix, Tables A-12 and A-13.

Livestock Production

General

Large numbers of livestock are raised in the United States, with the result that over 2 billion tons of manure are produced each year. Part of the total livestock waste production remains in the pastureland and rangeland, but large volumes accumulate in feedlots and buildings and must be collected, transported, and disposed of in an economical and inoffensve manner.

The term "livestock wastes" may mean any one of a number of things:

TABLE 4.3

Illustrative Characteristics of Fish, Seafood, and Fish Meal Processing Wastewater[a]

Item processed	BOD mg/liter	BOD kg/tonne[b]	Suspended solids mg/liter	Suspended solids kg/tonne[b]	Oil mg/liter	Oil kg/tonne[b]	Water use (m³/tonne)[c]
Groundfish							
Wetline process	600–1200	15	150–960	7	200–1500	13	5.0
Dryline process	100–1100	5	30–230	1	ND[d]–500	1	5.0
Lobster processing	840–1200	26	140–170	4	100–800	5	2.6
Crab processing	320–1000	40	135–660	19	100–900	21	5.4–7.0
Herring filleting	3200–5800	22	1150–5300	21	200–3000	10	6.0
Marinated filleting	6900–14,000	215	1500–4600	85	800–5000	83	2.5
Fish meal							
Bloodwater	190,000–315,000	—	4200–21,000	—	2500–3000	—	—
Stickwater	46,000–490,000	—	7600–21,500	—	100–500	—	—

[a]Adapted from (3). [Reproduced with permission of the Minister, Supply and Services Canada, from "Biological Treatment of Food Processing Wastewater—Design and Operations Manual," Report EPS 3-WP-79-7, Environment Canada, Ottawa, 1979.]
[b]Kilogram per tonne of raw seafood landed.
[c]Cubic meter of water per tonne of raw seafood landed.
[d]Nondetectable.

(a) fresh excrement, including both the solid and liquid portions, (b) total excrement but with bedding added to absorb the liquid portion, (c) the material after liquid drainage, evaporation of water, or leaching of soluble nutrients, (d) only the liquid, which has been allowed to drain from the total excrement, or (e) material resulting from aerated or anaerobic storage.

The characteristics of livestock wastes are a function of the digestibility and composition of the feed ration. The feces of livestock consist chiefly of undigested food, mostly cellulose fiber, which has escaped bacterial action. A portion of the other nutrients also escape digestion. Undigested proteins are excreted in the feces, and the excess nitrogen from the digested protein is excreted in the urine as uric acid for poultry and urea for animals. Potassium is absorbed during digestion but eventually almost all is excreted. Feces also contain residue from the digestive fluids, waste mineral matter, worn-out cells from the intestinal linings, mucus, bacteria, and foreign matter, such as dirt consumed along with the food. Calcium, magnesium, iron, and phosphorus are voided chiefly in the feces. Livestock wastes can contain feed spilled in the animal pens.

Municipal wastes, food processing wastes, and other industrial wastes are waterborne and result in liquid waste handling, treatment, and disposal systems. Livestock wastes are solid, semisolid, or liquid, depending upon how the production operation is designed and operated. The characteristics of livestock wastes are affected by decisions on how the wastes are to be handled. A broader set of waste management alternatives exists for livestock wastes than for other industrial wastes since opportunities exist for both liquid and solid waste handling, treatment, and disposal. Livestock wastes are generated as a semisolid, and there is logic in handling and disposing the wastes in this condition.

Individuals concerned with pollution control activities are interested in characteristics of these wastes in terms of BOD, COD, solids, and nutrient content. Suspended solids have little relevance to most livestock wastes since they usually exist as a solid or slurry rather than a dilute liquid waste. A number of ways have been used to report the characteristics of livestock wastes.

The pollutional characteristics can be reported in terms of milligrams per liter of the liquid slurry that results. Since the water content of the waste slurry will vary, depending on the quantity of water used in cleaning, spilled by the animals in drinking, and excreted by the animals and by any evaporation or rainfall that occurs between cleanings, values based on milligrams per liter can be expected to vary widely.

Livestock wastes are solid matter which contains some water rather than liquid waste containing some solids. Although the water content will vary, the solid content is a function of the ration fed and the species of animal and should be relatively constant per animal. Data can be presented as milligrams of a pollutional characteristic per milligram of total or volatile solids.

Other data relate the quantity of waste to the animal in terms of pounds per head of animal. This approach is the most realistic and valuable in estimating the gross wastes generated at a particular production unit.

Representative sampling is difficult with wastes as concentrated and hetero-genous as animal wastes. Until more experience is gained with analytical meth-ods on animal wastes and slurries, it is best to consider the available information as guidelines or reasonable estimates rather than accurate and precise data. Characteristics indicated in this book and in other literature should be used with an understanding of the variations in the data and their relative value. Although it is difficult to apply average livestock waste production values to a specific location and problem, knowledge of average values is useful to develop order-of-magnitude information concerning current and potential livestock production units. Summary data on animal wastes are in Table 4.4. These values represent an average of those that have been reported in the literature. Other data are presented in the Appendix.

Poultry

Studies on the physical composition of fresh poultry manure have shown that it contains 75–80% moisture, 15–18% volatile solids, and 5–7% ash, with an average particle density of 1.8 and a bulk density of about 65 lb/ft^3. Based upon 75% moisture, the manure contained 1375 Btu per pound of wet manure

TABLE 4.4

Summary of Animal Waste Characteristics[a]

Parameter[b]	Dairy cow	Beef feeder	Swine feeder	Sheep feeder	Poultry		Horse
					Layer	Broiler	
Raw manure (RM)[c]	82	60	65	40	53	71	45
Total solids	10.4	6.9	6.0	10.0	13.4	17.1	9.4
Volatile solids	8.6	5.9	4.8	8.5	9.4	12.0	7.5
BOD$_5$	1.7	1.6	2.0	0.9	3.5	—	—
COD	9.1	6.6	5.7	11.8	12.0	—	—
Nitrogen (total, as N)	0.41	0.34	0.45	0.45	0.72	1.16	0.27
Phosphorus (as P)	0.073	0.11	0.15	0.066	0.28	0.26	0.046
Potassium (as K)	0.27	0.24	0.30	0.32	0.31	0.36	0.17

[a]From J. R. Miner and R. S. Smith, eds., "Livestock Waste Management with Pollution Control," North Central Regional Research Publication 222, MWPS-19. Midwest Plan Service, Iowa State University, Ames, 1975.

[b]Characteristics are in terms of weight of parameter per day per 1000 liveweight units (pounds or kilograms).

[c]Feces and urine with no bedding.

(5). Manure excreted from a chicken per day represents about 5% of the body weight of the bird. Waste volume ranges from 0.05–0.06 gal per bird per day.

Essentially all of the broilers produced in the United States are grown on litter which absorbs moisture and provides a nesting material for the birds. The litter may be sawdust, peanut hulls, wood shavings, or other suitable material. About 1.3 billion cubic feet of used broiler litter must be disposed of every year. The frequency of removal and disposal of broiler manure and litter depends on the number of growing cycles that can be completed on one batch of litter without disease problems. Practices vary among producers, but broiler houses are cleaned about two to three times per year. The wastes are spread on land and used for fertilization of pasture for beef cattle.

Turkey production experiences a seasonal variation, with the major producers active during the summer and fall. Waste accumulation in both confinement houses and open feeding areas must be considered. Young turkeys are kept in houses on litter for the first 8 weeks and are then placed in open range areas at densities of from 400 to 2500 birds per acre for the remainder (12–16 weeks) of their growing cycle. The wastes in turkey houses usually are removed twice a year and are handled in the same way as broiler litter.

The amount of manure produced (wet basis) was 1.25 lb and 1.35 lb of feed consumed for male and female turkeys, respectively. The average production rate was 0.64 and 0.53 lb per bird per day on a wet basis for males and females. Very little difference in the nutrient content was observed between sexes. The nitrogen, phosphorus, and potassium content was 1.36, 0.49, and 0.71%, respectively, of the wet solids (6).

Duck production operations have greater water usage than do other poultry operations. The volume of wastes from the duck farms on Long Island ranged from 10 to 34 gal per duck, with an average of 18 gal per duck. More recent water usage is near the lower end of the range as duck producers have sought to minimize their waste flows. The characteristics of duck wastewater are variable, due to variable water flow rates. Sand, undigested food, and manure particles make up the wastes (Fig. 4.1). Duck production is seasonal, with the peak period being from April to October. The characteristics of wastewater from duck farms are noted in the Appendix, Table A-14.

Swine

The daily waste production from swine is a function of the type and size of animal, the feed, the temperature and humidity within the building, and the amount of water added in washing and leakage. The quantity of feces and urine produced increases with the weight and food intake of the animal. The feeding regime will affect the properties of the manure. Approximately 30% of the consumed feed is converted to body tissue, and the remainder is excreted as urine

Fig. 4.1. Production of ducks showing exterior duck runs and available flowing water.

and manure. Feed conversion efficiencies averaged 3.2 lb of commercial meal consumed per pound of liveweight gained. The data ranged from 2.6 to 3.8 lb of feed per pound of weight gain. The manure production ranges from 6 to 8% of the body weight of a hog per day. The total amount of wastewater to be handled in a swine operation can be affected more by the amount of wastage from the pressure waterers than any other factor. Volume of manure production can average about 1 gal per 100-lb animal per day. Wet manure can contain 5–9% total solids. Of these, 83% may be volatile.

There can be a linear relationship between the quantity of (a) feces, urine, and manure production; and (b) meal, water, and meal plus water consumed by growing finishing pigs. The feces-to-meal ratio ranged from about 0.6 for pigs fed on a floor feeding regime with a water-to-meal ratio of 2.5:1 to approximately 0.67 for a pipeline feeding regime with a water-to-meal ratio of 4:1 (7). The corresponding values for the urine-to-water ratio ranged from 0.6 to 0.7.

Because the land is the ultimate acceptor of swine wastes, nutrients in the wastes are of interest. Comparative data on nutrients characteristics are presented in the Appendix, Table A-15. Other data on swine waste characteristics are summarized in Tables A-21 and A-22.

Dairy Cattle

Two types of wastes result from a dairy farm, the solid waste from the animals and the liquid wastes from the cleaning of the milking parlor. At some farms, these wastes are kept separate for treatment and disposal. At other farms, the liquid wastes are added to the manure in a storage pit to make the manure more amenable to transport and disposal as a liquid. Because dairy cattle manures normally are handled as a solid or slurry, concentrations in terms of milligrams per liter have little meaning. Manure production from dairy cattle ranged from 73 to 143 lb, with an average of 86 lb per animal per day (5) and represented 7–8% of the body weight of the animal per day. Moisture content averages 80–88% in fresh mixed manure and urine. Urine makes up about 30% of the weight of the manure. Characteristics of dairy cattle manure are compared in the Appendix, Table A-16. Fresh manure contains from 80 to 85% volatile matter.

Dairy farms using stanchion barns will use bedding such as sawdust, wood shavings, straw, or chopped corn stalks for the animals. The mixture of the manure and bedding is removed daily and transported by solid-manure-handling equipment. Free-stall dairy barns use no bedding and have a more liquid material of combined feces and urine.

An average dairy cow will produce between 14 and 18 tons of feces and urine per year. Manure has about 300 to 400 lb of dry matter per ton. There are about 10 lb of nitrogen, 4 lb P_2O_5, and 9 lb K_2O in each ton of wet manure defecated by the cow.

Each dairy farm has a milking area or parlor where the animals are milked twice a day, generally on a 12-hr cycle. On small dairy farms, the milking areas may be used only a fraction of the day, while on the very large farms, the milking parlors are in almost continuous use. Milk sanitation regulations require that the milking parlor and milk transfer lines be cleaned after every milking session. A chlorine cleaning solution is used for the pipelines, and fresh water is used for the milking parlor itself. The volume of cleaning water is related to the size of the milking parlor and the total length of milk transfer lines rather than to the number of animals milked. The milking parlor wastes will contain the milk residue in the pipelines plus any manure and debris flushed from the parlor. Where the milking parlor wastes are treated separately from the cattle manure, the manure and other solids in the parlor should be swept out prior to cleaning the parlor and should be handled with the other solid wastes. The characteristics of milking parlor wastes are related to the waste management at the dairy farm. Comparative data on these wastes are presented in the Appendix, Tables A-17 and A-23.

Beef Cattle

Most of the cattle raised for meat production are produced in the feedlots of the Midwest and West, particularly in the grain-producing states of the Great Plains. Few beef cattle are raised in confinement housing, although there is an increase in this method, especially in the colder climates.

The wastes of beef cattle feedlots will be somewhat different from those of cattle on general farms or on pasture. Since the early 1960s, more efficient rations and feeding programs have helped maintain the economic position of the cattle feedlot operation. Increasingly, feeder cattle are fed high-concentrate rations in the feedlots. These rations lend themselves to mechanical processing and handling.

Cattle in feedlots are started on high-roughage rations and shifted to high-concentrate rations. A high-concentrate ration has about 75–85% digestible material and 5–7% minerals, resulting in about 60 lb of wet manure per day, which includes about 17 lb of urine from a 900-lb steer. Urine contains about 6% dry matter and is the major source of moisture in the manure. Urine serves as the major carrier of mineral wastes from the animal. Fresh manure has a moisture content of about 85%.

A large quantity of cattle are still fed on farms and ranches that produce and use large quantities of harvested roughage or pasture for feed. Because of the higher quantity of nondigestible matter in roughage, cattle on pasture or roughages may be expected to produce a greater quantity of waste per day than cattle fed a large quantity of concentrates.

Lignins and hemicellulose are relatively stable materials that are only slowly degraded by microorganisms. These materials form a large fraction of high roughage feeds and resultant wastes. Cattle feces also contain lignoprotein complexes, which are produced in the digestive tract of the animal and are similar to the humus found in soil. The total dry weight of cattle manure may contain up to 20–25% of these humuslike compounds (8).

Waste removed from beef cattle feedlots (Fig. 4.2) will have characteristics different from fresh manure. The characteristics change as the manure undergoes drying, microbial action, wetting by precipitation, and mixing and compaction by animal movement. The characteristics of beef cattle waste are compared in the Appendix, Tables A-18 and A-22.

Veal Cattle

Production of animals for veal requires the confinement and growth of young cattle to the 200- to 300-lb weight level. The calves are purchased when approximately 1 week old, weighing about 100 lb, and sold when 13 weeks old, weighing about 300 lb. The animals are kept in stalls with the wastes flushed by use of a gutter system. The animals are fed high-concentrate diets, milk, and milk products and produce a more liquid waste than do mature animals. Waste production has been estimated at 2 gal per day per head and 0.16 BOD per day per head, with a BOD concentration of about 10,000 mg per liter (9). Grab samples of wastes generated at a veal operation indicated a flow of 0.9 gal per day per animal, with a BOD of 28,000 mg per liter when the animals were

Fig. 4.2. Removal of accumulated manure at a beef feedlot.

maximum age. The BOD production was 0.22 lb of BOD per animal per day. Other characteristics of veal calf manure are shown in Table A-24.

Other Animals and Agricultural Products

The preceding data provide information on the characteristics of livestock for food production. Knowledge of the characteristics of wastes from other animals, such as those from pet stores, stables, veterinary facilities, animal farms, and pharmaceutical testing laboratories may be of interest. Data of the pollutional characteristics of pets and small animals are meager. Some available information is presented in the Appendix, Tables A-19 and A-20.

References

1. Ralls, J. W., Mercer, W. A., Graham, R. P., Hart, M. R., and Maagdenberg, H. J., Dry caustic peeling of tree fruit to reduce liquid waste volume and strength. *In* ''Proceedings of the Second National Symposium on Food Processing Wastes,'' pp. 137–167. Pacific Northwest Water Lab., U.S. Environmental Protection Agency, Corvallis, Oregon, 1971.
2. Beefland International Inc., ''Elimination of Water Pollution by Packinghouse Animal Paunch

and Blood," Final Rep. Proj. 12060 FDS. U.S. Environmental Protection Agency, Washington, D.C., 1971.

3. Environment Canada, "Biological Treatment of Food Processing Wastewater—Design and Operations Manual," Report EPS 3-WP-79-7. Environment Protection Service, Ottawa, Canada, 1979.

4. Riddle, M. J., and Murphy, K. L., "An Effluent Study of a Fresh Water Fish Processing Plant," Environment Canada. Rep. 3-WP-73-8. Ottawa, Ontario, 1973.

5. Sobel, A. T., Some physical properties of animal manures associated with handling. *In* "Proceedings of the National Symposium on Animal Waste Management," Publ. SP-0366, pp. 27–32. Am. Soc. Agric. Eng., St. Joseph, Michigan, 1966.

6. Berry, J. G., Sutton, A. L., and Larson, J. R., The production rate and composition of manure from growing turkeys *In* "Proceedings of the Agricultural Waste Management Conference" (R. C. Loehr, ed.), pp. 153–158. Cornell University, Ithaca, New York, 1974.

7. O'Callaghan, J. R., Dodd, V. A., O'Donoghue, P. A. J., and Pollock, K. A., Characterization of waste treatment properties of pig manure. *J. Agric. Eng. Res.* **16**, 399–419 (1971).

8. McCalla, T. M., Frederich, L. R., and Palmer, G. L., Manure decomposition and fate of breakdown products in soil. *In* "Agricultural Practices and Water Quality," DAST 26, 13040 EYX, pp. 241–255. Federal Water Pollution Control Administration, U.S. Dept. of the Interior, Washington, D.C., 1969.

9. Scheltinga, H. M. J., and Poelma, H. R., Treatment of farm wastes, *In* "Proceedings of a Symposium on Farm Wastes," pp. 138–142. Inst. Water Pollut. Contr., Newcastle-upon-Tyne, 1970.

5

Biological Treatment Fundamentals

Introduction

Biological treatment processes are the most common processes used for the treatment, stabilization, or disposal of agricultural wastes. Decisions concerning the most appropriate processes to be used depend upon the characteristics of the wastes and the objective to be achieved. The processes may treat wastes that are liquids, slurries, or solids; may be aerobic, anaerobic, or facultative; may be used to achieve effluent standards, odor control, or to generate energy; may be in controlled structures such as tanks or ponds; or may occur in the soil. Examples of biological treatment processes that can be used with agricultural wastes include oxidation ponds, aerated lagoons, oxidation ditches, anaerobic lagoons, anaerobic digesters, composting, and land treatment.

Because the processes are biological, an understanding of the processes must be based upon the fundamentals of microbiology and on the transformations in biological waste treatment units. If this understanding can be achieved, rational predictions of performance become possible and the capabilities of a process can be better utilized. Without an understanding of the fundamentals, the processes can be treated only as "black boxes" in which the performance is subject to parameters seemingly beyond our control. Lack of proper understanding means that successful design and operation of biological processes must be based only on prior performance, which may be difficult to translate to different wastes and environmental conditions.

Biochemical Reactions

In the biological systems, microorganisms utilize the wastes to synthesize new cellular material and to furnish energy for synthesis. The organisms also can use previously accumulated internal or endogenous food supplies for their respiration and do so especially in the absence of external or exogenous food sources. Synthesis and endogenous respiration occur simultaneously in biological systems, with synthesis predominating when there is an excess of exogenous food and endogenous respiration dominating when the exogenous food supply is small or nonexistent.

Regardless of the biological system utilized, the principles of energy, synthesis, and endogenous cellular respiration are basic. The rates at which these reactions occur are a function of the environmental conditions imposed by, and/or on, a given biological treatment process.

The general reactions that occur can be illustrated in Eq. (5.1).

$$\text{Energy-containing metabolizable wastes} + \text{microorganisms} \rightarrow$$
$$\text{end products} + \text{more microorganisms} \tag{5.1}$$

Equation (5.1) represents energy–synthesis reactions in which the wastes are metabolized for energy and the synthesis of new cells. The energy utilized in Eq. (5.1) is obtained during the metabolism of the wastes. Synthesis or growth is affected by the ability of the microorganisms to metabolize and assimilate the food, the presence of toxic materials, the temperature and pH of the system, and the presence of adequate accessory nutrients and trace elements.

The wastes must contain sufficient carbon, nitrogen, phosphorus, and trace minerals to satisfy nutritional requirements. In a biological system, the indispensable nutrient that is present in the smallest quantity needed for microbial growth will become the limiting factor. With most organic wastes, adequate nutrients are available, and the biological reactions proceed at a rate constrained only by environmental factors such as temperature, pH, and inhibitory compounds.

When growth becomes limited, the microorganisms die and lyse, releasing the nutrients of their protoplasm for utilization by still-living cells in an autoxidative or endogenous cellular respiration process:

$$\text{Microorganisms} \rightarrow \text{end products} + \text{fewer microorganisms} \tag{5.2}$$

Endogenous respiration proceeds in the presence as well as the absence of an external food source. The rate of cellular oxidation and endogenous respiration is related to the mean time the cells have undergone treatment, i.e., solids retention time.

In the presence of waste material (food), microbial metabolism will occur to produce new cells and energy, and the microbial solids will increase. In the absence of food, endogenous respiration will predominate, and a reduction of the net microbial solids will occur. The microbial mass will not be reduced to zero, even with a long endogenous respiration period, however. A residue of about 20–25% of synthesized microbial mass will remain. Even in a long-term biological treatment system there will be a minimum rate of solids accumulation. Any inert solids in the raw waste will increase the rate of solids buildup in the unit. Eventually these solids must be removed from the units.

When the organic matter is metabolized and converted into microbial cells, the waste is only partially stabilized. As indicated in Eq. (5.2), the microbial cells are capable of further degradation. Only when the microbial cells are oxidized or removed does a stabilized effluent result.

It is possible to design and operate a biological treatment unit to function in any portion of a synthesis–endogenous respiration relationship. The specific design will depend on the characteristics of the desired effluent. If the treated wastes are to be discharged to surface waters, a high-quality effluent will be required. This can be attained by operating well into the endogenous region and removing residual solids from the effluent before it is discharged. If the land is the ultimate disposal point, a high-quality effluent may not be necessary. In this case the treatment unit can be operated without separation of the solids since further degradation will take place on the soil.

Basic Biological Processes

Biological processes can be defined by the presence or absence of dissolved oxygen, i.e., aerobic or anaerobic, by their photosynthetic ability or by the mobility of the organisms, i.e., suspended or adherent growth. Common examples of processes that are utilized for waste treatment are shown in Table 5.1. Since the terms are not mutually exclusive, some processes can be defined in more than one manner.

Aerobic

As used with biological waste treatment processes, the term "aerobic" refers to processes in which dissolved oxygen is present. The oxidation of organic matter, using molecular oxygen as the ultimate electron acceptor, is the primary process yielding useful chemical energy to microorganisms in these waste treatment processes. Microbes that use oxygen as the ultimate electron acceptor are aerobic microorganisms.

TABLE 5.1

Common Biological Treatment Processes

Aerobic	Anaerobic
Activated sludge units	Anaerobic lagoons
Trickling filters	Digesters
Oxidation ponds	Anaerobic filters
Aerated lagoons	Photosynthetic
Oxidation ditch	Oxidation ponds
Suspended growth	Adherent growth (fixed film)
Activated sludge	Trickling filters
Aerated lagoons	Rotating biological contractors
Mixed digesters	Anaerobic filters
Oxidation ditch	Denitrification columns

Anaerobic

Some microorganisms are able to function without dissolved oxygen in the system. Such microorganisms can be called "anaerobic organisms" or "anaerobes." Certain anaerobes cannot exist in the presence of dissolved oxygen and are obligate anaerobes. Examples of these are the methane bacteria commonly found in anaerobic digesters, anaerobic lagoons, and swamps. Anaerobes obtain their energy from the oxidation of complex organic matter but utilize compounds other than dissolved oxygen as oxidizing agents. Oxidizing agents are defined broadly as electron acceptors. Oxygen is not necessary to have an oxidation reaction. Oxidizing agents other than oxygen that can be used by microorganisms include carbon dioxide, partially oxidized organic compounds, sulfate, and nitrate. The process by which organic matter is degraded in the absence of oxygen frequently is called fermentation.

Facultative

Only a few species of organisms are obligate anaerobes or aerobes. A large number of organisms can live in either the absence or presence of oxygen. Organisms that function under either anaerobic or aerobic conditions are facultative organisms. When oxygen is absent from their environment, they are able to obtain energy from degradation of organic matter by nonaerobic mechanisms but, if dissolved oxygen is present, they metabolize the organic matter more completely. Organisms can obtain more energy by aerobic oxidation than by anaerobic oxidation.

Biological waste treatment units may be designed to be either aerobic or anaerobic. There are occasions when anaerobic conditions occur in units that are designed to be aerobic. Examples of these conditions are when organic matter has settled to the bottom of oxidation ponds and streams, i.e., benthic deposits; when aerobic systems are overloaded because of an increase in the strength of the raw waste; and in the interior of activated sludge floc particles and trickling filter growths. The majority of the organisms in biological waste treatment processes are facultative organisms.

Photosynthetic

Photosynthesis is the utilization of solar energy by the chlorophyll of green plants for the incorporation of carbon dioxide and other inorganic constituents in the production of cellular material. In this process molecular oxygen is formed. The photosynthetic organisms of interest in biological treatment systems are algae and rooted or floating plants. Examples of such biological treatment systems include oxidation ponds, streams, reservoirs, lakes, and high-rate algal production systems to recover the nutrients in wastes.

Suspended Growth

Suspended growth refers to processes that have mixtures of microorganisms and the organic wastes. The microorganisms are able to aggregate into flocculant masses and to move with the liquid flow. Agitation of the liquid keeps microbial solids in suspension. Suspended growth processes may be either aerobic or anaerobic. Anaerobic suspended growth units can be agitated by mechanical mixing and gas diffusion. Activated sludge units, aerated lagoons, oxidation ditches, and well-mixed anaerobic digesters are suspended growth processes.

Adherent Growth

Microbial growth is adherent when the microorganisms grow on a solid support medium and the wastes flow over or come in contact with the organisms. The support media can be large stones, rocks, slag, corrugated plastic sheets, or rotating disks. Commonly, the organic wastes flow over or through the openings of the supporting media. Although the vast majority of adherent growth systems currently used for waste treatment are aerobic, a few are anaerobic. Examples of adherent growth units are trickling filters, rotating biological disks, and anaerobic filters.

Energy Relationships

Knowledge of the energy relationships of microbial cells permits an understanding of energy available for synthesis and respiration, of production of

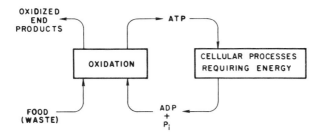

Fig. 5.1. Schematic pattern of energy transfer in the ATP–ADP system.

microbial cells in biological waste treatment units, and of the nature of the expected end products under certain conditions. All cells, whether animal, plant, or microbial, use similar fundamental mechanisms for their energy-transforming activities. These activities involve transferring chemical energy from food to the processes which utilize energy for the functions and survival of living cells. In both aerobic and anaerobic cells, the energy of the food material is conserved chemically in the compound adenosine triphosphate (ATP). ATP is the carrier of chemical energy from the oxidation of foods, either aerobic or anaerobic, to those processes of the cells which do not occur spontaneously and can proceed only if chemical energy is supplied. These processes are involved in the performance of osmotic, mechanical, or chemical work. In the context of this book, the food for the cells would be organic wastes of agriculture.

During the oxidation in the cells, ATP is formed from adenosine diphosphate (ADP). ATP is the high-energy form of the energy-transporting system, and ADP is the lower-energy form. A portion of the energy of the oxidation thus is conserved as the energy of the ATP. This process operates in a continuous dynamic cycle, receiving energy during the oxidation of foods and releasing energy during the performance of cellular work. A molecule of inorganic phosphate (P_i) is released when ADP is formed and incorporated in ATP when the ATP is formed. The principle of the cellular energy cycle is shown in Fig. 5.1. Although ATP is not the only energy-carrying compound in every cellular reaction, it is the common intermediate in the energy transformation in the cells.

The purpose of biological waste treatment is to stabilize or oxidize the organic wastes of man, industry, and agriculture. Oxidation is the process in which a molecule or compound loses electrons. Reduction is the process in which a molecule or compound gains electrons. Examples of these processes are noted in Fig. 5.2. In the first case ferrous iron is oxidized to ferric iron, with the release of an electron. In the second case carbon dioxide is reduced to methane. The carbon gains electrons and is reduced.

The general oxidation–reduction relationship can be indicated by

$$AH + B \rightarrow A + BH \tag{5.3}$$

Oxidation

$$Fe^{2+} \longrightarrow Fe^{3+} + e^-$$
(Ferrous)   (Ferric)

Reduction

$$CO_2 \longrightarrow CH_4$$
(C^{4+}) (C^{4-})

Fig. 5.2. Examples of oxidation and reduction.

where B is the electron (hydrogen) acceptor and is being reduced, and A is the compound being oxidized. Although all the reactions involve oxidation, they sometimes are referred to in terms of the type of hydrogen acceptor. Transformations in which oxygen is the hydrogen acceptor are called oxidation; when nitrate is the hydrogen acceptor, denitrification; when sulfate is the hydrogen acceptor, sulfate reduction; and with carbon dioxide as the hydrogen acceptor, the transformation is methane fermentation. Oxidation and reduction reactions do not occur independently but as coupled reactions. When a compound is oxidized, another compound must be reduced.

The reducing agent is an electron donor, and an oxidizing agent is an electron acceptor. Each electron donor has a characteristic electron pressure and each electron acceptor has a characteristic electron affinity (1). Electron donors may be arranged in a series of decreasing electron pressures. The tendency will be for electrons to flow from compounds having the highest electron pressure to compounds lower in the series.

A schematic diagram of the electron flow in the aerobic oxidation of an organic compound illustrates the oxidation–reduction sequence (Fig. 5.3). In each step, two electrons are passed along. The electrons flow from compounds

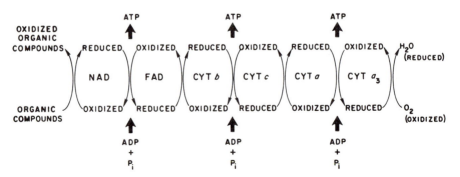

Fig. 5.3. Schematic diagram of the oxidation of organic compounds.

having the highest electron pressure, with the oxidized forms of the electron carriers serving as electron acceptors. The electron carriers noted in Fig. 5.3 are nicotinamide adenine dinucleotide (NAD); flavin adenine dinucleotide (FAD); cytochromes (CYT) b, c, a; and cytochrome oxidase (CYT a_3). The cytochromes are iron-containing enzymes. The iron is reduced and oxidized as the oxidation–reduction reactions occur. The respiratory sequence involving the cytochromes is the final common metabolic pathway by which all electrons derived from the oxidation of different organics flow to oxygen, the final oxidant, or acceptor of electrons in aerobic cells.

In anaerobic cells, the cytochrome pathway does not exist. The energy-conserving steps of the cytochrome system are not available to organisms that do not use oxygen as the terminal electron or hydrogen acceptor. An example of the difference in relative energy conversions by aerobic and anaerobic cells can be illustrated by the metabolism of glucose (Fig. 5.4). Anaerobic cells obtain energy from the conversion of glucose to lactate, which then leaves the cell as a metabolic waste. The energy available to anaerobic cells from this conversion is only about 7% of the amount that would be available if glucose were oxidized aerobically. Aerobic organisms can conserve for themselves a greater portion of the available energy from the metabolism of organic matter than can anaerobic organisms. Thus anaerobic organisms must process a greater quantity of food to obtain the same amount of energy.

This information is useful in predicting the products and the efficiency of aerobic and anaerobic biological treatment processes. The synthesis of organisms per molecule of ATP should be the same for most bacteria since, once the substrate energy is converted to ATP or biological energy, the growth of organisms follows generally similar biochemical pathways. Both aerobic and anaerobic bacteria have essentially the same composition and both contain ATP. The concept of ATP as an energy resource permits the formulation of a general relationship between substrate and growth (2). Because the energy recovery per unit of food is so small for anaerobic organisms, it follows that the amount of microbial cells synthesized per unit of food metabolized will be significantly less than that for aerobic organisms.

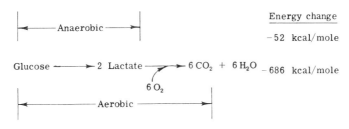

Fig. 5.4. Aerobic and anaerobic metabolism of glucose.

Figure 5.4 shows the end products of aerobic metabolism in the most oxidized state, carbon dioxide and water. The end products of anaerobic metabolism are in a partially oxidized state and, if oxygen were available, would have an oxygen demand.

For a given organic loading, aerobic conditions will produce a more oxidized end product or effluent than will anaerobic conditions and will permit synthesis of a greater quantity of microbial cells. These additional cells are an asset because it is thus possible to have a larger amount of active microbial solids to increase the removal of organic wastes. The greater synthesized microbial cells in an aerobic unit will increase the sludge disposal problem, however. Anaerobic conditions will produce smaller quantities of microbial cells for ultimate disposal, but because of the decreased rates of synthesis, there may be problems in maintaining adequate microbial solids in anaerobic units. The end products of anaerobic units are in the partially oxidized state. Some of the end products, e.g., methane, carbon dioxide, hydrogen, and nitrogen gases, can be exhausted to the atmosphere without problems. Others can produce disagreeable odors, e.g., mercaptans, amines, or volatile acids, and will exert an oxygen demand when released to the environment.

Microorganisms

Biological waste treatment processes contain a mixture of microorganisms capable of metabolizing organic wastes. Within limits, they can adjust to varying organic loads and environmental influences, such as temperature and pH, that may be imposed. Extreme temperature, high concentrations of metal ions, or toxic chemicals can decrease or eliminate the activity of the microorganisms. The microorganisms in various biological treatment systems include bacteria, fungi, algae, protozoa, rotifers, crustacea, bacteriophage, worms, and insect larvae, depending upon environmental conditions.

Bacteria

Bacteria are the most important group of microorganisms in waste treatment systems. The diverse biochemical activities of bacteria, as a group, enable them to metabolize most organic compounds found in municipal, industrial, or agricultural wastes. Aerobes and facultative bacteria are active in all aerobic treatment units. Facultative and obligate anaerobic bacteria are active in anaerobic treatment units. Bacteria are single-cell microorganisms that metabolize soluble food. Insoluble foods are converted into soluble food by microbial enzymes. The bacteria exist in a variety of forms, usually some modification of a cylinder or ovoid, with dimensions on the order of a few micrometers. They exist in waste treatment processes in agglomerations of various arrangements and species.

A microbe is a complex organized system. A representative formula for bacterial cells is $C_5H_7O_2N$ or $C_{75}H_{105}O_{30}N_{15}P$. The composition of bacteria is not constant and varies according to the stage of growth and the particular substrate utilized. Storage of reserve materials can occur during the growth phase and will alter the representative composition of bacterial protoplasm. This empirical formula expresses only the average proportions of the principal constituents in a bacterial cell. The cell contains many other elements in small amounts.

Bacteria can be divided into groups, depending upon their source of carbon used for synthesis of protoplasm. Organisms that use organic carbon as their carbon source are heterotrophic organisms, while organisms that utilize carbon dioxide for cellular carbon are autotrophic organisms. Heterotrophs are the most numerous and important group of organisms in the common biological waste treatment processes.

The microbial species having the fastest growth rate and the ability to utilize most of the available organic matter will be the predominant species. Shifts in microbial predominance occur in waste treatment systems as environmental conditions such as temperature, pH, available dissolved oxygen, ultimate hydrogen acceptor, or available food vary in the system.

A useful characteristic of some bacteria is their ability to flocculate. Such flocculation permits the removal of microbial solids in a subsequent solids separation unit and assists in the production of a good quality effluent.

Fungi

Fungi are nonphotosynthetic, multicellular, aerobic, branching, and filamentous microorganisms that metabolize soluble food. Both bacteria and fungi can metabolize the same kinds of organic material. The environmental conditions will determine which group of organisms will predominate. Fungi will predominate at low pH levels, low moisture content, in low-nitrogen wastes, and when certain nutrients are missing. The composition of fungal cells can be represented empirically by $C_{10}H_{17}O_6N$.

Fungi are not active in anaerobic systems. Since fungal cells contain less nitrogen than bacterial cells, fungi may compete more favorably in wastes having a lower nitrogen content than required for bacterial synthesis. Many fungi grow well at pH levels of 4 to 5, levels at which it is difficult for bacteria to compete.

The filamentous nature of the fungi make them less desirable in biological waste treatment units because they do not settle well. Under the normal environmental conditions that exist in most waste processes, the fungi will not predominate. Fungi will be of secondary importance in common, properly operating aerobic biological treatment units.

Filamentous organisms may predominate when food processing wastes that are nutritionally unbalanced, such as sugar processing or cheese processing wastes, are treated in aerobic treatment processes. These wastes have a high carbon-to-nitrogen (C/N) ratio. Adding nutrients such as nitrogen changes the nutritional conditions, results in a predominance of bacteria and nonfilamentous organisms, and results in a sludge with better settling properties. A nutritionally balanced substrate is desirable for efficient performance of biological treatment or stabilization processes.

Algae

Algae are photosynthetic autotrophs. The composition of algal cells can be represented by $C_{106}H_{180}O_{45}N_{16}P$. Since the nutritive requirements of algal species are different, this formula is an empirical average. Algae obtain their energy from sunlight and utilize inorganic materials such as carbon dioxide, ammonia or nitrate, and phosphate in the synthesis of additional cells. In photosynthesis, molecular oxygen is formed. It is released to the environment and utilized by bacteria as the bacteria metabolize available organic matter. The design and management of oxidation ponds attempts to balance and exploit both groups of organisms.

Algae obtain carbon dioxide from the following sources in water or wastewater: (a) absorption from the atmosphere, (b) respiration of aerobic and anaerobic heterotrophic organisms, and (c) bicarbonate alkalinity. As the carbon dioxide is removed from a wastewater by growing algae, the pH will increase, pH values as high as 10 are not uncommon in active algal systems such as oxidation ponds and similar units. Although algal growth can be controlled by carbon limitation, carbon from alkalinity and bacterial carbon dioxide production provides an ample amount of carbon for algal growth. Carbon in natural systems rarely limits algal growth.

Algae are of consequence only where sufficient sunlight can penetrate the liquid. Algae will not predominate where there is high turbidity, as in activated sludge units and aerated lagoons, where sunlight is excluded, or where the liquid is dark in color.

In the absence of sunlight photosynthesis ceases, and the endogenous respiration of the algae continues in the same manner as it does with bacteria. The algae thus present an additional oxygen demand on the unit in which they exist.

After 0.5–1 year of aerobic decomposition, an average of 50% of the initial nitrogen and phosphorus remained in the undecomposed algal fraction, while the other 50% was regenerated. Under anaerobic conditions 40% of the nitrogen and 60% of the phosphorus were regenerated (3). Nutrients from dead algal cells are released to surface waters for a long period of time. To accomplish a high degree of organic carbon, nitrogen, and phosphate removal in biological treatment units utilizing algae, the algal cells must be removed from the unit effluent before discharge.

Protozoa

Protozoa are single-celled organisms that can metabolize both soluble and insoluble foods. The protozoa found in aerobic treatment systems include flagellates, free-swimming ciliates, and stalked ciliates which are attached to solid particles by stems. Protozoa reduce the concentration of bacteria and nonmetabolized particulate organic matter in a treatment system and assist in producing a higher-quality and clearer effluent.

Activated sludge units free of protozoa produced effluents of high turbidity. The turbidity was caused by the presence of large numbers of dispersed bacteria. As a result, effluent BOD and nonsettleable solids were high. The addition of ciliated protozoa to these units increased effluent quality and decreased bacterial numbers (4). The succession of protozoa types in aerobic systems can be related to the degree of treatment in the system.

Protozoa generally have more complex nutritional requirements than do bacteria or fungi. Because of their utilization of particulate organic matter and their need for dissolved oxygen, protozoa will exist in well-stabilized systems in which the soluble food has been converted to microbial cells and in which the oxygen supply exceeds the oxygen demand. Since they are sensitive to dissolved oxygen changes, they can serve as indicators of the status of aerobic biological waste treatment.

Protozoa have been observed in anaerobic treatment of sewage solids and in systems treating animal wastes, especially ruminant wastes. The role of protozoa in these systems is unknown, but it is postulated as the same as in aerobic systems, i.e., metabolism of particulate material and bacteria, and clarification of the resultant effluent.

Rotifers

Multicellular organisms such as rotifers, which can metabolize solid food, are found in highly stabilized systems having dissolved oxygen at all times. Rotifers metabolize solid particles, some of which the protozoa cannot use, and also assist in producing a nonturbid effluent.

Crustacea

Crustacea are multicellular organisms with hard shells. They grow in well-stabilized systems, using smaller organisms as their major source of food. In doing so, they assist in producing a clarified effluent and are indicative of a high-quality effluent from aerobic treatment systems. *Daphnia* are crustacea commonly found in aerobic treatment systems producing a low BOD effluent such as oxidation ponds. These invertebrates feed by filtering algae, protozoa, bacteria, and organic detritus from the water.

Summary

The predominance of the various forms of microorganisms in biological systems may at times be indicative of the performance and environmental conditions in the systems. Microscopic examination of the biological system can be utilized as a tentative guide to the quality of the effluent, the degree of treatment that has been accomplished, and changes occurring in the systems.

A schematic of the relative predominance of the types of organisms in an

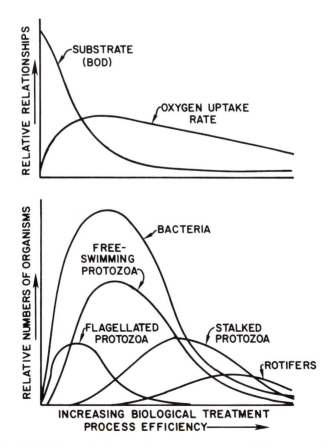

Fig. 5.5. Relative predominance of microorganisms in aerobic treatment systems.

aerobic treatment unit is shown in Fig. 5.5. With experience, the observer can relate the presence and predominance of various types of microorganisms in a process with trends in effluent quality and process performance.

For instance, when there is a large quantity of substrate (BOD) in an aerobic unit, the oxygen uptake rate will be relatively high, and bacteria, free-swimming protozoa, and flagellated protozoa will be the predominant microorganisms. Because of the high BOD and type of microorganisms in the unit, the treatment efficiency will be low, and the solids will not settle rapidly.

When the treatment efficiency of the unit is high, the oxygen uptake rate is lower, and stalked protozoa and rotifers begin to increase in numbers. Large numbers of stalked protozoa usually occur in a well-stabilized system, i.e., a system with low BOD and sludge of good settling characteristics. Because of their need for dissolved oxygen, rotifers and crustacac will be found in well-stabilized systems.

Thus, the relative types and numbers of microorganisms in an aerobic process can be a routine tool to follow the performance of the process.

Biochemical Transformations

A number of changes take place in biological units. Some of the transformations affect the constituents of the wastes undergoing treatment, thus affecting the quality of the unit effluent. Others affect the properties and quantity of microbial solids. Many of the important transformations in biological waste treatment systems are discussed in subsequent sections. These are by no means the only ones, but they are fundamental transformations in a variety of treatment systems.

Carbon

The oxidation of organic carbon-containing compounds represents the mechanism by which heterotrophic organisms obtain the energy for synthesis. The process is called respiration. The general relationships were noted in Eq. (5.1). In aerobic treatment systems, organic carbon is transformed via many steps to synthesized microbial protoplasm, $C_5H_7O_2N$, and carbon dioxide:

$$\text{Organic carbon} + O_2 \rightarrow C_5H_7O_2N + CO_2 \qquad (5.4)$$

The uptake of oxygen and formation of carbon dioxide represent the effects of respiration.

In anaerobic systems, molecular oxygen cannot be the terminal electron acceptor, and all of the respired carbon will not be transformed to carbon dioxide. Under anaerobic conditions, organic carbon is converted to microbial solids, carbon dioxide, methane, and other reduced compounds. Anaerobic metabolism leading to the formation of methane occurs in a series of steps. For simplicity these can be summarized as the conversion of complex organics to simpler compounds:

$$\text{Organic carbon} \rightarrow \text{microbial cells} + \text{organic acids, aldehydes, alcohols, etc.} \quad (5.5)$$

and the conversion of the simpler compounds to gaseous end products:

$$\text{Organic acids} + \text{oxidized organic carbon} \rightarrow$$
$$\text{microbial cells} + \text{methane} + \text{carbon dioxide} \quad (5.6)$$

Little stabilization of organic matter occurs in the first step [Eq. (5.5)]. Stabilization of the organic matter occurs in the second step [Eq. (5.6)], in which the carbon compounds, carbon dioxide (CO_2) and methane (CH_4), are released to the atmosphere and removed from the substrate. The oxygen demand of the waste is thus reduced. At standard conditions, the production of 5.6 ft^3 of methane results in the stabilization of 1 lb of ultimate oxygen demand.

Nitrogen

Nitrogen is an important nutrient in biological systems. Nitrogen is about 12% of bacterial protoplasm and 5–6% of fungal protoplasm. In waste matter, nitrogen will be present as organic and ammonia nitrogen, the proportion of each depending upon the degradation of organic matter that has occurred. In biological systems, organic nitrogen compounds can be transformed to ammonium nitrogen and oxidized to nitrite and nitrate nitrogen:

$$\text{Organic N} \rightarrow \text{ammonium N} \rightarrow \text{nitrite N} \rightarrow \text{nitrate N} \quad (5.7)$$

The oxidation of ammonia to nitrite and nitrate is termed "nitrification" and occurs under aerobic conditions. A more basic definition of nitrification is the biological conversion of inorganic or organic nitrogen compounds from a reduced to a more oxidized state. In waste treatment the term usually is used to refer to the oxidation of ammonia. A residual dissolved oxygen concentration of about 2 mg per liter has been found necessary for optimum nitrification. Autotrophic bacteria, such as *Nitrosomonas*, which obtain energy from the oxidation of ammonia to nitrite, and *Nitrobacter*, which obtain energy from the oxidation of nitrite to nitrate, are organisms that in combination can accomplish the complete oxidation of nitrogen.

Ammonia nitrogen is the main soluble nitrogen end product in anaerobic units. The release of ammonia nitrogen to aerobic treatment units or to receiving streams creates an added oxygen demand to these systems. The oxidation of 1 lb of ammonia nitrogen to nitrate nitrogen will require 4.57 lb of oxygen. The oxygen demand of ammonia is significant and requires consideration when evaluating the effect of discharging wastes to the environment and when evaluating the design of adequate biological treatment processes.

Denitrification is the process by which nitrate and nitrite nitrogen are reduced to nitrogen gas and gaseous nitrogen oxides under anoxic conditions. This process requires the availability of electron donors (reducing agents). The necessary donors can be organic material (such as methanol), the addition of untreated wastes, unmetabolized organic matter, or the endogenous respiration of microbial cells.

Denitrification offers the opportunity to reduce the nitrogen content of waste effluents by having a fraction of the nitrogen exhausted to the atmosphere as an inert gas.

Phosphorus

The sources of phosphorus in wastewaters include organic matter, phosphates originating in cleaning compounds used for process cleanup, and the urine of man and animals. The organic phosphorus is transformed to inorganic phosphorus during biological treatment.

Condensed phosphates constitute a substantial portion of the phosphorus in municipal sewage. The form of phosphates in wastewaters is of interest since phosphate removal techniques generally are evaluated on their ability to remove orthophosphates. The hydrolysis of condensed phosphates to orthophosphate [Eq. (5.8)] is affected by environmental conditions such as temperature and microbial concentration.

$$\text{Tripolyphosphate } (P_2O_{10}^{5-}) + H_2O \rightarrow \text{orthophosphate } (PO_4^{3-}) + H^+ \qquad (5.8)$$

The rate of hydrolysis of condensed phosphates in the following systems decreases in the order given: activated sludge, untreated wastewater, algal cultures, and natural waters.

Aerobic biological treatment will convert condensed phosphates to orthophosphates. Anaerobic treatment will result in other changes. A primary step in anaerobic treatment is the liquefaction of organic matter, and inorganic phosphorus compounds will be released from organic compounds. The effluent from an anaerobic unit can contain a greater concentration of soluble phosphorus compounds than the influent. The release of such effluents to other parts of a waste treatment facility or to the environment can complicate and/or negate phosphorus removal processes at the facility.

Sulfur

Microbial transformations of sulfur are similar to those of nitrogen. Both sulfide and ammonia are decomposition products of organic compounds. Both are oxidized by autotrophic bacteria, as are other incompletely oxidized inorganic sulfur and nitrogen compounds. Sulfate and nitrate are reduced by microorganisms under anaerobic conditions.

Inorganic unoxidized sulfur compounds and elemental sulfur are oxidized by photosynthetic and chemosynthetic bacteria as well as by certain heterotrophic microorganisms. Under anaerobic conditions, sulfide is the reduced end product, and under aerobic conditions, sulfate is the oxidized end product.

All organisms contain sulfur and are involved in the transformations of sulfur to some degree. The assimilation of sulfur into cellular protoplasm is the primary reaction of heterotrophic organisms. With other organisms, sulfur transformations can provide the energy for metabolism, and sulfur compounds can be hydrogen donors or acceptors. Because of these reactions, certain bacteria are designated as sulfur bacteria. These bacteria are autotrophic, can utilize sulfur or incompletely oxidized inorganic sulfur compounds as reducing agents, i.e., direct or indirect hydrogen donors, and can assimilate carbon dioxide as their sole source of carbon.

Food and Mass

The primary purpose of biological waste treatment is to oxidize the organic content of the waste, i.e., the food for the microorganisms. The waste concentration decreases as the microbial mass increases. In aerobic systems, approximately 0.7 lb of cell mass is synthesized for every 1.0 lb of food, as BOD, that is oxidized. Following extensive endogenous respiration, or aerobic digestion of the cells, the 0.7 lb of cells will be reduced to about 0.17 lb of residual cellular material that remains for ultimate disposal. The actual residual cellular solids in a system will be somewhere between the latter two values, depending upon how the aerobic system is operated, i.e., the degree of endogenous respiration that takes place. Changes similar to these also occur in anaerobic systems.

Engineers generally use the volatile suspended solids concentration of a biological treatment unit as an estimate of the concentration of active microorganisms in the unit. While this parameter is an imperfect measure of the active mass, it has been a useful design and management parameter. Other parameters have been explored as better measures of both biomass and bioactivity in treatment units. These include dehydrogenase enzyme activity to measure overall rates of cellular oxidation reactions, specific enzymes involved in intermediary metabolism, and DNA concentration. ATP is a specific measure of microbial activity and can be used to estimate viable microorganism concentrations in a biological treatment unit.

Oxygen

Oxygen plays a critical role in biological systems since, when it serves as an ultimate hydrogen acceptor, the maximum energy is conserved for the microorganisms. Minimum dissolved oxygen concentrations of from 0.2 to 0.6 mg per liter are necessary to maintain aerobic systems. The dissolved oxygen concentrations in aerobic treatment units should be kept above about 1.0 mg per liter if oxygen limitations are to be avoided.

pH

Biological activity can alter the pH of a treatment unit. Photosynthesis, denitrification, organic nitrogen breakdown, and sulfate reduction are examples of biological reactions that can cause an increase in pH. Sulfate oxidation, nitrification, and organic carbon oxidation are examples of biological reactions that can cause a decrease in pH. The relative changes in pH will be affected by the buffer capacity of the liquid and amount of substrate utilized by the microorganisms.

Nutrient Needs

To achieve satisfactory biological treatment of wastes, the wastes must contain sufficient carbon, nitrogen, phosphorus, and trace minerals to sustain optimum rates of microbial synthesis. In most wastes, nutritional balance is not a problem, since there usually is more than enough nitrogen, phosphorus, and trace minerals with respect to the carbon used in cell synthesis. These excess nutrients can be a cause of eutrophication in surface waters when the treated effluent is discharged. Methods to control or manage these excess nutrients are becoming required before discharge of the treated effluents.

Certain wastes, such as some food processing wastes, may have a deficiency of specific nutrients, which need to be added in proper amounts to accomplish satisfactory biological waste treatment. Knowledge is required of the amount of nutrients that are needed, both to assure that adequate nutrients are available and to avoid excess nutrients in the resultant effluent. Besides being uneconomical, added nutrients appearing in the effluent have the potential of causing environmental quality problems. Inadequate quantities of nutrients, such as nitrogen and phosphorus, tend to decrease the rate of microbial growth, decrease the rate of BOD removal, and impair the settling characteristics of the sludge.

The common approach of avoiding nitrogen or phosphorus limitations is to add nutrients to obtain a BOD:N:P ratio of 100:5:1. This ratio is satisfactory if one wishes to assure no nutrient deficiency but is of little use if the purpose is to have low nitrogen and phosphorus levels in the effluent. The preceding ratio was designed to assure adequate nutrients in high-rate biological treatment. Studies with nutrient-deficient wastes established that 3–4 lb of N per 100 lb of BOD_5 removed and 0.5–0.7 lb of P per 100 lb of BOD removed would avoid nutrient-deficient conditions (5). This results in a BOD:N:P ratio of 100:3:0.6. Other studies with food processing wastes have noted that a BOD:N ratio of 100:2 or 100:1.5 was satisfactory in treating cannery wastes, without a decrease in process efficiency (6).

The actual nutritional needs will be related to the manner in which the biological treatment process is operated. A high-rate process will have a high rate of microbial synthesis and a higher nutrient requirement. However, for a stationary- or declining-growth biological treatment system, such as most treatment systems, a lower rate of microbial synthesis and nutrient requirement will prevail. With the long solids retention time, a matter of days in the common treatment systems, endogenous respiration of the microbial cells will release nutrients to the system. These nutrients will be used in the synthesis of new microbial cells. Approximately 0.11 lb of nitrogen will be released from the oxidation of 1 lb of microbial cells.

Certain trace elements also are needed for satisfactory microbial metabolism. These elements act as catalytic or structural components of larger molecules. Of particular interest are copper, molybdenum, zinc, selenium, cobalt, and manganese. Because only trace amounts of these elements are needed, agricultural wastes or residues usually contain adequate amounts. The separate addition of these and other trace elements to waste treatment systems rarely is necessary.

Oxygen Demand Measurements

Biochemical Oxygen Demand

The biochemical oxygen demand (BOD) test is one of the most widely applied analytical methods in waste treatment and water pollution control. The test attempts to determine the pollutional strength of a waste in terms of microbial oxygen demand and is an indirect measure of the organic matter in the waste. It evolved as an estimate of the oxygen demand that a treated or untreated waste will have on the oxygen resources of a stream. The acceptable BOD test is described in "Standard Methods" (7).

Experience with a number of organic wastes has indicated that the change in the BOD of the waste can be characterized by a first–order equation

$$\frac{dC}{dt} = -kC \tag{5.9}$$

where C is the waste concentration and k is a proportionality constant referred to as the BOD rate constant. Expressing the waste concentration in terms of the amount of oxygen required to biologically oxidize the waste, Eq. (5.9) can be written as

$$Y = L(1 - e^{-kt}) \tag{5.10}$$

in which Y is the oxygen or BOD exerted in time t and L is the ultimate amount of oxygen to biologically oxidize the carbonaceous waste or the ultimate first-stage BOD (Fig. 5.6a). The BOD test is conducted as a batch aerobic experiment in BOD bottles. Other oxygen demand determinations can be conducted in manometric respirometers and similar large containers.

Microorganisms can oxidize both carbon-containing compounds (carbonaceous demand) and nitrogen compounds (nitrogenous demand). The nitro-

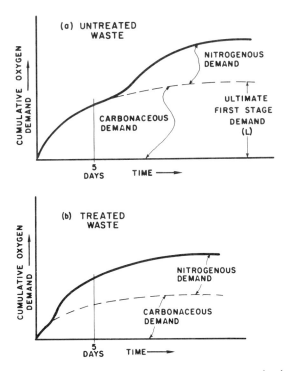

Fig. 5.6. Generalized oxygen demand patterns for heterogeneous untreated and treated wastes.

gen-oxidizing bacteria are autotrophs, normally not in large concentrations in untreated wastewaters. These organisms can be present in well-oxidized waste-waters, such as aerobically treated wastewater effluents from activated sludge and trickling filter plants, and in streams. If a low concentration of nitrifying organisms is present in a BOD bottle, a lag period can exist before the nitrifiers are present in large enough numbers to exhibit a noticeable nitrogenous demand. In wastewaters containing a number of organic compounds, such as agricultural wastewaters, a two-stage oxygen demand frequently can be observed if the oxygen demand is measured over a long enough period.

The BOD test is standardized at a 5-day period at 20°C using prescribed quality dilution water to permit comparison of results. The test does not provide an absolute measure of the oxygen-consuming organic matter in the waste. Unless otherwise noted, results reported as BOD indicate that the test has been conducted under standard conditions. BOD values are affected by time and temperature of incubation, presence of adequate numbers of microorganisms capable of metabolizing the waste, and toxic compounds. In practice a constant temperature of 20°C does not occur and water is not in a state of rest for 5 days in the dark. The bacteria used in the test are the real link with what actually occurs in practice.

The k value in Eqs. (5.9) and (5.10) changes with temperature, increasing with increasing temperatures. Relationships are available to determine k at temperatures other than 20°C:

$$k_T = (k_{20})\theta^{(T-20)} \tag{5.11}$$

Experiments have shown that θ can vary from 1.016 to 1.077. The value commonly used for θ is 1.047. The maximum or total oxygen demand is not affected by temperature, since it is a function only of the quantity of organic matter available.

With heterogeneous wastes, the 5-day period frequently occurs before the beginning of significant nitrification. With adequately treated wastes, the car-bonaceous demand is caused primarily by microbial cells and is low. In addition, a significant population of nitrifiers could be present. Under these conditions, the nitrogenous demand can occur early and before the 5-day test period (Fig. 5.6b). Because of these conditions, a treated wastewater may exhibit both a car-bonaceous and nitrogenous demand while an untreated wastewater may exhibit only a carbonaceous demand. When this occurs, evaluation of the BOD removal at a wastewater treatment facility will be in error since the oxygen demand of the influent and effluent samples does not measure the same materials.

The maximum oxygen demand (OD_m) will occur if all of the unoxidized organic and inorganic matter is oxidized completely. In most biological treat-ment units, this means that the organic carbon in a waste is oxidized to carbon dioxide and all of the organic and ammonia nitrogen is oxidized to nitrate. Mathematically, this can be expressed as

$$OD_m = 2.67C + 4.57N \qquad (5.12)$$

where C is the organic carbon concentration, and N is the sum of the organic and ammonia nitrogen expressed as nitrogen. In practice not all of the organic carbon in a waste may be oxidized, not all of the organic nitrogen in the waste may be converted to ammonia, not all of the ammonia may be oxidized, and a portion of the synthesized microbial cells will not be completely oxidized. Thus Eq. (5.12) will not determine the actual oxygen demand of a waste in a treatment system or a stream. It can, however, be used to estimate the maximum oxygen demand that could occur.

The BOD rate constant k is a function of the oxidizability of the waste material. Wastes having a high soluble organic content, such as milk wastes, will exhibit a higher k than wastes having a high particulate organic content, such as a cellulosic waste. Microorganisms must first solubilize particulate organic material before it can be metabolized, whereas they can metabolize soluble wastes more rapidly.

The first-order equations expressed in Eqs. (5.9) and (5.10) represent a best estimate of the rate of oxygen demand and have been the subject of considerable discussion over the years. Because of the heterogeneous nature of most waste mixtures, the different waste constituents can have varying reaction rates. For a specific waste, the oxygen demand may be better described by a second-order equation or a composite exponential equation. Heavy reliance should not be placed on the monomolecular or first-order oxygen demand relationship, since it is only an estimate of the complex reactions that are taking place in the biological reactor and frequently reflects only the carbonaceous demand. A nitrogenous demand and, in some wastes, an initial chemical oxygen demand also can occur.

The small quantity of oxygen present in the BOD test bottle, 2–3 mg, means that high-strength wastes, such as many food processing wastewaters and animal wastes, must be diluted prior to analyzing the waste. Prior to BOD analysis, animal wastes may require dilutions of 1:100 to 1:1000 or more. The difficulty in diluting wastes that are neither physically or chemically uniform decreases the precision of the standard BOD test, which is estimated to have a precision of $\pm$ 20% (8).

In spite of the problems associated with the BOD test, it remains an important analytical tool in water pollution control work since it is one of the few analyses that attempts to measure the effect of a waste under conditions approximating natural stream conditions.

Chemical Oxygen Demand

The lengthy analytical time of the BOD test as well as a desire to find a more precise measure of the oxygen demand of a waste led to the development of the chemical oxygen demand (COD) test. The COD test is a wet chemical combustion of the organic matter in a sample. An acid solution of potassium dichromate is used to oxidize the organic matter at high temperatures. Various COD procedures, having reaction times of from 5 min to 2 hr, can be used (7, 9).

The use of two catalysts, silver sulfate and mercuric sulfate, are necessary to overcome a chloride interference and to assure oxidation of hard-to-oxidize organic compounds, respectively. Animal wastes and certain food processing wastes such as those from sauerkraut, pickle, and olive processing can contain high chloride concentrations and will require the use of the mercuric sulfate in COD analyses or a chloride correction factor. Compounds such as benzene and ammonia are not measured by the test. The COD procedure does not oxidize ammonia, although it does oxidize nitrite.

BOD and COD analyses of a waste will result in different values because the two tests measure different material. COD values are always larger than BOD values. The differences between the values are due to many factors such as chemicals resistant to biochemical oxidation but not to chemical oxidation, such as lignin; chemicals that can be chemically oxidized and are susceptible to biochemical oxidation but not in the 5-day period of the BOD test, such as cellulose, long chain fats, or microbial cells; and the presence of toxic material in a waste, which will interfere with a BOD test but not with a COD test.

In spite of the inability of the COD method to measure the biological oxidizability of a waste, the COD method has value in practice. For a specific waste and at a specific waste treatment facility, it is possible to obtain reasonable correlation between COD and BOD values. Examples of the BOD and COD values of agricultural wastes are noted in Chapter 4. The method is rapid, more precise ($\pm 8\%$) (8), and in most circumstances provides useable estimates of the total oxygen demand of a waste.

Changes in both the BOD and COD values of a waste will occur during treatment. The biologically oxidizable material will decrease during treatment, whereas the nonbiological but chemically oxidizable material will not. The nonbiologically oxidizable material will exist in the untreated wastes and will increase because of the residual cell mass resulting from endogenous respiration. The COD/BOD ratio will increase as the biologically oxidizable material becomes stabilized.

The COD/BOD ratio can be used to estimate the relative degradability or oxidizability of a waste. A low COD/BOD ratio would indicate a small nonbiodegradable fraction. A waste with a high COD/BOD ratio such as animal wastes has a large nonbiodegradable fraction that will remain for ultimate disposal after treatment. Wastes that have been treated, such as waste-activated sludge or waste-mixed liquor from oxidation ditches, have a high COD/BOD ratio, indicating that most of the organic matter has been metabolized and that further treatment might not be economically rewarding.

As with BOD rates, COD data must be used with caution and judgment. Both can be used, separately and together, to estimate the oxidizability of a waste and its effect on a stream.

Total Organic Carbon

Total organic carbon (TOC) is measured by the catalytic conversion of organic carbon in a wastewater to carbon dioxide. No organic chemicals have been found that will resist the oxidation performed by the equipment now in use. The time of analysis is short, from 5 to 10 min, permitting a rapid estimate of the organic carbon content of wastewaters. The relationship of TOC values of wastes and effluents to pollution control results requires further understanding. A TOC value does not indicate the rate at which the carbon compounds degrade. Compounds analyzed in the TOC test, such as cellulose, degrade only slowly in a natural environment. Values of TOC will change as wastes are treated by various methods.

BOD and COD utilize an oxygen approach. TOC utilizes a carbon approach. There is no fundamental correlation of TOC to either BOD or COD. However, where wastes are relatively uniform, there will be a fairly constant correlation between TOC and BOD or COD. Once such a correlation is established, TOC can be used for routine process monitoring.

Total Oxygen Demand

The total oxygen demand (TOD) of a substance is defined as the amount of oxygen required for the combustion of impurities in an aqueous sample at a high temperature (900°C), using a platinum catalyst. The oxygen demand of carbon, hydrogen, nitrogen, and sulfur in a wastewater sample is measured by this method. The interpretation of TOD values to treatment plant efficiency or to stream quality requires further investigation but generally can be related to BOD and COD values. A portion of both TOC and TOD values will represent non-biodegradable matter. TOD and TOC methods are rapid and can be incorporated in wastewater and treatment plant control systems.

Temperature

General

Temperature is an important factor affecting waste treatment, since it influences the physical properties of the liquid under treatment, the rates of the biological and chemical treatment processes, and the waste assimilation capacity of land or water. Most investigations have concluded that, within specific temperature ranges, an equation of the form

$$k = ae^{bT} \qquad (5.13)$$

can be used to express the relationship of a reaction rate k and temperature T (°C). The values of a and b are a function of the specific process and reaction. Reaction rates or coefficients are determined or known at a specific temperature and can be estimated at other temperatures by

$$k_{T_2} = k_{T_1}e^{c(T_2 - T_1)} = k_{T_2}\theta^{(T_2 - T_1)}) \qquad (5.14)$$

where T_1 is the temperature at which the reaction rate or coefficient is known (k_{T_1}) and T_2 is the temperature at which the rate or coefficient is desired.

Equation (5.14) is only an approximation, since the temperature variation is theoretically exponential in nature. However, over a limited range, the relationship can be utilized for practical purposes. The temperature effect is the alteration of the rates of specific enzyme activity. In general, only the overall effect of temperature is of engineering and design interest, and Eq. (5.14) has proven adequate to describe the temperature effects on biological systems. The value of θ will be different for different treatment systems and different physical, chemical, and biological conditions. Table 5.2 summarizes values of θ for a number of biological processes.

For simplicity, it is assumed that θ is not a function of temperature. The error introduced in assuming θ to be independent of temperature is in the range of 10–15%, which may be acceptable in most engineering work. Where more precise results are necessary, specific studies may be necessary.

TABLE 5.2

Temperature Coefficients (θ) for Biological Waste Treatment Processes

	Temperature coefficients	
Treatment	Range	Average
Activated sludge	1.0–1.04	1.03
Aerated lagoon	1.085–1.1	—
Anaerobic lagoon	1.08–1.09	1.085
Trickling filter	1.035–1.08	—
Aerated lagoon	—	1.035
Aerated lagoon	—	1.05

Because temperature has an effect on many fundamental factors, e.g., viscosity, density, surface tension, gas solubility, diffusion, and enzyme activity, it is illogical to assume that θ should be constant over the entire temperature range affecting a treatment process. A number of investigations have demonstrated the interrelationship of θ with temperature. An example is shown in Fig. 5.7 (10). After the optimum temperature ranges for the specific microorganisms, the reaction rate decreases and θ becomes negative.

If $\theta = 1.072$, a reaction rate is doubled for an increase in temperature of 10°C. This value, 1.072, is in the same range for the coefficients for many processes and reactions and has led to the common generality that, for the most

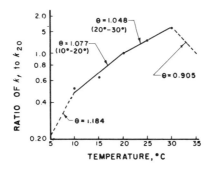

(a) TEMPERATURE COEFFICIENTS
FOR
CARBONACEOUS OXYGEN DEMAND CONSTANTS

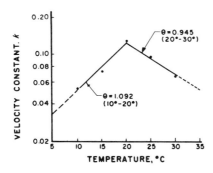

(b) TEMPERATURE COEFFICIENTS
FOR
NITROGENOUS OXYGEN DEMAND CONSTANTS

Fig. 5.7. Variation of carbonaceous and nitrogenous temperature coefficients with temperature (10).

part, biological reaction rates are doubled for each 10°C rise in temperature. This estimation generally is useful for only a given temperature range, around 20°C. Its use is less applicable at extreme temperature conditions.

In a biological treatment system, the reduction of the temperature of the biological unit by 10°C will require about double the active organisms in the unit to achieve equivalent process efficiencies. This can be accomplished by increasing the mixed liquor suspended solids (MLSS) concentration in the unit. Such an increase in MLSS may affect the viscosity of the liquid in certain cases and may reduce the solids settling rate. Although temperature variations of 10°C rarely occur abruptly, an uncovered unit can experience considerable temperature differences between winter and summer. Systems with solids recycle permit the operator to compensate for lower temperatures with increased microbial solids in the system. Systems without solids recycle do not have this flexibility. A knowledge of the effect of temperature on a biological system permits better design and operation of the system.

Process Equations

General

The proper utilization of biological treatment processes depends upon an understanding of the kinetics of the processes and the effects of environmental factors on the kinetics. If organic matter, i.e., waste, is added at a constant rate to a continuous flow biological treatment unit, the unit eventually will reach equilibrium conditions. Until equilibrium conditions occur, the microorganisms will respond to the waste addition and synthesize new organisms until the microbial mass is in equilibrium with the available food supply, i.e., waste. At equilibrium, the net microbial concentration is related to the available substrate and to the decay rate, or endogenous respiration of the organisms. At equilibrium the unit is a food-limited unit.

In this explanation, the substrate, i.e., the carbonaceous material, BOD, or COD, is assumed to be the limiting material. The same relationship would be observed for cases where some other nutrient, i.e., nitrogen, phosphorus, or a trace metal, would limit the maximum reaction rates. For agricultural wastes, nutrients other than carbon rarely are the limiting nutrient in actual waste treatment units. Certain food processing wastes may be deficient in nitrogen or phosphorus. In such cases, sufficient nutrients should be added to these wastes so that the waste treatment process will not be nutrient deficient.

Biological waste treatment generally occurs in a continuous flow, mixed reactor either with or without solids recycle. Mass balance equations for these reactors have been developed and are used to design such reactors so as to achieve the desired performance. These equations are used to determine (a) effluent substrate concentrations, (b) net microbial growth rate, (c) the solids retention time, (d) reactor microorganism concentration, (e) oxygen requirements, and (f) net sludge production. The following sections identify the relevant equations and indicate the effect of process design factors on treatment system performance.

Continuous Growth

The microbial cell concentration is a function of the substrate in the system. With a given and constant substrate input and a desired substrate effluent or removal efficiency, a continuous, nonrecycle process will contain a specific quantity of microbial cells. For this system the only factor that a design engineer can vary to produce different removal efficiencies and effluent quality is the hydraulic retention time (HRT). When a removal efficiency or an effluent substrate concentration is specified, there is only one set of conditions that will meet the specified removal efficiency.

The basic relationship for predicting microbial growth in a continuous flow system is

Net microbial growth per unit time
  = (waste utilized per unit time) × (microbial cell yield coefficient)
  − (organism decay coefficient) × (microbial mass in the system)   (5.15)

The net microbial growth produced per unit time can be designated as dX/dT and the waste utilized per unit time as dS/dt. Equation (5.15) can be expressed as

$$\frac{dX}{dt} = Y\frac{dS}{dt} - K_D X \qquad (5.16)$$

where Y is the microbial cell yield coefficient, K_D is the organism decay coefficient, and X is the microbial mass in the system. Equation (5.16) can be rearranged to yield

$$\frac{1}{\text{SRT}} = YU - K_D \qquad (5.17)$$

where SRT equals $X/(dX/dt)$ and U equals $(ds/dt)/X$.

The term $(dX/dt)/X$ is the net specific growth rate (time^{-1}) of the microorganisms in the continuous flow reactor and is the reciprocal of the solids retention time (SRT). Thus, the SRT of a biological treatment unit and the net specific growth rate of the microorganisms in the unit are interrelated. The SRT of a unit must be equal to or greater than the time it takes for the microorganisms to reproduce. Above this minimum SRT, as the SRT increases, the net specific growth rate decreases. This in turn means that there will be a smaller net microbial mass generated for subsequent disposal. The total sludge requiring disposal will be the sum of the net microbial growth in the system and the inert organic and inorganic solids carried in the unit.

The specific substrate utilization rate $(dS/dt)/X$ or U is the rate at which the biodegradable waste (S) is removed from the unit per unit time per unit of microbial mass metabolizing the waste. It commonly is known as the food-to-microorganism ratio (F/M). Mathematically, U can be expressed as

$$U = \left(\frac{S_0 - S}{X} \right) \left(\frac{Q}{V} \right) \qquad (5.18)$$

where S_0 = the influent waste concentration, S = effluent waste concentration, X = microbial mass in the reactor, V = the volume of the reactor, and Q = the volumetric flow rate through the reactor.

These parameters, SRT and U, and Eq. (5.17) are key factors in the understanding, design, and operation of biological treatment systems. The solids retention time (SRT) is sometimes referred to as MCRT (mean cell residence time) or θ_c by various authors. SRT is the time that the microbial mass is retained in the biological system.

Fundamentally, SRT should be determined using the quantity of active microorganisms in the system. However, measuring the active microbial mass in biological treatment systems is difficult. Fortunately, other parameters can be used to determine the SRT of a system. Assuming that complete mixing results in a uniform distribution of the active microorganisms and other solids in a biological treatment system, SRT can be determined by using the quantity of other forms of solids in the system, i.e., volatile suspended solids, total suspended solids, or total solids. In practice, the SRT of the system can be determined by

$$\text{SRT} = \frac{\text{weight of solids in the system}}{\text{weight of solids leaving the system/time}} \qquad (5.19)$$

The solids in the system are those in the aeration unit and in any secondary solids separation unit used to clarify the mixed liquor from the aeration basin.

The actual SRT of a biological treatment system must be greater than the minimum time it takes for the microorganisms to reproduce in the system. If this does not occur, the microorganisms will be removed from the system at a faster rate than they can multiply, and failure of the system will result. The critical SRT or cell washout time can be written as

$$\frac{1}{\text{SRT}_c} = YU_{\max} - K_D \tag{5.20}$$

$U_{\max}$ is the maximum substrate (waste) removal rate per unit of microorganisms. SRT_c is the minimum microbial residence time at which a stable microbial cell population can be maintained. Because the yield coefficient Y is smaller for anaerobic metabolism than for aerobic metabolism, the value of SRT_c for anaerobic processes is larger than for aerobic processes.

The concept of a minimum SRT is important in all biological waste treatment systems, especially in anaerobic treatment systems. In aerobic treatment systems, the energy relationships are such that microorganisms can reproduce rapidly, a matter of hours or less, depending upon the specific system and its management. It is extremely rare to note an aerobic treatment system that failed because its SRT was less than the minimum time necessary for microbial reproduction.

In anaerobic systems, the microorganisms reproduce less rapidly, and a longer minimum SRT is required to accommodate the slower net growth rate. An example of the relationship between the SRT and the efficiency of substrate removal is shown in Fig. 5.8 (11). Values for these parameters that have been obtained for agricultural and other waste treatment systems are presented in Tables 5.3 and 5.4.

Nonrecycle Systems

Many biological treatment systems used with agricultural wastes can be considered as completely mixed, nonrecycle biological treatment systems (Fig.

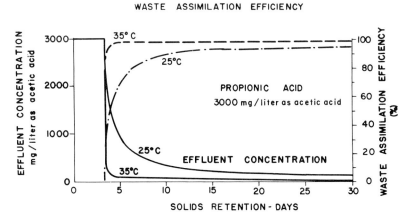

WASTE ASSIMILATION EFFICIENCY

Fig. 5.8. Process efficiency as related to SRT in an anaerobic treatment unit (11).

TABLE 5.3

Ranges of Kinetic Coefficients for Aerobic and Anaerobic Treatment Processes

		Treatment	
Coefficient	Units	Aerobic	Anaerobic
Y	lb volatile suspended solids/lb BOD_{ult}	0.3–0.4	0.03–0.15
K_D	day^{-1}	0.02–0.06	0.01–0.04
U	lb BOD/day/lb volatile suspended solids	4–24	4–20
SRT_c	days	0.1–0.3	2–6

5.9). Examples of such systems include aerated lagoons, oxidation ditches, and mixed anaerobic digestion units. A characteristic of these systems is that no solids are recycled back to the biological units to increase the microbial solids concentration. The microbial solids concentration in these units is directly related to the concentration of waste (food) in the unit.

In such systems, the hydraulic retention time HRT is equal to the SRT because the systems are considered completely mixed and the liquids and solids stay in the system the same length of time. HRT is the theoretical time that a volume of liquid remains in the system:

$$\text{HRT} = \frac{\text{volume of system}}{\text{volume of liquid leaving the system/time}} = \frac{V}{Q} \tag{5.21}$$

The degree of stabilization can be regulated by controlling either the SRT or HRT of such a system.

A substrate balance for a nonrecycle system (Fig. 5.9) will yield

$$S_0 - S_i = X_1 U(\text{HRT}) \tag{5.22}$$

Therefore, for a specific influent waste concentration, the amount of substrate removed and stabilization efficiency is a function of the liquid retention time, the microbial cells in the system, and the substrate removal rate per microorganism.

The waste removal rate U is a system loading rate and has been used as a design loading rate, i.e., quantity of substrate added per unit of microorganisms per unit time. Such rates are generally in terms of pounds of BOD or COD per day per pound of mixed liquor suspended solids.

A mass balance for the nonrecycle system will yield Eq. (5.17), assuming that the microbial mass entering the system, X_0, is small compared to the microbial mass, X_1, in the system. For simplicity, the term YU can be replaced by μ, the microbial growth rate. The term $\mu - K_D$ is the net growth rate of the microbial solids in the system, which is important since it estimates the ac-

TABLE 5.4
Kinetic Coefficients Determined in Specific Waste Treatment Systems[a]

Waste	Y	K_D (day^{-1})	SRT$_c$ (days)	Temperature (°C)	Reference
Packinghouse waste	0.76 lb VSS/lb BOD	0.17	—	35	(12)
Synthetic milk waste	0.37 lb VSS/lb COD	0.07	—	20–25	(12)
Synthetic carbohydrate and protein wastes	—	—	0.52	10	(13)
	—	—	0.21	20	(13)
	—	—	0.14	30	(13)
Ammonia oxidation	0.05 lb VSS/lb N	—	1.5	19	(14)
			0.7	27	(14)
Nitrite oxidation	0.02 lb VSS/lb N	—	1.2	19	(14)
Pear processing	0.49 lb VSS/lb COD	—	0.115	20	(15)
Peach processing	0.46 lb VSS/lb COD	—	0.055	15	(15)
Apple processing	0.56 lb VSS/lb COD	—	0.03	10	(15)
Potato processing	0.6–0.8 lb VSS/lb BOD	—	—	—	(16)
Field extended aeration system	0.54 lb VSS/lb BOD	0.014	—	—	(17)
Municipal wastewater Summary of data from 6 plants	0.336 lb VSS/lb COD	0.16	—	20–25	(18)

[a]VSS, volatile suspended solids.

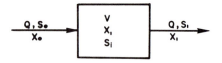

Q = FLOW
S = SUBSTRATE (WASTE) CONCENTRATION
X = MICROBIAL CONCENTRATION
V = REACTOR VOLUME

Fig. 5.9. Completely mixed, nonrecycle biological treatment system.

cumulation of microbial solids that ultimately must be removed from the system. Equation (5.17) also verifies an earlier statement that synthesis and respiration occur simultaneously in biological systems, with synthesis predominating when food is in excess ($\mu > K_D$) and respiration predominating when food, or essential nutrient, is limited ($K_D > \mu$).

The net microbial solids production is inversely related to SRT:

$$\frac{dx}{dt} = \frac{X}{SRT} \tag{5.23}$$

Thus, a biological unit with a long SRT will have less net microbial solids for wasting than will a unit with a shorter SRT. The amount of microbial solids to be wasted can be varied by controlling the SRT of the unit. A long SRT biological unit is less effective in removing inorganic nutrients because net microbial solids production and incorporation of the nutrients into microbial cells is less.

The total quantity of solids that will accumulate or require removal will include the excess microbial solids and the relatively stable organic and inorganic solids contained in the untreated waste which are not altered in the treatment process. The total quantity of excess solids can be estimated by

$$\frac{dX_s}{dt} = \frac{dX}{dt} + \frac{dX_i}{dt} \tag{5.24}$$

where dX_s/dt represents the rate of total sludge solids produced, dX/dt represents the rate of net microbial solids produced, and dX_i/dt represents the rate of stable inorganic and organic solids accumulation.

The value for dX_i/dt varies with the type of waste. Agricultural wastes can have values of dX_i/dt ranging from close to zero for wastes such as milk processing wastes to large values with wastes having large quantities of nonbiodegradable material such as cattle wastes.

The rate at which solids will accumulate in a biological treatment system is dependent upon the waste loading rate and the rates at which solids are lost from the system, inert solids enter the system, volatile waste solids and microbial solids are oxidized in the system, and the waste is being converted into microbial cells.

As noted in Chapter 1, in many situations a specific effluent substrate concentration must be achieved to meet effluent limitations or water quality criteria. In a nonrecycle biological treatment system, SRT, HRT, and U are related directly, and both SRT and U are directly related to the effluent waste concentration (Eq. 5.22) and the process treatment efficiency. Therefore, controlling SRT or U in a nonrecycle system will control the process efficiency and the effluent substrate concentration.

In aerobic systems, the preceding equations can be used to estimate the oxygen requirements. The quantity of oxygen required will be a summation of that required for substrate removal (synthesis) and that required for respiration. For a process such as in Fig. 5.9

$$V \frac{dO}{dt} = aQ(S_0 - S_i) + bK_D X_1 V \tag{5.25}$$

where dO/dt is in terms of milligrams per liter per unit of time, a is a constant used to convert substrate units to oxygen units utilized in synthesis (units of oxygen per unit of BOD or COD), and b is a constant used to convert cell mass units to oxygen units (unit of oxygen per unit of cell mass). Cell mass generally is expressed as volatile suspended solids, although this parameter is an inadequate measure of the active microorganisms. The rate of oxygen utilization is

$$\frac{dO}{dt} = \frac{(S_0 - S_i)a}{\text{HRT}} + bK_D X_1 \tag{5.26}$$

The oxygen requirement varies directly with the substrate to be removed and inversely with the HRT of the system. The longer the retention time of the biological system, the lower the rate of oxygen demand in the system and the greater the total oxygen utilization. The constants a and b and Eq. (5.26) normally do not include the oxygen required for nitrification. The oxygen requirement for nitrification is a function of the amount of ammonia that is nitrified, and the equations can be modified by including the following term:

$$\frac{4.57(\Delta NH_3)}{\text{HRT}} \tag{5.27}$$

where ΔNH_3 is the ammonia oxidized in the unit during aerobic treatment.

Solids Recycle

For a given substrate removal rate, a larger quantity of waste can be treated if the mass of organisms in the system is greater. Systems with solids recycle (Fig. 5.10) can increase the concentration of active organisms in a biological unit and can provide engineers and operators with additional alternatives to obtain satisfactory performance and to decrease the size of the biological unit. The solids separator, generally a sedimentation unit, plays an important role in such systems, since the quantity of recycled solids and the effluent quality depend on the efficiency of the solids separator. The microbial solids must be separated easily; otherwise, the system will approximate a system without recycle (Fig. 5.9), the advantages of recycle will be lost, and the system may not produce an effluent of the desired quality.

A substrate, microbial mass, and oxygen requirement balance can be made for the recycle system in the same manner as was done for the nonrecycle system. A microbial mass balance on the entire system operating at equilibrium and assuming the influent solids concentration to be negligible illustrates that

$$\mu - K_D = \frac{QX_2}{VX_1} = \frac{1}{(HRT)(r)} = \frac{1}{SRT} \tag{5.28}$$

where r is the mixed-liquor-to-effluent cell ratio X_1/X_2 and is numerically greater than 1.0. The relationship between the cell mass in the system, X_1, and the substrate removed, $S_0 - S_i$, for the system in Fig. 5.10 can be expressed by substituting $(HRT)(r)$ for SRT to yield

$$X_1 = \frac{Yr(S_0 - S_i)}{1 + K_D(SRT)} \tag{5.29}$$

which indicates that the microbial mass in the system will be a function of both

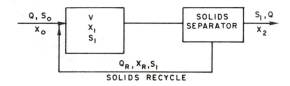

Q = FLOW
S = SUBSTRATE (WASTE) CONCENTRATION
X = MICROBIAL CONCENTRATION
V = REACTOR VOLUME

Fig. 5.10. Completely mixed biological treatment system with recycle.

the substrate removal and the cell ratio. The microbial mass in a biological system with recycle is greater than one without recycle, all other factors being equal. The effluent substrate concentration can be expressed as

$$S_i = S_0 - \left(\frac{X_1[1 + K_D(\text{SRT})]}{Yr} \right) \tag{5.30}$$

The oxygen requirements for a system with recycle also can be described by Eq. (5.31), which results from combining several previous equations:

$$\frac{dO}{dt} = (S_0 - S_i) \left(\frac{a}{\text{HRT}} + \frac{bK_D r}{U(\text{SRT})} \right) \tag{5.31}$$

For a biological system with recycle, the oxygen requirement is related to both HRT and SRT. The oxygen requirement for substrate removal for synthesis is a function of HRT, while the oxygen requirement for respiration is related to the microbial cell residence time.

The HRT of a system with recycle would be less than a system without recycle to achieve a specific substrate removal efficiency. When design values are used in these equations, the oxygen demand rate for a system with recycle will be greater, because of a shorter HRT and a larger quantity of microbial solids in the system, than for a system without recycle used to treat the same wastes.

As indicated by Eq. (5.28), the SRT of the system is a function of both the HRT and the mixed-liquor-to-effluent cell ratio of the system. This permits the system to have a long SRT to obtain a high treatment efficiency, low S_i with a short HRT, and a smaller-size biological unit. Solids recycle also offers the opportunity to maintain a SRT greater than the minimum required for cell growth, with an HRT which could be less than the minimum SRT.

A system with cell recycle offers greater flexibility for design and operation than a system without recycle. When a treatment process is subject to variable waste loads, the microbial solids in the biological unit can be varied by changing the recycle ratio to adjust to the waste loads and produce a consistent effluent. This requires a qualified operator, a consideration when deciding to use a system with or without recycle.

A system without recycle will require a longer HRT and a larger biological unit than will a system with recycle to obtain the same quality of waste effluent. A large biological unit can dampen the surges caused by variable waste load and may provide adequate time for the microbial solids to adjust to an increased waste concentration in the unit. For many agricultural wastes, the decision to use a system with or without solids recycle will hinge on the availability of an operator to manage a system with recycle. Because many facilities producing agricultural wastes are located where land costs are not excessive, treatment systems without recycle, such as oxidation ponds, aerated lagoons, and oxidation ditches are common.

Application to Agricultural Wastes

While fundamental to all biological systems, the mathematical relationships and equations previously noted rarely have been applied to systems treating agricultural wastes. The use of these relationships offers an opportunity to those who wish to develop better biological processes for agricultural wastes and to better understand the processes currently handling these wastes. Because agricultural wastes are organic wastes, some type of biological treatment will be utilized prior to ultimate disposal on the land or to surface waters.

The design engineer can utilize the mathematical relationships to obtain satisfactory performance. The actual design of a waste treatment facility will include an estimation of possible overloads to the system, temperature effects, operator ability and interest, differences in waste characteristics, and allowance for incalculable risks.

Some of the specific differences that are involved with agricultural wastes include the solids concentration in the raw waste, long retention time, non-homogeneity in the biological system, and the oxygen requirements. With municipal and certain industrial wastes, the quantity of solids in the untreated waste is small compared to the quantity of solids in the treatment system. Such an assumption cannot be made for agricultural wastes, especially food processing and animal wastes. These wastes contain a high concentration of solids, which must be considered in all mass balance and mathematical relationships. These wastes frequently have a large nonbiodegradable fraction, which will not be removed in a biological treatment unit and will remain for ultimate disposal. In most of the previous mathematical relationships, the effect of an influent nonbiodegradable fraction was not considered. The total solids for disposal will include the net increase in microbial solids and the nonbiodegradable solids.

One of the basic assumptions made in developing the basic equations was that the biological treatment process was completely mixed. This is not always the case with processes treating agricultural wastes. The solids content and particle size of some of the wastes preclude complete mixing, and solids sedimentation and accumulation in the quiescent areas of the biological unit can occur. In evaluating the performance of biological units treating agricultural wastes, this possibility should be closely checked.

The oxygen requirements are based upon maintaining an excess of dissolved oxygen, generally at least 1–2 mg per liter. When lesser amounts of dissolved oxygen exist in the system, the system is oxygen limited and the microbial reaction rates and treatment efficiency decrease. In long detention time units having a residual dissolved oxygen content, a population of nitrifying organisms can be established, and an oxygen demand caused by nitrification can result.

For most agricultural wastes, especially those disposed of on the land, a residual dissolved oxygen concentration is not required. The minimum input oxygen requirement would be to avoid the odors that are produced under highly reduced conditions.

Notation

X Microbial cell concentration in a biological treatment unit, mass/volume

μ Net specific growth rate, unit of cells produced per unit of existing cells per unit time, time^{-1}

S Substrate concentration, mass/volume

U Substrate removal rate, unit of substrate removed per unit of existing cells per unit time, time^{-1}

Y Microbial yield coefficient, unit of cells produced per unit of substrate removed, mass/mass

Q Waste flow rate to a treatment process, volume/time

S_0 Influent waste concentration, mass/volume

S_i Effluent waste concentration, mass/volume

V Reactor volume, volume

X_1 Mixed liquor microbial cell concentration, generally measured as suspended solids or volatile suspended solids, mass/volume

HRT   Mean hydraulic residence time in a biological reactor, V/Q, time

X_0 Microbial cell concentration in the influent wastes, generally measured as suspended or volatile suspended solids, mass/volume

K_D Microorganism decay rate, unit of cells decreased per unit of existing cells per unit time, time^{-1}

SRT   Mean microbial solids residence time in the biological unit, time

SRT_c Minimum microbial solids residence time, time

X_s Total sludge solids concentration in the process, inert plus microbial solids, mass/volume

X_i Stable inorganic and organic solids in the biological process, mass/volume

a A constant to convert substrate units to oxygen units, unit of oxygen per unit of substrate

b A constant to convert cell mass units to oxygen units, unit of oxygen per unit of cell mass

O Oxygen requirement, mass/volume

ΔNH_3 Ammonia oxidized to nitrate during an aeration time of HRT under aerobic conditions, mass/volume

X_2 Microbial cell concentration in the effluent of a secondary solids separator in a biological treatment process with solids recycle, mass/volume

Q_r Recycle solids flow rate, volume/time

r Mixed-liquor-to-effluent microbial cell ratio, X_1/X_2

References

1. Lehninger, A. L., "Bioenergetics." Benjamin, New York, 1965.

2. McCarty, P. L., Thermodynamics of biological synthesis and growth. *Adv. Water Pollut. Res.* **2,** 169–200 (1964).
3. Foree, E. G., Jewell, W. G., and McCarty, P. L., The extent of nitrogen and phosphorus regeneration from decomposing algae. *Adv. Water Pollut. Res., Proc. Int. Conf. 5th 1970,* 1971.
4. Curds, C. R., Cockburn, A., and Vandyke, J. M., An experimental study of the role of ciliated protozoa in the activated sludge process. *J. Inst. Water Pollut. Contr. (G.B.)* **61,** 312–324 (1968).
5. Helmers, E. N., Frame, J. D., Greenberg, A. E., and Sawyer, C. N., Nutritional requirements in the biological stabilization of industrial wastes. III. Treatment with supplementary nutrients. *Sewage Ind. Wastes* **24,** 496–507 (1952).
6. Parker, C. D., and Skerry, G. P., Cannery waste treatment in lagoons and oxidation ditch at shepparton, victoria, australia. *In* "Proceedings of the Second National Symposium on Food Processing Wastes," pp. 251–270. Pacific Northwest Water Lab., U.S. Environmental Protection Agency, Washington, D.C., 1971.
7. "Standard Methods for Examination of Water and Wastewater," 15th ed. Am. Public Health Assoc., New York, 1980.
8. Analytical reference service sample type VII. Water oxygen demand. U.S. Dept. of Health, Education and Welfare, R. A. Taft Sanit. Eng. Cent., Cincinnati, Ohio, 1960.
9. Jeris, J. S., A rapid COD test. *Water Wastes Eng.* **4,** 89–91 (1967).
10. Zanoni, A. E., Secondary effluent deoxygenation at different temperatures. *J. Water Pollut. Contr. Fed.* **41,** 640–659 (1969).
11. Lawrence, A. W., and McCarty, P. L., Kinetics of methane fermentation in anaerobic treatment. *J. Water Pollut. Contr. Fed.* **41,** R1–R17. (1969).
12. Gates, W. E., A rational model for the anaerobic contact process. *J. Water Pollut. Contr. Fed.* **39,** 1951–1970 (1967).
13. Garrett, M. T., and Sawyer, C. N., Kinetics of removal of soluble B.O.D. by activated sludge. *Proc. Purdue Ind. Waste Conf.* **7,** 51–77 (1952).
14. Knowles, G., Downing, A. L., and Barrett, M. J., Determination of kinetic constants for nitrifying bacteria in mixed culture, with the aid of an electronic computer. *J. Gen. Microbiol.* **38,** 263–278 (1965).
15. Esvelt, L. A., and Hart, H. H., Treatment of fruit processing waste by aeration. *J. Water Pollut. Contr. Fed.* **42,** 1305–1326 (1970).
16. Guttormsen, K., and Carlson, D. A., Status and research needs of potato processing wastes. *In* "Proceedings of the First National Symposium on Food Processing Wastes," pp. 27–38. Pacific Northwest Water Lab., Federal Water Quality Administration, U.S. Environmental Protection Agency, Corvallis, Oregon, 1970.
17. Middlebrooks, E. J., and Garland, C. F., Kinetics of model and field extended aeration wastewater treatment units. *J. Water Pollut. Contr. Fed.* **40,** 586–599 (1968).
18. Eckhoff, D. W., and Jenkins, D., "Activated Sludge Systems—Kinetics of the Steady and Transient States," SERL Rep. 67-12. College of Engineering, University of California, Berkeley, 1967.

6

Ponds and Lagoons

Introduction

Ponds and lagoons are among the simplest treatment systems in current use. They have found wide use with municipal and agricultural wastes. The major types of ponds and lagoons can be classified as facultative, aerobic, anaerobic, or aerated. Typical characteristics of the pond and lagoons are noted in Table 6.1.

Facultative ponds are the most common ponds in use. The term "facultative" describes ponds which have aerobic conditions in the upper layers and anaerobic processes occurring in the bottom layers, especially in the settled solids. The ponds are frequently termed "oxidation ponds" or "waste stabilization lagoons." Design loadings rarely exceed 50 lb of BOD per acre per day, and effluents may have BOD concentrations in the 20–40 mg per liter range. Particulate matter is the main source of the BOD in the effluent since the influent wastes have been converted to bacterial or algal protoplasm.

In aerobic ponds, organic matter is decomposed solely through aerobic oxidation, with the oxygen obtained by mixing and photosynthesis. Few natural ponds in practice are designed to operate in this manner. Aerobic ponds are designed with a large surface-area-to-volume ratio. Oxygen is introduced by liquid recirculation, wind, or mechanical mixing and by photosynthesis. Continuous mechanical movement of the liquid must be employed if photosynthetic ponds are to be aerobic and not facultative. Loadings of 100–200 lb of BOD per

TABLE 6.1

Typical Characteristics of Common Ponds and Lagoons Used for Waste Treatment

Characteristic	Facultative	Aerobic	Anaerobic	Aerated
Description	Natural surface aeration and photosynthesis provide oxygen in the upper aerobic layer; lower layer contains no dissolved oxygen; anaerobic activity in the bottom settled solids	Aerobic conditions maintained throughout depth by photosynthesis; high concentration of algal cells may be in effluent	Anaerobic conditions occur throughout the lagoon due to high organic loading	Aerobic conditions maintained throughout most of the lagoon by surface or diffused aeration equipment, settling of mixed solids required before effluent is discharged
Purpose	Treatment of wastewater to achieve point source discharge standards; usually built in series for this purpose; can be used for aerobic storage of wastewater prior to land treatment	Production of algal cells as a source of protein; waste treatment	Preliminary treatment of wastewaters; followed by an aerobic unit or by land treatment	Treatment of wastewaters; can be used following anaerobic ponds or as the only treatment process
Mixing	Natural mixing by wind action and gas evolution by algae and anaerobic action	Natural mixing by wind action and gas evolution by algae	No mixing	Surface or diffused aeration equipment used for mixing; complete mixing is not accomplished
Hydraulic detention time[a] (days)	Long, frequently greater than 30	5–10	Depends on whether used primarily for settling or treatment; 5 to 30+	1–10
Depth (m)	1–2	0.3–1	2–5	2–5

[a]Depends on climate.

acre per day can be employed, with effluents containing 10–20 mg per liter BOD if the algal cells are removed.

Anaerobic ponds are systems in which the concentration of organic wastes applied per unit area is sufficient to bring about complete depletion of dissolved oxygen by limiting photosynthesis, by high bacterial oxygen demand, or by both. In these ponds, up to 75% of the applied BOD can be accounted for as methane and carbon dioxide released to the atmosphere. Methane formation is the primary biological process for carbonaceous BOD removal in these units. Anaerobic ponds are deep, have a small surface-area-to-volume ratio, and may be loaded in excess of 400 lb of BOD per acre per day. Effluents have a high BOD concentration.

Aerated lagoons are biological treatment units in which the oxygen demand is met by mechanical aeration equipment. The continuous oxygen supply permits the aerated lagoon to treat more wastewater per unit volume per day. The design of an aerated lagoon can be approximated by the kinetics of a completely mixed biological reactor. The lagoons normally operate without solids recycle and can achieve from 50 to 90% BOD removal, depending upon loading, required effluent quality, and whether the solids in the effluent are removed prior to discharge.

Oxidation Ponds

Oxidation ponds are relatively shallow, diked structures with a large surface area to maintain aerobic conditions. In areas where the land is relatively flat, inexpensive, and available, oxidation ponds will be more economical than other types of aerobic biological treatment. Their predominant use has been in rural areas with adequate sunlight, wind action, and available land.

The term "aerobic" does not completely describe the biochemical reactions taking place in the pond. While ample dissolved oxygen may exist in the upper portion of the pond, there may be little or no dissolved oxygen in the lower depths. The solids layer on the bottom of the pond is devoid of oxygen, and anaerobic conditions prevail. Oxidation ponds are so designed and loaded that these anaerobic conditions have little noticeable effect on the quality of the effluent from the ponds. To control the biological processes involved and to design oxidation ponds adequately, attention must be given to the factors that affect the process: temperature, light, organic loading, pond size and shape, and hydraulic considerations.

Oxidation ponds approach natural purification more closely than any other treatment process; however, the objectives of natural purification and waste treatment are different. The goal of natural purification is to recycle nutrients and organic matter. Treatment has the goal of removing pollutants. Oxidation ponds tend to recycle rather than remove nutrients.

Biochemical Reactions

Bacteria and algae are the key microorganisms in an oxidation pond. Heterotrophic bacteria are responsible for the stabilization of the organic matter in the pond. A portion of the entering BOD settles and undergoes anaerobic fermentation in the bottom sludges. This fermentation will reduce the sludge volume if temperatures are adequate but will release the products of fermentation to the liquid layer.

As the soluble organic wastes entering the pond and released from the bottom sludges are metabolized by the bacteria, end products such as carbon dioxide, ammonium and nitrate ions, and phosphate ions become available for the growth of algae. Solar energy furnishes the energy for the growth of the algae. As the autotrophic algae produce new protoplasm, oxygen is an end product and can be used by the heterotrophic bacteria. A simplified sketch of the microbial relationships in an oxidation pond is noted in Fig. 6.1. The generation of oxygen will be in proportion to the carbon converted into algal protoplasm. If BOD removal is a primary objective of an oxidation pond, the design of the pond must provide for carbon removal by methane fermentation or conversion of carbonaceous material to algae, with removal of the algal cells from the effluent.

Bacteria are responsible for the oxidation and reduction processes that take place in the ponds. Algae play an important role in using excess carbon dioxide and creating available oxygen. At low algal densities, oxygen production may be

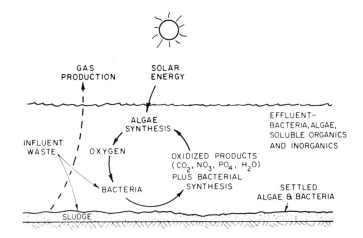

Fig. 6.1. Schematic sketch of biological interactions in an oxidation pond.

insufficient to satisfy more than a small fraction of the BOD in the pond. In most oxidation ponds, there are enough algal nutrients from bacterial metabolism and in the influent wastes to support an adequate algal concentration. These algae can produce oxygen in excess of the bacterial oxygen demand. Such excess oxygen production accounts for the supersaturated oxygen concentrations observed in many ponds.

Surface reaeration cannot meet the oxygen theoretically needed for the oxidation of organic matter in an oxidation pond. Photosynthetically produced dissolved oxygen in the pond can be a barrier against oxygen input by reaeration. As the concentration of dissolved oxygen increases, the rate of reaeration decreases until at saturation no reaeration can occur. In a pond containing growing algae, the dissolved oxygen usually approaches or exceeds saturation during daylight hours. Photosynthesis is the major source of oxygen in all aerobic and facultative ponds.

Satisfactory performance of an oxidation pond depends upon the balance between the bacteria and algae. An excess of bacterial activity over algal activity, such as caused by high waste loads or inhibition of algal metabolism, will lead to oxygen depletion, to odor nuisance, and poor quality effluent from the pond. Excess algal activity, such as caused by excess algal nutrients and environmental conditions favorable to algal growth, will lead to excessive algal cells in the pond effluent.

The concept that organic wastes are stabilized or oxidized in an oxidation pond is valid only in the sense that the original waste organics are converted into a more stable organic form, algal cells. Oxidation ponds are organic matter generators since algal cells are produced. Mixing, temperature, and radiation are the important factors affecting the growth and concentrations of algae in oxidation ponds. In properly functioning oxidation ponds, more algal cells are produced than are generated solely from the carbon dioxide released from bacterial metabolism, since the carbon dioxide in the pond waters also can be used. The net result is an increase in organic matter in the system. The removal of carbon dioxide from the pond waters results in an increase in the pH of the waters. In active oxidation ponds, the pH has been known to rise above 9.0.

The production of algal cells can be illustrated by Eq. (6.1).

$$106CO_2 + 90H_2O + 16NO_3 + PO_4 + \text{light energy} \rightarrow$$
$$C_{106}H_{180}O_{45}N_{16}P_1 + 154.5O_2 \tag{6.1}$$

The equation indicates that the weight of oxygen produced per unit weight of algae synthesized is approximately 2.0. The oxygen-generating capacity of algae varies from species to species, and the growth of 1 gm of algae in wastewater is associated with 1.3–1.8 gm of oxygen production. A value of 1.6 is commonly used in estimating oxygen production. The oxygen demand in an oxidation pond should be equal to or less than the photosynthetic oxygen production if anaerobic conditions are not desired. However, anaerobic reactions play a major role in the stabilization of BOD in an oxidation pond. Anaerobic conditions in an oxidation pond, such as those that might occur in the lower portions and bottom of the pond, are not undesirable unless odors, nuisance conditions, or lowered efficiency result.

It can be observed that 1 gm of carbon will produce about 1.9 gm of organic matter, 1 gm of nitrogen will produce about 12 gm of organic matter, and 1 gm of phosphorus will produce about 75 gm of organic matter as algal cells. These relationships indicate the importance of nitrogen and especially phosphorus in the eutrophication of surface waters, since the same relationship will hold in surface waters other than oxidation ponds.

Carbon, nitrogen, and phosphorus are the key nutrients in oxidation ponds. Rapidly growing algae are about 55% carbon as C, 9% nitrogen as N, and 1% phosphorus as P. Based on these percentages or using Eq. (6.1), it is possible to estimate the quantity of algae that a given waste will support and to estimate which of the nutrients will limit algal growth. In low-rate ponds and ponds with long retention times, algal growth and oxygen production are related to the carbon dioxide available. Because of the small amounts of nitrogen and phosphorus required, it is rare that these elements become limiting factors, especially with municipal and most agricultural wastes. Certain food processing wastes may be incapable of supporting satisfactory algal growth due to a deficiency of a critical nutrient. Examples might be wastes from sugar or starch production, from certain canning operations where a low nitrogen-to-carbon ratio may exist, or from sugar beet wastes, which may have a shortage of phosphorus.

Sulfur transformations are the major cause of odors from facultative oxidation ponds. Under anaerobic conditions, organic sulfur is transformed to hydrogen sulfide.

$$\text{Organic sulfur} \rightarrow H_2S \uparrow \qquad\qquad (6.2)$$

The emission of hydrogen sulfide and other reduced sulfur compounds is dependent upon the sulfide concentration, the pH of the pond, and the opportunity for gas–liquid transfer. Mixing enhances hydrogen sulfide emission.

Photosynthetic sulfur bacteria can occur in certain oxidation ponds. These bacteria use reduced sulfur as an electron donor and remove hydrogen sulfide from solution (Eq. 6.3). As a result, no sulfide odors occur.

$$CO_2 + H_2S + \text{light energy} + \text{photosynthetic sulfur bacteria} \rightarrow$$
$$\text{sulfur bacteria synthesis} + H_2SO_4 \qquad\qquad (6.3)$$

The bacteria are biological deodorizers. They are purple or red and will occur where adequate sunlight and reduced sulfur exist. They are most commonly found in ponds treating wastewaters with a high sulfur content, such as beet sugar wastewaters and oil refinery wastewaters.

Well-designed oxidation ponds have demonstrated the ability to handle fluctuating loads. Because of large liquid volume and long detention times, the active bacterial mass in the pond will be low. Long detention times compensate for the low bacterial mass, and high degrees of BOD removal can be achieved.

Oxygen Relationships

The net oxygen transferred into an oxidation pond is the sum of the algal oxygen production and the quantity added from or lost to the atmosphere. Because sunlight is the energy source for algae, algal growth is a function of the penetration of sunlight into the pond, which in turn is a function of the turbidity of the water. The efficiency of light conversion by algae grown in wastewater has been found seldom to exceed 10–12% and is usually much less. Oxidation ponds are designed with a shallow water depth, generally 3–5 ft, to permit maximum light penetration and algal growth. The light available at locations having no ice cover is more than adequate to support the required photosynthetic activities. Ice cover and cold weather will reduce both light and temperature to the point that oxygen production by the algae is negligible or zero.

Usually there is no measurable oxygen production under ice, because of the cold temperatures and because the ice reduces light transmission. Depending upon the thickness and clarity of the ice, light transmission may range from 2 to 60% of that transmitted without the ice (1).

Although the pond waters are cold under the ice, some microbial activity will occur. Because of the limited light penetration, anaerobic conditions occur from the onset of the ice cover until ice breakup. During mixing that may occur in the spring, the anaerobic end products, such as hydrogen sulfide, are emitted from the pond and are the source of the odors generally noted during that time of the year. After ice breakup, aerobic conditions in the upper layers are restored in a short time. During cold weather, oxidation ponds serve more as solids sedimentation units than as biological treatment units.

The depth of an oxidation pond is determined by the depth of light penetration since the efficiency with which algae convert light energy to cellular materials depends upon the intensity of light. In laboratory studies with common algal species, a light intensity of 30 ft-c was the lowest at which a measurable overall algal efficiency was obtained (2).

A pond 4 ft deep, allowing light equivalent to 15/kcal/liter/day, will produce about 50 lb of oxygen per day, with an increase of pH to 9.5. Successful pond designs currently are limited to loading rates of 40–50 lb of BOD per acre per day. Extensive mixing would be required to utilize the maximum photosynthetic oxygenation rate of 100–200 lb/acre/day noted in completely aerobic ponds.

The oxygen produced by algae is contained in the upper layer of the pond and must be mixed throughout the pond to be of benefit. Wind is the general source of mixing, although liquid recirculation has been tried. With active algal growth, supersaturated oxygen concentrations may exist in the top layer of the pond, and some oxygen loss to the atmosphere will occur. If unsaturated conditions exist in the top layers, the wind action will assist in transferring the oxygen from the atmosphere to the pond. Good mixing increases the capacity of the pond to handle increased pollution loads.

The beneficial effect of algae occurs only when the energy from the sun is available. During nighttime, the respiration of the algal cells will represent an oxygen demand on the system. Normal oxidation ponds may produce supersaturated oxygen conditions during daylight hours, which generally are sufficient to keep the ponds aerobic during the night.

Oxygen production in wastewater oxidation ponds can vary from 2.1 gm/m²/hr at midmorning to 0.03 gm/m²/hr in the early evening (1). A schematic relationship of oxygen production and oxygen demand is illustrated in Fig. 6.2. Wind action and oxygen transfer from the atmosphere may be sufficient to

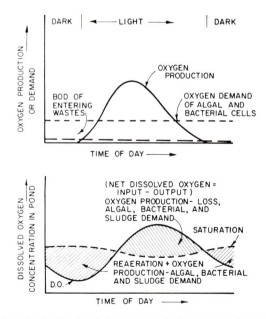

Fig. 6.2. Schematic sketch of typical oxygen production, oxygen demand, and oxygen concentration in an oxidation pond.

maintain aerobic conditions during nighttime. The rate of oxygen demand of algal cells without light energy is essentially that of endogenous respiration. Diurnal pH as well as oxygen relationships occur in oxidation ponds.

Temperature Relationships

Temperature affects the rate of metabolism of microorganisms in an oxidation pond, the loading rate of a pond, and its size. The maximum rate of metabolism will occur in late summer and early fall when the temperature of the liquid is at a maximum. No attempt is made to control the temperature in the oxidation pond, and it will fluctuate in response to ambient temperatures. The design of an oxidation pond will reflect the expected temperature conditions. Feasible pond loading rates, in terms of pounds of BOD per acre per day, are less in colder climates than in warmer climates.

The temperature effect on the waste degradation rate in oxidation ponds has been shown to be similar to that of other microbial systems, i.e.,

$$K/K_{20} = \theta^{(t-t_{20})} \tag{6.4}$$

where K represents the rate reactions and t the liquid temperature in terms of °C. A θ value of 1.072 was found to fit available data. The equation was valid between 3 and 35°C. These temperatures represent the practical limit caused by decreased activity as the temperature approached freezing and approached the thermal inactivation level of many algae species, respectively (3).

The fate of the settled organic matter on the pond bottom is a function of the conditions existing at the bottom. If the temperature is near 4°C or if the pH is below 5.5, decomposition of the organic matter is slow, and the organic matter accumulates. If the bottom temperature is higher, acid decomposition occurs and if the environmental conditions are correct, methane fermentation may occur.

The pattern of solids accumulation on the bottom of oxidation ponds, anaerobic ponds, and incompletely mixed aerated lagoons is cyclic, increasing the winter when there is little solids degradation and decreasing in the summer when the accumulated volatile solids are degraded. There will be a net solids accumulation in these ponds and lagoons due to a buildup of inert solids in the influent wastes and undecomposed organic matter.

Gas production in terms of cubic feet produced per pound of BOD applied was found to be maximum when the pond water temperature was greater than 18°C (2). At less than this temperature, gas production was less than the theoretical amount expected from solids decomposition. The gases from these solids contained 60–75% methane. At temperatures less than 18°C methane fermentation slowed and organic matter accumulated. These solids decompose when the temperature becomes greater than 18–19°C (Fig. 6.3). Gas production from

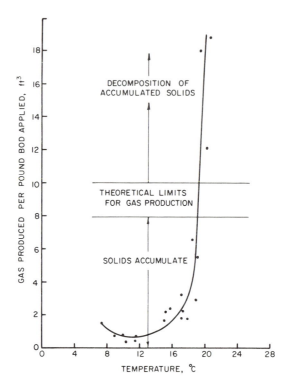

Fig. 6.3. Gas production as a function of temperature and BOD loading in a pond or lagoon (2). [Reprinted by permission from *International Journal of Air and Water Polution* **7**, W. S. Oswald, C. G. Golueke, R. C. Cooper, H. K. Kee, and J. C. Bronson. Water reclamation, algal production and methane fermentation in waste ponds. Copyright 1963, Pergamon Press, Ltd.]

solids decomposition will occur when the water temperature is greater than 15°C (Fig. 6.4).

Facultative ponds should be designed to include solids settling and decomposition. Design for optimum algal growth can be secondary since in these ponds, the algae in the surface layers serve only to maintain an aerobic zone and to control odors.

Loading Rate

The environment within an oxidation pond is unable to be controlled by the design engineer. Only physical factors such as size and loading rate can be controlled. A wide fluctuation in environmental conditions is natural, and the design must be such that the process will operate satisfactorily within these conditions. Two types of loading rates are important, the hydraulic and the oxygen demand (BOD) loading rates. The hydraulic loading rate is determined by the detention period and the total pond volume. The BOD loading rate is

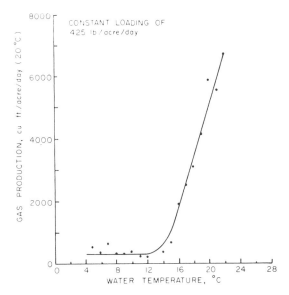

Fig. 6.4. Mean gas production as a function of water temperature at a constant loading in a pond or lagoon (2). [Reprinted by permission from *International Journal of Air and Water Pollution* **7**, W. S. Oswald, C. G. Golueke, R. C. Cooper, H. K. Kee, and J. C. Bronson. Water reclamation, algal production and methane fermentation in waste ponds. Copyright 1963, Pergamon Press, Ltd.]

determined by the rate of bacterial activity in decomposing the organic matter and the rate at which algae can produce the oxygen.

Rational design of an oxidation pond includes determining the ultimate influent BOD, desired effluent quality, quantity of oxygen that must be produced for BOD satisfaction, quantity of algae that must be synthesized to produce the required oxygen, and solar energy required to sustain algal growth under actual conditions. The required surface area of the oxidation pond would be equal to the total energy required per day to sustain algal growth, divided by the energy in the algal cells at a given conversion of solar energy per surface area. An oxidation pond loading rate should be related to the ratio of BOD load to the oxygen production by algae.

The aforementioned approach is feasible, although it has rarely been applied to the design of oxidation ponds. Empirical loading rates, approximating this approach, have been used for design. The loading rate for oxidation ponds is usually expressed in terms of pounds of BOD per surface area per day in recognition of the fact that the primary energy input is sunlight entering through the pond surface.

Pond performance is affected by environmental conditions such as temperature, solar radiation, and wind speed. Since these vary at a given geographical location, the design criteria reflect the influence of prevailing climatological conditions. The allowable organic loading rate on a pond is a function of the rate at which biological processes can satisfactorily decompose the organic matter without creating nuisance conditions. Design loading rates range from 20 to greater than 50 lb of BOD_5 per acre per day. The lower figure is common in northern areas where cooler temperatures and ice coverage is common. Higher loadings are used in southern and warmer areas (Table 6.2).

During the summer, considerably higher loading rates than the noted average can be treated in oxidation ponds. The practice is to design oxidation ponds for conditions when oxygen production will be the lowest. In most of the United States the policy has been to reduce allowable loadings to the range of about 20–50 lb of BOD per acre per day so that problems of the transition period may be minimized. Loading rates up to 100 lb of BOD per acre per day can be used in tropical and semitropical climates to obtain BOD and COD removal greater than 90 and 80%, respectively.

The relatively low organic loadings used for design require large surface areas with resultant long liquid detention times. For example, a waste with a BOD_5 of 300 mg per liter and a flow of 1 million gallons per day (mgd) would require a pond having a surface area of 50 acres, assuming a permissible loading rate of 50 lb of BOD per acre per day. If the pond averaged 5 ft in depth, the liquid detention time would be about 80 days. If land is available, oxidation ponds can be satisfactory for high-volume, low BOD concentration wastes.

A minimum liquid depth of about 3 ft has been found desirable in an oxidation pond to reduce the growth of aquatic plants and mosquito propagation. Because of the necessity to maintain this depth, oxidation ponds may be unsatisfactory for relatively low-volume wastes having high BOD concentrations. Some agricultural wastewaters, especially those from animal production and food processing operations, fall in the latter category. Consider a waste having a BOD_5 of 1000 mg per liter and a flow of 0.2 mgd. The surface area required

TABLE 6.2

Oxidation Pond Design Criteria in the United States[a]

	Region		
Design criteria	North	Central	South
Organic loading (lb BOD/acre/day)			
Mean	26	33	44
Range	16.7–40	17.4–80	30–50
Detention time (days)			
Mean	117	86	31
Range	30–180	25–180	20–45

[a]Data from Canter et al. (4).

would be 33 acres, assuming a loading rate of 50 lb of BOD_5 per acre per day. If the average evaporation plus seepage loss from the pond exceeded 0.23 in. per day, the loss from these sources would be greater than the liquid input per day, and maintenance of a satisfactory liquid depth would be difficult. If the organic loading were 20 lb of BOD_5 per acre per day, the surface area required would be 83 acres, and an average evaporation plus seepage loss of only 0.1 in. per day would be greater than the liquid input per day. Greater difficulties would exist for higher-strength, low-volume wastes. The required surface area, average liquid detention time, and the ability to maintain an adequate liquid depth should be evaluated critically for high-strength, low-volume wastes.

Lagoons should not receive significant amounts of surface runoff. Provision should be made for excluding surface water from the ponds.

Effluent Quality

The objective of using oxidation ponds is to treat the incoming wastes without causing nuisance conditions and to produce an effluent of acceptable quality. In adequately designed oxidation ponds, the effluent will contain a negligible amount of the influent waste, and on this basis high removal rates are possible. The effluent will contain bacterial and algal cells, which will exert an eventual oxygen demand in the waters receiving the effluent. In active oxidation ponds, the algal concentration is considerably larger than that of the bacteria. The oxygen demand of an oxidation pond effluent essentially will be in proportion to the concentration of algae in the effluent.

In low-rate oxidation ponds in which the applied organic loading is low, the algal disposal problem may be small since the algae are in low concentrations. Many of the algae are consumed by crustacea such as *Daphnia* or settle slowly and become part of the complex material at the lagoon bottom. Where anaerobic conditions exist, digestion occurs. A large portion of the suspended algae is entrained in pond currents and discharged into the receiving streams.

Removal of phosphates, nitrates, and other plant nutrients from the pond effluent will be high only if the algal cells are removed from the effluent. Controlled algal production and separation from oxidation pond effluents can be a feasible process for the removal of nitrogen and phosphorus in certain locations where tertiary treatment is required.

Sludge

The influent solids do not settle uniformly throughout the pond. The pattern of deposition is a function of the size and density of the solids and the liquid velocities in the pond. Most of the solids settle near the inlet. A solids deposition

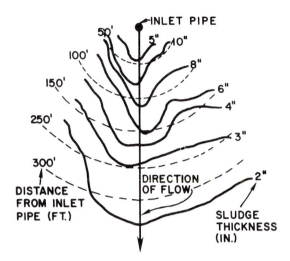

Fig. 6.5. Solids deposition pattern in an oxidation pond treating food processing wastes. Solid lines represent sludge thickness (in inches); dashed lines represent distance (in feet) from inlet pipe.

pattern that occurred in an oxidation pond treating food processing wastes is noted in Fig. 6.5.

The solids that settle in an oxidation pond can help seal the pond bottom. In an animal wastewater storage pond, a solids layer on the pond bottom restricted the movement of water (5). The initial sealing was caused by physical entrapment of suspended particles in soil. This was followed by a secondary mechanism of microbial growth that sealed the surface and prevented water movement.

Bacteria Removal

In addition to the removal of solids and BOD, oxidation ponds also remove many of the pathogens that enter with the wastewater. Fecal coliform organisms are the indicator organisms commonly used to estimate pathogen removals (Chapter 13). Coliform removals can range from 50 to 90% or more, depending upon factors such as temperature and detention time. In a five-pond system with a total detention time of 29 days, the fecal coliform organism concentration was reduced from an average of 5×10^7 per 100 ml in the untreated wastewater to an average of 20 per 100 ml in the discharge from the last pond (6). The mean temperature in the pond system was 26°C, and the mean BOD reduction was 92%.

Chlorination (Chapter 13) also has been used to reduce the bacterial content of pond effluents to meet specific discharge standards. Removal of the solids from the lagoon effluent improves the chlorination efficiency by reducing the chlorine demand. Sulfide exerts a significant oxygen demand. For sulfide concentrations of 1.0–1.8 mg per liter, a chlorine dose of 6–7 mg per liter was required to produce the same chlorine residual as a chlorine dose of 1.0 mg per liter for conditions of no sulfide (7). Adequate disinfection can be obtained with combined chlorine residuals of between 0.5 and 1.0 mg per liter after a contact time of about 50–60 min or less.

Design and Operation

The oxidation pond site should be convenient to the source of the waste but where prevailing winds will not cause any odors to drift to residential areas. The pond should be located in areas where wind action is likely to be maximum. Ravines and sheltered valleys are not appropriate locations.

The topography of the area and the soil type will influence the site selection. The site should be on impervious soils or on soils suitable for sealing. The pond bottom should be compacted with clay or other material to avoid excess seepage, which can cause difficulties in maintaining adequate water depth and possible contamination of groundwater.

Pond side slopes are influenced by the nature of the soil and size of the installation. For outer slopes, a ratio of 3:1 or greater, horizontal to vertical, has been found satisfactory. Flat inner slopes have the disadvantage of providing shallow areas conductive to emergent vegetation. Wave action is more severe for larger installations, warranting flatter slopes to minimize erosion.

Liquid depth should be more than 2–3 ft since shallower depths result in growth of aquatic weeds, which in turn increase mosquito breeding. Vegetation-free banks are important in the control of mosquito breeding. Vegetation on the banks should be kept cut and removed and should not be allowed to float on the water to provide harborage for the mosquito larvae. Covering the banks of a pond at the waters edge with solid material such as plastic, asphalt, cement, etc., to prevent growth of vegetation is useful to control mosquito breeding and to minimize erosion. Mosquito larvae are seldom found in the open water of a pond. The shape of the ponds should be uniform and essentially circular or rectangular, with few areas where floating material may accumulate. The ponds should have adequate freeboard above the maximum water line, generally 3 ft for ponds having large surface areas. Weeds, straw, and grass clippings should be excluded from the pond since they will increase the organic load in the pond.

Temporary odor problems, such as caused during the spring transition from anaerobic to aerobic conditions, can be controlled by the addition of nitrates. The nitrates will act as hydrogen acceptors in the absence of oxygen to prevent reduced sulfur-containing compounds from being generated. Sodium or ammonium nitrate are common sources of hydrogen acceptors. Mechanical aeration also can be used during such periods.

Continuous odor problems indicate that the oxygen demand in the oxidation pond exceeds the supply. Solutions to the problem include reducing the BOD loading or using mechanical aeration to convert the oxidation pond to an aerated lagoon.

The effluent from oxidation ponds should discharge to surface waters only if the discharge will not result in violation of state water quality standards. The effluent may receive further treatment or be disposed of by land treatment systems.

Application to Agricultural Wastes

Oxidation ponds rarely are used as the sole treatment process to meet effluent limitations. Where they are used, they are part of an overall treatment system and may be used as a final or polishing treatment step. Multiple ponds in series or parallel generally are preferred for large flows or waste loads.

Oxidation ponds are not recommended for animal manures or the runoff from feedlots. This material has a high oxygen demand (BOD), and very large areas will be required. Anaerobic ponds are used with these wastes, especially as storage prior to land treatment or other treatment processes.

Oxidation ponds have been used with more dilute wastes such as those from milking centers and food processing operations. Resultant effluent quality has been variable. Without proper care, oxidation ponds can become overloaded and cause odors, and their effluent may not meet permit discharge requirements. The following examples illustrate how they have been used.

An evaluation of six lagoons treating milking center wastewater noted that lagoons designed at a BOD loading rate of 28 lb per acre per day were aerobic and emitted few odors (8). The effluent BOD from these lagoons varied seasonally and ranged from 6 to 80 mg per liter. The effluent from such lagoons can be applied to adjacent cropland to capture the nutrients, for supplemental irrigation, and for disposal.

The performance of three facultative lagoons treating milking center wastewater was determined over a 12-month period. The pollutant removals were the following: BOD_5, 96%; suspended solids, 90%; COD, 92%; fecal coliform and fecal streptococci, 99% (9). A design standard of 28 lb BOD per acre per day (31.1 kg per hectare per day) appeared reasonable for these lagoons, although the results indicated that a loading of 36 lb of BOD per acre per day (40 kg/ha/day) could be feasible.

Meat packing wastes have been treated in a combination of anaerobic–aerobic ponds. Oxidation ponds treating the effluent from the anaerobic ponds achieved a BOD reduction of 59%, with a detention time of 18 days. The loading rate averaged 130 lb of BOD per acre per day (10).

High-rate oxidation ponds for poultry processing wastes were loaded at the rate of 214 lb of BOD per acre per day. BOD removal efficiencies of 70–96% were obtained. The treatment facilities consisted of primary sedimentation, an equalization tank and two oxidation ponds operated in parallel (11). A two-stage oxidation pond system has been used to obtain 80–95% overall BOD removal for poultry processing wastes (12). The first stage was a high-rate pond 8 ft deep loaded at a rate of 935 lb of BOD per acre per day, with a detention time of 7.9 days. The first pond was anaerobic but achieved 72.5% BOD removal. The second stage was a shallow photosynthesis pond, 2 ft deep and loaded at 74 lb of BOD per acre per day. BOD removal efficiencies in the second stage were 9% when the algae were not removed from the effluent and 70% when the algae were filtered out.

A three-stage oxidation pond successfully treated bean cannery waste during short (10-week) packing seasons (13). The first and second stages were heavily loaded and were anaerobic. The loading on the total pond area was 260 lb of BOD per acre per day. BOD reduction was about 78%, and suspended solids removal about 77%. Temporary odorous conditions were corrected by adding sodium nitrate to the pond influent and aerating the influent. The effluent was spray-irrigated on a ryegrass field at a rate of 67 lb of BOD per acre per day, without odors, ponding, or contamination of an adjacent stream.

The effluent limitations established for agricultural wastes are in terms of mass units, i.e., quantity of pollutants per unit of production, such as kilograms of BOD per 100 kg of raw material or final product for canned fruits and vegetables. Because the limitations are in mass units, industry has a variety of options to consider in meeting them. The industry can (a) continue its normal wastewater discharge and achieve an effluent concentration that will result in the limitation being met, (b) reduce the wastewater flow, achieve a higher effluent concentration and meet the limitations, (c) use additional technology, such as filtration, to remove pollutants that remain in the pond effluent, or (d) use alternative disposal processes, such as spray irrigation, during periods when the limitations are not being met. Oxidation ponds can continue to be used by industry to meet effluent limitations using the preceding alternatives.

Aerobic Ponds

Oxidation ponds have more treatment capacity in terms of loading rate than that recommended by most state regulatory agencies, especially if it is realized that the entire volume of the pond is available for treatment. The engineering challenge is to utilize this additional capacity.

The work by Oswald and co-workers over the years has established the basic factors involved in oxidation ponds (14, 15). These studies identified the role of photosynthesis and algal age in conventional ponds treating sewage. The loading rate to which a conventional pond can be subjected is a function of the active algal cells in the pond capable of participating in the necessary photo-synthetic reactions and of the settling in the pond of some of the influent organic matter. As noted earlier, a conventional pond is characterized by both aerobic and anaerobic decomposition and by a low concentration of active algal cells.

By reducing pond detention times and using shallow depths to foster actively growing algal cells, 95% BOD removals were obtained with BOD loads of up to 225 lb/acre/day in summer and 110 lb/acre/day in the winter (14). These removals could be obtained only when the large quantities of algae were removed from the pond effluent. If the algae were not removed, the BOD removal was 7%.

The high-rate aerobic oxidation pond is able to transform the unstable influent organic matter into algal protoplasm. The treatment process is incomplete unless the algal cells are removed from the effluent, again emphasizing the need for heterotroph–autotroph balance unless solids separation is contemplated with an oxidation pond.

This high-rate process is feasible in climates with continuously mild temperatures and continuous sunlight, where adequate management of the pond is available, and where the separated algal cells can be disposed of adequately and perhaps profitably. The use of the concentrated algal cells as animal feed has been contemplated.

Anaerobic Lagoons

General

Anaerobic lagoons are similar to aerobic oxidation ponds only in that both are impoundments used for holding and treating liquid wastes. Anaerobic lagoons are units that are loaded such that surface reaeration and photosynthetic activity cannot maintain aerobic conditions. Many lagoons currently labeled anaerobic lagoons are overloaded aerobic lagoons. An anaerobic lagoon bears only a superficial resemblance to aerobic lagoons, has a different purpose, and should be designed on a different basis.

The purpose of anaerobic lagoons is the destruction and stabilization of organic matter, not water purification. The lagoons can be used as primary sedimentation units to reduce the load on subsequent treatment units. They differ from primary sedimentation units in that the settled solids are not routinely removed but are left in the unit to degrade. A gradual buildup of solids occurs, the rate of buildup being a function of the solids loading rate, the characteristics of the raw waste, and the rate of the solids stabilization. Periodic solids removal is necessary. The biodegradable fraction of the solids undergoes anaerobic decomposition. Considerable gas may be evolved, with a resultant decrease in BOD and COD of the lagoon contents.

Size

There is no need for a large surface area to promote surface aeration and/or to obtain adequate light energy for photosynthesis. Anaerobic lagoons require less land area than for an equivalent aerobic lagoon since they are more heavily loaded. The depth of the lagoon is not restricted by light penetration. Anaerobic lagoons should be built with a small surface area and as deep as possible. In practice, anaerobic lagoons have been from 5 to 15 ft in depth. The small surface area promotes anaerobic conditions and decreases the needed land area. Long liquid detention times are not required. Detention periods as short as 3–5 days have been successful. Longer times also have been successful.

In anaerobic lagoons there is a relatively solid-free liquid layer above a layer of settled solids. A floating scum layer may occur, depending on the type of waste. With a small surface area, the scum can form an effective floating cover to minimize surface reaeration and to provide some insulation of the lagoon contents during cold weather.

Loading

Anaerobic lagoons are comparable to single-stage, unmixed, unheated digesters. Loading values should be based on pounds per volume per time, as is done for other anaerobic systems. Loadings of 300–2000 lb of BOD_5 per acre per day have been reported for anaerobic lagoons, illustrating the difference in loading rates between anaerobic lagoons and oxidation ponds. Loadings of 0.36–10.4 lb of VS per 1000 ft^3 per day have been reported for lagoons treating a variety of animal wastes. Higher loading rates also can be successful. The advantage of anaerobic lagoons lies with wastes that are highly concentrated and in high loading rates. A minimum loading rate of 15–20 lb of BOD per 1000 ft^3 per day should be satisfactory for most wastes.

The design of anaerobic lagoons should be on the basis of the BOD or solids loading rate since these factors more clearly represent the factors affecting the microbial reactions in the lagoon. Loadings and designs based on number of animals or pounds of food processed are subject to differences in animal feed; types of animal housing; process efficiency; and management efficiency, such as amount of waste feed in the manure, liquid and solid wastes separation at the source in food processing operations, and other in-plant waste control measures.

Inadequacies and failures of anaerobic lagoons at a variety of loading rates have been reported. The establishment and maintenance of conditions suitable for optimum biological metabolism plays a larger role in the success or failure of anaerobic lagoons than does the loading rate. Attention should be paid to control of pH, alkalinity, temperature, and mixing (Chapter 9).

Unbalanced conditions frequently occur during the start-up of an anaerobic lagoon and when environmental factors abruptly change, such as when the excess solids are removed from the lagoon or when the lagoon contents warm in the spring. Under these conditions, it is important to control the environment in the lagoon until a proper population of methane-generating bacteria is established.

Since anaerobic conditions are not inhibited if adequate buffer capacity, i.e., alkalinity, is present, additional alkalinity can be added until an optimum environment is created. Lime has been a common chemical for this purpose, although other chemicals such as sodium bicarbonate, ammonium carbonate, and anhydrous ammonia can be used. The additional alkalinity should be mixed throughout the lagoon contents to avoid localized pH variations that may continue to inhibit optimum biological reactions. Figure 6.6 (16) illustrates initial start-up conditions in an anaerobic lagoon treating beef cattle feedlot wastes. The lagoon was placed in operation in early August. The data indicate that, even with warm temperatures, attention to proper environmental conditions was vital to proper lagoon performance. Alkalinity, as lime, was added to the lagoon as noted until equilibrium was achieved.

When an anaerobic lagoon is in biochemical balance, odors are at a minimum and rarely are more than those of the surrounding area. An odor problem results when the lagoon is out of balance. Nuisance conditions can arise as a unit is placed into operation, after it is cleaned, after intermittent mixing, and as

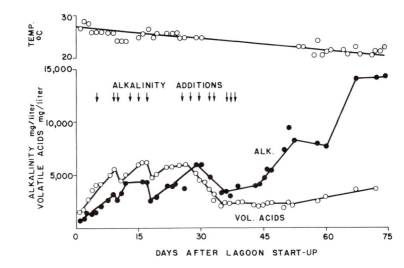

Fig. 6.6. Start-up conditions in an anaerobic lagoon treating beef cattle feedlot wastes (16).

warm temperatures are established in the spring. Every attempt should be made to minimize the time a lagoon is out of balance.

Oxygen and oxidized material such as nitrates should not be added to anaerobic lagoons, since they will be used as hydrogen acceptors in preference to oxidized organic matter and carbon dioxide, with the result that anaerobic metabolism can be inhibited if enough quantity of such oxidized material is added. Whenever chemical additives are required, they should be added in the most reduced forms.

Mixing

When the solids in the anaerobic lagoon are actively digesting, considerable quantities of gas are evolved. The evolution of the gas serves to mix the contents of the lagoon, making the organic material more readily available to the active organisms. Gasification rarely mixes the lagoon contents completely and is limited to the time of the year when the liquid temperature is warm, generally above 15°C.

Gas Production

Conversion of the wastes into gas, discharge of the lagoon contents, and seepage into the ground are the only ways by which waste material is removed from an anaerobic lagoon. Seepage through the lagoon bottom and sides should be minimized by adequate sealing to avoid pollution of subsurface waters. Of the remaining methods, gasification removes the greatest quantity of waste material from the lagoon. The theoretical quantity of gas produced is 8–10 ft^3 per pound of biodegradable solids added to the unit. High rates of gas production in anaerobic lagoons occur only when the temperature of the lagoon contents are warm.

The gases generated will consist of 60–70% methane, with the remainder being carbon dioxide and inert gases.

Temperature

Temperature is one of the most important factors affecting the performance of anaerobic lagoons. Because the lagoon is built with the ground for insulation and is generally uncovered, it is subject to temperature fluctuations that can affect the biological system. Anaerobic lagoons function better in warmer climates and are less effective in the colder climates. Maximum decomposition and gas production takes place when the temperature is higher than 17–19°C (Fig. 6.3).

When the temperature in the lagoon is low, an anaerobic lagoon becomes little more than a sedimentation tank. Solids increase on the bottom of the tank during cold weather and undergo decomposition when the warmer temperatures occur. A cover of scum and grease helps maintain the temperature in the lagoon. In addition to providing insulation, it reduces the emission of odors.

Effluent Quality

Anaerobic lagoons offer a possible approach for the treatment of concentrated organic wastes. BOD reductions in anaerobic lagoons can be respectable, 60–90%; however, due to the high loading rates, the effluent from anaerobic lagoons is unlikely to be suitable for discharge to surface waters, even with high BOD removals. The effluent will contain significant concentrations of oxygen-demanding material, solids, and nutrients. The quality of the effluent is decreased during start-up operations and when low temperatures exist in the lagoon.

Anaerobic lagoons are followed by an aerobic unit, generally an oxidation pond or an aerated lagoon if an effluent suitable for discharge to a stream is desired. The degree of treatment that occurs in an anaerobic lagoon may be satisfactory if the contents are to be discharged to the land for disposal, although odors may be a problem during distribution of the contents on the land.

Solids Removal

The rate of decomposition of solids entering an anaerobic lagoon depends on environmental factors such as the temperature of the lagoon and the degree of mixing that takes place. At low temperatures, the quality of the settled solids is similar to those that entered the lagoon. Little decomposition takes place.

The amount of decomposition will depend on the biodegradability of the entering solids. Studies with beef cattle manures have demonstrated a volatile solids reduction of between 40 and 55%. In an anaerobic lagoon treating similar wastes, about 50% of the entering wastes can be expected to accumulate in the period between lagoon cleanings, even under optimum conditions.

The period between solids removals can be estimated if the volatile solids content of the raw waste and the volatile solids reduction in the lagoon can be estimated, and the resultant solids concentration on the bottom of the lagoon is approximated. Since no decomposition of inert or nonvolatile solids takes place in the lagoon, the introduction of excessive amounts of inert matter should be minimized since it results only in a more rapid filling of the lagoon and greater frequency of solids removal from the lagoon.

All solids should not be removed from an anaerobic lagoon when it is cleaned. Solids removal from an anaerobic lagoon in cold weather can affect the resultant BOD and volatile acid levels in the liquid layer of the lagoon (Fig. 6.7).

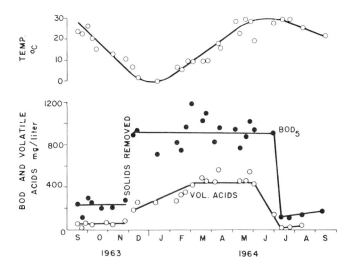

Fig. 6.7. Changes in the characteristics of the liquid effluent from an anaerobic lagoon due to solids removal (17).

As depicted in Fig. 6.7, it was not until the middle of the following summer that conditions in an anaerobic lagoon returned to what they were before cleaning.

The settled solids that are removed periodically from active anaerobic lagoons have undergone degradation, stabilization, and concentration. They are less pollutional than the entering untreated solids. However, their quality is such that they should not enter receiving waters. Land disposal offers an acceptable method of disposal for these solids.

Application to Agricultural Wastes

Anaerobic lagoons have been used with a variety of agricultural wastes as a part of an overall treatment system. The discharge from an anaerobic lagoon is either applied to land for irrigation and as a fertilizer or is treated in aerobic processes. The following provide examples of how such lagoons have been used.

Anaerobic lagoons have been used with oxidation ponds to treat milking parlor wastes. The milking parlor waste from an 80-cow dairy was treated in an anaerobic lagoon loaded at 9 lb of BOD_5 per day per 1000 ft³. BOD reductions averaged 85% for the summer with liquid temperatures of 85°F and 20% in the winter when the liquid temperature was 35°F. The average BOD of the effluent from the anaerobic unit during summer was about 100 mg per liter (18). The anaerobic unit acted as a sedimentation unit during the winter. The settled solids were degraded during the spring, with the result that the BOD of the effluent from the anaerobic unit during the spring was greater than that entering the unit during this period.

An anaerobic–aerated lagoon system provided overall BOD removals of 99%, suspended solids removals of 98%, and grease removals of 98% from packinghouse wastes. The anaerobic lagoon had a water depth of 15 ft, a BOD design loading of 15 lb of BOD per 1000 ft^3 per day and averaged 65% BOD removal. The actual loading was 12.3 lb of BOD per 1000 ft^3 per day (18). A full-scale study resulted in an anaerobic lagoon loaded at about 29.3 lb of BOD per 1000 ft^3 per day, or 195% of design capacity. Liquid detention time was about 5 days. Removals that occurred in this anaerobic lagoon averaged 82% for BOD, 68% for COD, 78% for grease, 69% for volatile solids, and 59% for total suspended solids (19). The organic ammonia was converted to ammonia nitrogen in the lagoons, resulting in an average concentration of 120 mg per liter in the effluent, an increase of about 200%. The pH remained about 7.0 throughout the year. The temperature ranged from 60 to 75°F, with an annual average of 69°F.

Hog processing wastes were successfully treated in an anaerobic pond having an average design detention time of 5 days. A cover of grease kept the temperature above 80°F at all times, with a maximum of 94°F during the summer (20). The pond was loaded at 14.7 lb of BOD$_5$ per 1000 ft^3 per day and achieved a 78% BOD$_5$ removal and a 90% suspended solids removal.

Anaerobic ponds have been used with success in the meat-packing and meat processing industry, because these wastes contain high solids and BOD concentrations. In many cases such ponds are used to pretreat the wastes before discharge to a municipal plant or to an aerobic unit for further treatment. State regulatory agencies in certain states have accepted design loadings of 15 lb of BOD per 1000 ft^3 per day for anaerobic ponds and have allowed for 60% BOD removal when the anaerobic ponds are followed by some sort of aerobic treatment.

Beet sugar wastes have been treated in an anaerobic unit followed by either a facultative pond or an algae pond. In this instance, the anaerobic lagoon was loaded with screened and settled flume water up to 2060 lb ultimate BOD per acre per day. A portion of the oxygen demand in the anaerobic lagoon was met by a floating aerator capable of transferring about 300 lb of oxygen per day into the lagoon water when the dissolved oxygen was zero (21). At loadings above about 1000 lb of ultimate BOD per acre per day, odors caused by sulfides became an increasingly severe problem. Although the anaerobic lagoon accepted high loadings and accomplished high removals, the effluent had a high BOD and was unsuitable for discharge without further treatment.

Full-scale systems of anaerobic lagoons, followed by either aerated or aerobic lagoons, have been used to obtain 80–99% removal of BOD, suspended solids, and grease (Table 6.3) (22). Nutrient removals were not as great. Organic nitrogen removals ranged from 46 to 94%, ammonia nitrogen increased rather

TABLE 6.3
Combined Anaerobic Lagoon–Aerobic Lagoon Treatment of Meat-Packing Wastes[a]

Plant number	BOD	Suspended solids	Grease	Nitrogen				Phosphorus	
				Organic	Ammonia (NH_3-N)[b]	Total Kjeldahl	Nitrate	Total (P)	Soluble (P)
				Reduction (%)					
1	98	95–99	95–99	93–95	+	26–46	—	16–49	+–34
2	82–99	77–95	99	84–90	+	0–32	—	+–56	+–72
3	89–97	83–91	94–98	71–85	+	+–17	—	+–28	+–48
5	99	99	99	46	+	—	—	58	73
6	99	90	97	91	+	46	—	48	25
7	98	94	87	90	+	57	—	26	+
				Final effluent concentration (mg/liter)					
1	30–250	12–144	13–39	8–20	62–120	—	0–1.2	9–20	6–18
2	200–400	106–124	15–24	14	84	—	0	7.7	4–9
3	40–90	110–127	15–20	18–20	59–25	—	0–3.8	7–13	3–9
5	17–20	16–68	4–17	4–5	50–60	—	0.5–0.8	6–10	5–8
6	30	308	61	34	86	—	0	3–15	7–10
7	20	52	—	7	37	—	—	—	—

[a] Data from Crandall et al. (22). Top portion of table represents the precentage reduction of the noted parameters in the combined lagoon system; bottom portion represents the final effluent concentrations (milligrams per liter) of the parameters.

[b] + = increase.

than decreased, and soluble phosphate changes ranged from an increase to removals of over 50%. Soluble nitrogen and phosphorus will result from microbial metabolism in the anaerobic unit and are not removed effectively in an aerobic lagoon. Control of these nutrients will require additional, specific treatment processes.

Minimum volumes for the size of anaerobic lagoons to be used with animal manures have been recommended (Table 6.4). The volumes noted are for a mild climate, such as that through the middle of the United States. In colder climates larger volumes are needed, due to lower degradation rates, and in warmer climates smaller volumes will suffice. In addition to the volumes shown, a freeboard of 2–3 ft is needed.

Aerated Lagoons

General

An aerated lagoon differs from an oxidation pond in that aerobic conditions are maintained by mechanical or diffused aeration equipment. Aerated lagoons consist of an aeration device in a square or rectangular basin. Usually, the lagoon is an earthen basin with some protection on the banks for wave action caused by the aeration unit (Fig. 6.8).

A major advantage of the aerated lagoon is the continuous oxygen transfer caused by the aeration equipment. The continuous oxygen supplied by the aeration equipment in an aerated lagoon permits this unit to treat more wastewater per unit volume per day. Aerated lagoons have been used to improve the effluent

TABLE 6.4

Minimum Recommended Volumes for
Anaerobic Lagoons Treating
Animal Wastes[a]

Animal	Minimum volume (cubic meter/1000 kg of animal)[b]
Dairy Cattle	72
Beef Cattle	63
Swine	62
Laying Hen	147
Broiler	187

[a]Adapted from (23).
[b]Cubic meter = 35.3 cubic feet.

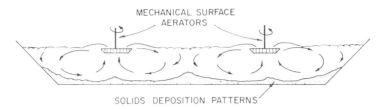

Fig. 6.8. Schematic of an aerated lagoon with surface aerators.

quality of overloaded oxidation ponds, to alleviate nuisance conditions in such units, to permit a more economical use of land, and to treat wastes without prior treatment.

The principal variables which must be evaluated for design are rate of biological reaction in the aerated lagoon, effect of temperature on the reaction rate, oxygen requirements, synthesis and oxidation of solids in the lagoon, mixing requirements, and pH and nutrient balance.

Operation and Design

The microbial characteristics of an aerated lagoon are similar to those of an activated sludge unit rather than an oxidation pond. Algae will not be important and generally are not found in the lagoon mixed liquor because of the turbulence and turbidity. The aeration equipment will be required to satisfy the entire oxygen demand.

The aerated lagoon is a dilute, well-mixed, biological treatment unit operating without solids recycle and having detention times on the order of 1–10 days, depending upon loading and desired effluent quality. Mixing is sufficient to distribute oxygen throughout the unit but may not be sufficient to keep all of the solids in suspension. The aeration equipment should be designed to deliver the required quantity of oxygen and to provide a minimum of solids deposition in the unit. The aeration equipment normally runs 24 hr a day.

For aerated lagoons, the loading rate should be expressed in terms of the pounds of oxygen demand per unit volume per day (lb of BOD per 1000 ft^3 per day) rather than pounds per surface area per day. The total volume will serve as a biological reactor, and the amount of surface area is of lesser importance. Because the biological reaction rates are affected by temperature, design detention time to achieve a given effluent quality should be expressed as a function of temperature.

A common approach to the design of aerated lagoons has been to consider them as completely mixed with respect to both dissolved oxygen and uniform solids concentrations. Mathematical models developed with this assumption have been applied to the design of aerated lagoons. These assumptions represent an ideal situation that may be only approximated in practice. Figure 6.8 represents a typical aerated lagoon and indicates that solids will settle out where velocities are low. The velocity profile is determined by the size, pumpage, and placement of the aerator, the liquid depth; and the size of the lagoon. The settled organic solids will become anaerobic and release soluble organic end products and gases to the aerobic upper layers. Because both aerobic and anaerobic areas exist in the aerated lagoon, it may more properly be referred to as a "facultative lagoon." If the aerated lagoon is not completely mixed, empirical relationships have been used to approximate the BOD removal in these units.

If the aerated lagoon is well mixed such that there is reasonably uniform distribution of dissolved oxygen and suspended solids in the lagoon, the design can be estimated by the kinetics of a completely mixed biological reactor. If BOD removal is assumed to be a first-order reaction, Eq. (6.5) expresses the relationship between the influent and effluent BOD.

$$S_0 = S_e(1 + Kt) \qquad (6.5)$$

where S_0 = BOD_5 in the influent (milligrams per liter), S_e = BOD_5 in the effluent (milligrams per liter), K = a BOD removal constant (day^{-1}), and t = liquid retention time (days). The magnitude of K is a function of temperature and the biodegradability of the influent organic matter.

If the suspended solids concentration in the aerated lagoon is known and it is assumed that the lagoon is completely mixed and that the suspended solids can be used to represent the active mass of microorganisms, Eq. (6.6) can be used to calculate the BOD removal.

$$S_0 = S_e(1 + kxt) \qquad (6.6)$$

In this situation, x = the mixed liquor suspended solids, milligrams per liter, and k = a BOD removal constant, day^{-1} · (mg/liter)$^{-1}$.

In Eq. (6.5), S_e should represent the BOD_5 in the effluent caused by any influent BOD_5 that remains untreated. In practice, the effluent BOD_5 measures the oxygen demand of the untreated BOD_5, any microbial solids in the effluent, and any nitrogenous demand. For this reason and because aerated lagoons are not always completely mixed, field results may not agree with results obtained by theoretical relationships.

The effluent solids concentration from an aerated lagoon may be higher than that for an oxidation pond or comparable aerobic unit since there is little natural opportunity for solids separation. The effluent suspended solids concentration can be reduced by baffling a portion of the aerated lagoon to permit a quiescent zone where suspended solids may settle. A tube settling unit can be installed in the effluent end of the lagoon to reduce the solids content of the effluent. An aerated lagoon can be followed by an unmixed oxidation pond, a quiescent settling pond, or a final clarifier to produce a low solids effluent.

BOD removals of 50–70% result if the suspended solids are not separated from the effluent, whereas 90–95% removals can be obtained if an effluent solids separation unit is used. Figure 6.9 illustrates the relationship that was obtained for the effluent BOD and suspended solids resulting from the treatment of potato processing wastes in a completely mixed aerated lagoon (24). Similar results can be expected from mixed aerated units where natural solids separation does not occur. Where very high BOD removals are required, effluent solids removal is required.

The aerated lagoon does not need operational control from a biological standpoint. Biological equilibrium will be established, based on the actual loading conditions, and will automatically adjust to various changes in loads.

Temperature

The liquid temperature in an aerated lagoon depends upon the mixing that takes place and the rate at which heat is lost. The temperature of an aerated

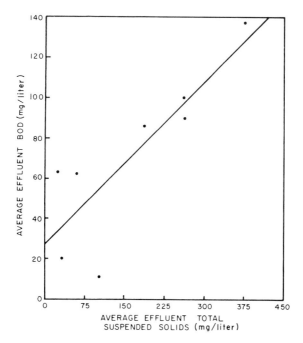

Fig. 6.9. Effluent BOD and suspended solids relationships for an aerated lagood treating potato processing wastes (24).

lagoon can be estimated from a heat balance incorporating the influent liquid temperature T_i, the effluent liquid temperature T_e, the mean lagoon temperature T_w, and the ambient air temperature T_a. If heat loss takes place primarily through the lagoon surface, the heat loss will be proportional to the temperature differential between the air and liquid and the surface area of the lagoon, A.

$$(T_i - T_e)Q = fA(T_w - T_a) \tag{6.7}$$

In this equation, all temperatures should be in degrees Celsius, area in square feet, and Q in million gallons per day. The proportionality factor f or overall heat transfer coefficient has been found to have a value of 8×10^{-6} and 12×10^{-6} (25, 26).

The following equation was developed to predict aerated lagoon temperatures (26).

$$T_L = \frac{8.34QT_i + 145A(T_A - 2)}{145A + 8.34Q} \tag{6.8}$$

T_L is the average weekly aerated lagoon temperature (°F), T_i the average weekly influent temperature (°F), T_A the average weekly air temperature, Q the wastewater flow (gallons per day), A the surface area of the aerated lagoon (square feet), and 145 the average heat exchange coefficient (Btu/ft²/day/°F). The term $(T_A - 2)$ approximates the equilibrium temperature of the aerated lagoon when exposed to air and assuming no liquid influent or effluent. The results obtained from Eq. (6.8) were relatively insensitive to the magnitude of the heat exchange coefficient.

Ice accumulation can occur on floating aerators during subzero temperatures. The accumulation can result in increased power usage, as the aerators settle deeper in the lagoon, and in overturning or sinking of the aerators. Electronic overload switches, electric heaters, and extra bouyancy can minimize such problems.

Aerated lagoons can function satisfactorily and are capable of over 80% BOD removal, even in cold climates (27), if attention is given to the fundamentals of the process when it is designed. Diffused aeration and aeration guns were more satisfactory in cold climates than were surface aerators.

Odor Control

With many agricultural wastes, the objective of using an aerated lagoon is odor control rather than meeting an effluent discharge limitation. Under such conditions, the aeration equipment need not meet the total oxygen demand in the lagoon nor mix the lagoon contents. A minimum of 1 hp per 1000 ft² of lagoon surface area was adequate surface agitation to accomplish odor and scum control in an aerated lagoon treating animal wastes (28).

Data from several lagoons treating animal wastes have led to the conclusion that satisfactory odor control will be achieved if sufficient aeration is provided to satisfy 50% of the COD (29). A surface-area-to-power input ratio of about 125 m^2 per kilowatt (1000 ft^2 per horsepower) also should be provided.

Aerated lagoons being used for odor control should minimize suspension of the settled solids. Aerators which promote surface pumpage rather than deep mixing are desirable.

Depth

Selection of the depth of aerated lagoons is related to the desirability for mixed conditions and heat conservation. By increasing the lagoon depth and therefore decreasing the surface area, heat loss is reduced. However, as depth is increased, it is more difficult to obtain complete mixing. For most aerated lagoons, a depth of 3–5 m is common. A greater depth is satisfactory when only surface aeration and odor control are desired.

Aeration Units

Because an aerated lagoon is not completely mixed, the dissolved oxygen concentration will vary throughout the unit. If the lagoon is large and the recirculation time through the aeration zone of the aerator is great, then a concentration gradient will develop, with the highest concentrations existing near the aerator and decreasing with increasing distance from the aerator.

The average liquid detention time in an aerated lagoon is much longer than in an activated sludge unit, and the solids concentration is much lower. Oxygen uptake rates are typically in the less-than-10 mg per liter range as opposed to the 25–50 mg per liter range for conventional activated sludge units.

Two aeration systems can be used for aerated lagoons: (a) diffused air units, in which compressed air is forced through perforated plastic pipe or an aeration gun near the bottom of the lagoon, and (b) surface aerators, which mechanically create turbulence at the liquid surface, such as with a series of blades that are partially submerged. An example of an aeration gun is illustrated in Fig. 6.10, while examples of surface aerators are presented in Figs. 6.11 and 6.12.

Mechanical surface aerators pump large quantities of water, causing oxygen to be dissolved in the water. The rate of pumping and type of air–water interface that is created influence the performance of the units. Low-speed units provide maximum pumping, but low-turbulence and high-speed units provide less pumping and greater turbulence. Performance must be determined in large-scale units or under actual conditions.

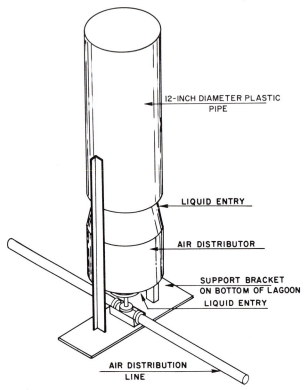

Fig. 6.10. Schematic of aeration gun used to aerate lagoons and lakes using compressed air. (Courtesy of Aero-Hydraulic Corp.)

Early types of mechanical aerators were attached to fixed supports in an aeration basin. These units had the disadvantage of being sensitive to variations in water levels, which could alter the oxygen transfer characteristics of the aerators. The development of floating surface aerators has removed this problem and has allowed the aerator to perform uniformly as the liquid level in the lagoon fluctuated. The simplicity and ease of maintenance of surface aerators, especially floating units, has caused them to be successful with agricultural wastewaters.

The aeration equipment is expected to mix the lagoon as well as transfer the necessary oxygen. In large systems with a high oxygen demand rate, if the aeration units can transfer enough air to meet the oxygen demand of the system, adequate mixing normally will occur. In low-rate systems or small systems, the mixing requirements may control, and the aeration equipment will be designed to keep the solids in suspension. Under these conditions, adequate oxygen will be available to meet the oxygen demand of the microorganisms. If it is desired to

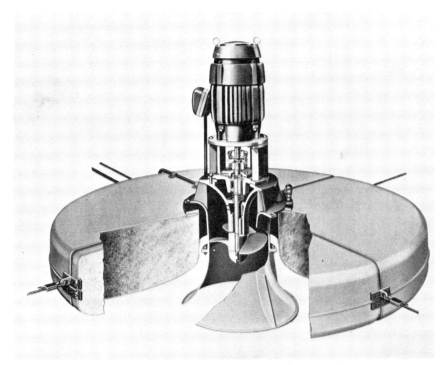

Fig. 6.11. Cross section of floating surface aerator. (Courtesy of Envirex. Inc., a Rexnord company.)

have complete mixing, the liquid velocity at the farthest point in the aeration unit will determine the suspension of the particles and the ultimate power requirements. A velocity of about 0.5 ft per second will maintain microbial solids in suspension.

Power

The power required for aerated lagoons is that needed to transfer the oxygen and to mix the lagoon contents. If only odor control is desired, the power required will be considerably less than that needed to meet the total oxygen demand and to provide for complete mixing. The actual power required to maintain solids in suspension is a function of the nature of the solids, the lagoon depth and geometry, and the mixing efficiency of the aeration equipment.

The power requirements for mixing usually are specified in terms of the horsepower required to either maintain specified minimum fluid velocities in the lagoon or to maintain a homogeneous, suspended solids concentration in the lagoon mixed liquor. Values of horsepower per 1000 gal have meaning only for a

Fig. 6.12. Surface aerators in an aerated lagoon treating brewery wastes. (Courtesy of Envirex, Inc., a Rexnord company.)

specific aeration unit since the pumpage of liquid from various designs of surface aerators can vary by factors of 2 to 10. A more meaningful factor to keep solids in suspension would be the pumpage that an aerator produces per 1000 gal or 1000 ft^3 of basin volume, since it is the pumpage that distributes the oxygen and keeps the solids in suspension. The transfer of oxygen is related to the quantity of fluid passing through a rotor or surface aerator.

In an aerated lagoon with detention times greater than about 1-day detention time, the horsepower for aeration is a function of the oxygen demand and tends to be represented by a first-order reaction of horsepower and detention time. The horsepower for mixing is a function of the volume of the lagoon and will vary directly with the detention time, since for a given flow rate, the volume increases directly with detention time. Figure 6.13 illustrates these relationships.

Where mechanical surface aeration is used to supply sufficient mixing to attempt uniform dissolved oxygen concentrations throughout the basin but not necessarily suspension of all the solids, power levels of 0.015 to 0.02 hp per 1000 gal of basin volume have been suggested (30). Evaluation of two full-scale aerated lagoon systems treating municipal wastes indicated that from 0.2 to 0.33 hp per 1000 ft^3 (0.027–0.044 hp per 1000 gal) of lagoon capacity was required if adequate mixing was maintained (31). The mechanical aeration units were capa-

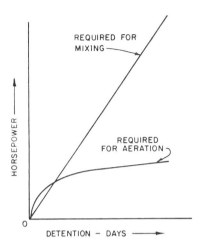

Fig. 6.13. Power relationships required for aeration and mixing in an aerated lagoon.

ble of transferring from 1.75 to 1.80 lb of oxygen per horsepower hour at 20–25°C and zero dissolved oxygen in the lagoon contents. The oxygen transfer relationships of mechanical aerators under process conditions range from 1.5 to 2.0 lb of oxygen per horsepower hour. Data from municipal waste aerated lagoons demonstrated that power levels of 0.03 to 0.05 hp per 1000 gal would completely mix the lagoons (26) and keep the suspended solids in suspension. Below 0.03 hp per 1000 gal of lagoon capacity, complete mixing was not accomplished.

Surface aerators rarely can be selected without a good knowledge of the process operating conditions, basin geometry, and aerator spacing as well as the need for the aerator to both mix and aerate. Where it is expected that these factors as well as capital and operation costs are critical, pilot and full-scale studies are valuable to select proper aeration equipment.

Application to Agricultural Wastes

Aerated lagoons have been used for the treatment of agricultural wastes. The following applications illustrate typical results.

Pea processing wastes have been treated in an aerated lagoon having a detention time of 5.5 days. Four 50-hp surface aerators were used, and the BOD loading rate ranged from 7.0 to 10.3 lb/1000 ft³/day. Reductions in total BOD and COD by the aerated lagoon averaged 76 and 59%, respectively (32). Dissolved BOD and COD in the waste was reduced by 95 and 82%, respectively.

Aerated lagoon treatment of pear, peach, and apple processing wastes provided greater than 70% BOD removal. Suspended solids in the effluent were the principal source of effluent BOD and COD (33). Nutrient addition was necessary to achieve successful treatment. Nutrient-deficient sludge did not settle. Surface aerators accomplished adequate mixing at 0.04 hp per 1000 gal and transferred 2 lb of oxygen per horsepower hour under operating conditions.

A series aerated lagoon system achieved 96% BOD reduction with wine wastes (34). Aerators having a capacity of 60 hp were installed in the first lagoon, which was loaded at 1200 lb of BOD per acre per day. Three 5-hp aerators were used in the second lagoon. The overall BOD removal decreased 5–10% during the winter. BOD removal of 75% was accomplished in the first lagoon, and an additional 84% in the second lagoon.

Aerated lagoon treatment of duck wastewater has produced an average of 85% BOD removal. The removals on the 26 farms using this treatment ranged from 60 to 95%. The aerated lagoons were designed to have an average detention time of 5 days. Because of variable flows, the actual detention times varied considerably from the average. The aerators used in the lagoons had power relationships that varied from 0.008 to 0.04 nameplate hp per 1000 gal lagoon capacity.

Many other studies could be cited to illustrate the applicability of aerated lagoons to agricultural wastes. The success of aerated lagoons depends upon utilizing biological treatment fundamentals in design, obtaining realistic information on the characteristics of the waste to be treated, and seeing that the actual waste loads sent to the unit are those for which the unit is designed. Aerated lagoons can perform satisfactorily and help industry meet the effluent limitations.

References

1. Bartsch, A. F., and Allum, M. O., Biological factors in treatment of raw sewage in artificial ponds. *Limnol. Oceanogr.* **2**, 77–84 (1957).
2. Oswald, W. J., Golueke, C. G., Cooper, R. C., Gee, H. K., and Bronson, S. C., Water reclamation, algal production and methane fermentation in waste ponds. *Int. J. Air Water Pollut.* **7**, 627–648 (1963).
3. Hermann, E. R., and Gloyna, E. F., Waste stabilization ponds. III. Formulation of design equations. *Sewage Ind. Wastes* **30**, 963–975 (1958).
4. Canter, L. W., Englande, A. J., and Mauldin, A. F., Loading rates on waste stabilization ponds. *J. Sanit. Eng. Div., Am. Soc. Civ. Eng.* **95**, SA6, 1117–1129 (1968).
5. Chang, A. C., Olmstead, W. R., Joanson, J. B., and Yamashita, G., The sealing mechanism of wastewater ponds. *J. Water Pollut. Contr. Fed.* **46**, 1715–1721 (1974).
6. Mara, D. D., Silva, S. A., and de Ceballos, B. S. Design verification for tropical oxidation ponds. *J. Env. Eng. Div. Am. Soc. Civ. Eng.* **105**, 151–155 (1979).
7. Johnson, B. A., Wight, J. L., Bowles, D. S., Reynolds, J. H., and Middlebrooks, E. J., "Waste Stabilization Lagoon Microorganism Removal Efficiency and Effluent Disinfection with Chlorine," EPA-600/2-79-018. Municipal Environmental Research Laboratory, U.S. Environmental Protection Agency, Cincinnati, Ohio, 1979.

8. Zall, R. R., Guest, R. W., and Weaver, D. E., Lagoons—a treatment for milking center waste. *J. Milk Food Technol.* **36**, 37–41 (1973).

9. Bland, R. R., Martin, J. H., and Loehr, R. C., Treatment of milking center wastewater in facultative ponds, *In* "Livestock Waste: A Renewable Resource," pp. 221–224. Amer. Soc. Agric. Engrs., St. Joseph, Michigan 1980.

10. Sollo, F. W., Pond treatment of meat packing plant wastes. *Proc. Purdue Ind. Waste Conf.* **15,** 386–398 (1961).

11. Anderson, J. S., and Kaplovsky, A. J., Oxidation pond studies of evisceration wastes from poultry establishments. *Proc. Purdue Ind. Waste Conf.* **16,** 8–21 (1961).

12. Nemerow, N. L., Baffled biological basins for treating poultry plant wastes. *J. Water Pollut. Contr. Fed.* **41,** 1602–1621 (1969).

13. Nemerow, N. L., and Scott, S. D., Lagoons plus spray irrigation for effective bean cannery waste treatment. *Proc. Purdue Ind. Waste Conf.* **27:** 555–563 (1972).

14. Oswald, W. J., Fundamental factors in stabilization pond design: Advances in biological treatment. *Int. J. Air Water Pollut.* **5,** 357–393 (1963).

15. Oswald, W. J., Gotaas, H. B., Golueke, C. G., and Kellen, W. R., Algae in waste treatment. *Sewage Ind. Wastes* **29,** 437–455 (1957).

16. Loehr, R. C., Treatment of wastes from beef cattle feedlots—field results. *In* "Proceedings of the Agricultural Waste Management Conference" (R. C. Loehr, ed), pp. 225–241. Cornell University, Ithaca, New York, 1969.

17. Loehr, R. C., and Ruf, J. A., Anaerobic lagoon treatment of milking parlor wastes. *J. Water Pollut. Contr. Fed.* **40,** 83–94 (1968).

18. Wymore, A. H., and White, J. E., Treatment of a slaughterhouse waste using anaerobic and aerated lagoons. *Proc. Purdue Ind. Waste Conf.* **23,** 601–618 (1969).

19. Baker, D. A., and White, J. E., Treatment of meat packing waste using PVC trickling filters. *In* "Proceedings of the Second National Symposium on Food Processing Wastes," pp. 287–312. Pacific Northwest Water Lab., U.S. Environmental Protection Agency, Denver, Colorado, 1971.

20. Niles, C. F., and Gordon, H., Operation of an anaerobic pond on hog abattoir wastewater. *Proc. Purdue Ind. Waste Conf.* **26,** 612–625 (1971).

21. Oswald, W. J., Goleuke, C. G., Cooper, R. C., and Tsugita, R. A., Anaerobic-aerobic ponds for treatment of beet sugar wastes. *In* "Proceeding of the Second National Symposium on Food Processing Wastes," pp. 547–598. Pacific Northwest Water Lab., U.S. Environmental Protection Agency, Denver, Colorado, 1971.

22. Crandall, C. J., Kerrigan, J. E., and Rohlich, G. A., Nutrient problems in meat industry wastewaters. *Proc. Purdue Ind. Waste Conf.* **26,** 199–212 (1971).

23. Midwest Plan Service, "Livestock Waste Management with Pollution Control," North Central Regional Research Publication 222. Iowa State University, Ames, Iowa, 1975.

24. Richter, G. A., Aerobic secondary treatment of potato processing wastes. *In* "Proceedings of the First National Symposium on Food Processing Wastes, pp. 39–71. Pacific Northwest Water Lab., U.S. Environmental Protection Agency, Portland, Oregon, 1970.

25. Timpany, P. L., Harris, L. E., and Murphy, K. L., Cold weather operation in aerated lagoons treating pulp paper mill wastes. *Proc. Purdue Ind. Waste Conf.* **26:** 776–787 (1972).

26. Malina, J. F., "Design Guides for Biological Wastewater Treatment Processes," Final Rep., Proj. 11010 ESQ. U.S. Environmental Protection Agency, Washington, D.C., 1971.

27. Pick, A. R., Burns, G. E., Van Es, D. W., and Girling, R. M., "Evaluation of Aerated Lagoons as a Sewage Treatment Facility in the Canadian Prairie Provinces," Proc. Int. Symp. Water Pollut. Contr. Cold Climates, Water Pollut. Contr. Res. Ser. 16100EXH. U.S. Environmental Protection Agency, Washington, D.C., 1971.

28. Humenik, F. J., Overcash, M. R., and Miller, T., Surface aeration: design and performance for

lagoons. *In* "Managing Livestock Wastes," pp. 568–571. Amer. Soc. Agric. Eng., St. Joseph, Michigan 1975.

29. Humenik, F. J., Overcash, M. R., Barker, J. C., and Westerman, P. W., Lagoons: state of the art. *In* "Livestock Waste: A Renewable Resource," pp. 211–216. Am. Soc. Agric. Engrs., St. Joseph, Michigan 1980.

30. Eckenfelder, W. W., and Wanielista, M. P., Theoretical aspects of aerated lagoon design, *Pap. Sanit. Eng. Div. Spec. Conf. Am. Soc. Civ. Eng.* (1965).

31. McKinney, R. E., and Benjes, H. H., Evaluation of two aerated lagoons. *J. Sanit. Eng. Div. Am. Soc. Civ. Eng.* **91,** SA6, 43–56 (1965).

32. Dostal, K. A., "Aerated Lagoon Treatment of Food Processing Wastes," Final Rep. Proj. 12060. Water Quality Office, U.S. Environmental Protection Agency, Washington, D. C., 1968.

33. Esvelt, L. A., and Hart, H. H., Treatment of fruit processing wastes by aeration. *J. Water Pollut. Contr. Fed.* **42,** 1305–1326 (1970).

34. Tofflemire, T. J., Smith, S. E., Taylor, C. W., Rice, A. C., and Hartsig, A. L. Unique dual lagoon system solves difficult wine waste treatment problems. *Water Waste Eng.* **19,** F1–F5(1970).

7

Oxygen Transfer

Introduction

Knowledge of the fundamentals that affect oxygen transfer is important to the design and operation of all aerobic waste treatment processes. Such processes are used to produce effluents suitable for point source discharge and to stabilize the wastes and reduce odors prior to subsequent use or disposal.

In aerobic processes, oxygen is transferred to a liquid primarily by mechanical equipment. The equipment serves to (a) provide the oxygen needed by the microorganisms to oxidize the organic matter and (b) keep the biological solids in suspension. The aeration equipment must be able to introduce dissolved oxygen as rapidly as it is utilized by the microorganisms as they oxidize the organic matter.

There basically are two types of aeration equipment: (a) mechanical aerators and (b) diffused air bubble aerators. With mechanical aeration, the rapid movement of impellers continually creates new surface areas through which the oxygen can be transferred to the liquid. Diffused air aerators introduce air bubbles into the liquid. Oxygen transfer occurs from the bubble to the liquid.

Aeration can be a major energy-consuming process in a waste treatment system. With the increasing cost of energy, there is heightened interest in a better understanding of the factors affecting oxygen transfer and in improving the design and operation of aeration systems to reduce the energy demand while meeting treatment objectives.

159

Transfer Relationships

Basic Relationships

The rate at which oxygen is transferred to a liquid is related to (a) the solubility of oxygen, (b) the characteristics of the liquid being aerated, (c) the rate at which oxygen is being utilized by the microorganisms, and (d) other factors, such as the type of equipment and the liquid temperature.

Aeration is a gas–liquid mass transfer process which occurs when a driving force results from unsaturated conditions. In the liquid phase, the driving force is a concentration gradient.

The transfer of oxygen into a liquid occurs in several phases (Fig. 7.1):

1. The oxygen molecules are transferred from the bulk of the gas used for aeration, usually air, to the liquid interface. This occurs very rapidly and establishes a saturated oxygen layer at the gas–liquid interface.
2. The oxygen molecules pass through the gas–liquid interface and the liquid film by molecular diffusion. This is a relatively slow process and can be the limiting factor controlling the rate of oxygen transfer.
3. The oxygen is dispersed throughout the liquid by diffusion and convection.

Any factors, such as chemicals that may be in a wastewater, that impede the transfer of oxygen through the liquid film will have an adverse effect on the rate of oxygen transfer. The turbulence provided by the aeration equipment keeps the liquid film thickness very small and disperses the oxygen throughout the bulk of the liquid. The oxygen concentration in the liquid at the interface is the equivalent of the saturation concentration.

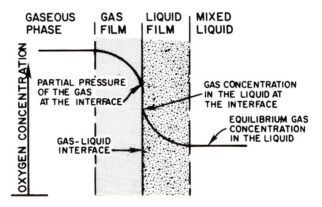

Fig. 7.1. Schematic representation of gas transfer from a gas to a liquid.

The mass transfer of oxygen, or other dissolved gas, in a liquid depends on the concentration gradient and the surface area through which the transfer occurs. Assuming that diffusion through the liquid film is the rate-limiting step, the mass transfer can be expressed as

$$N = D_L A \frac{\Delta C}{\Delta L} \tag{7.1}$$

where N = mass transfer per unit time, A = surface area through which diffusion occurs, D_L = diffusion coefficient through the liquid film, ΔC = difference in oxygen concentration at the interface and in the liquid, and ΔL = thickness of the liquid film.

Gas transfer commonly is considered in concentration rather than mass units. This can be obtained by dividing Eq. (7.1) by the volume (V) of the liquid being aerated:

$$\frac{dC}{dt} = \frac{N}{V} = \left(\frac{D_L A}{V} \right) \frac{\Delta C}{\Delta L} \tag{7.2}$$

In practice, neither the surface area, the thickness of the liquid film, nor the diffusion coefficient are able to be measured. These factors are combined into an overall gas transfer coefficient, $K_L a$. The resulting equation used to estimate the rate at which oxygen must be supplied by an aeration system is

$$\frac{dC}{dt} = K_L a(C_S - C_L) \tag{7.3}$$

where

dC/dt = rate of change of dissolved oxygen concentration with time, mass/volume-time

$K_L a$ = overall gas transfer coefficient, time^{-1}

C_S = oxygen saturation concentration for given atmospheric composition and pressure, mass/volume

C_L = actual oxygen concentration in the liquid at time t, mass/volume

The numerical value of the overall transfer coefficient $K_L a$ is a function of the gas–liquid interfacial area, the liquid turbulence, and the diffusivity of the oxygen in the liquid and is dependent upon the temperature of the liquid and the type and concentration of organics in the wastewater.

The saturation value of dissolved oxygen, C_S, is affected by temperature, dissolved inorganic material, and the air pressure at the depth the air is introduced into the wastewater. Oxygen is only slightly soluble (7–10 mg per liter) in water at ambient temperatures and pressures.

Most efficient oxygen transfer and aerator design, operation and performance occurs when (a) there is a long contact time between the gas and the liquid, (b) there is a large driving force, i.e., large oxygen concentration differential in the liquid, (c) there is a large contact area between the gas and the liquid, and (d) there are few interferences to the transfer of the gas through the interface and liquid film to the liquid.

Equation (7.3) indicates that the dissolved oxygen deficit is a driving force that attempts to achieve saturation. In a biological treatment unit, the oxygen utilization rate of the microorganisms (R_r) also will exist and Eq. (7.3) must be modified to include both factors. The microorganisms in the aerobic process will utilize the dissolved oxygen in the liquid as a hydrogen acceptor (Fig. 7.2).

$$\frac{dC}{dt} = K_L a(C_S - C_L) - R_r \tag{7.4}$$

The microbial oxygen utilization rate in a biological treatment system is dependent upon the available food, number of active organisms, and temperature. In low-rate, conventional aerobic treatment systems, R_r can be in the range of 5–50 mg O_2/liter/hr. The uptake rate R_r can be in the range of 50–100+ mg of O_2 per liter per hour in a high-rate system. On a mass basis, R_r can range from 5 to 40 mg of O_2 per hour per gVSS for conventional units such as an oxidation ditch and activated sludge process.

The oxygenation capacity of an aeration unit or system is a measure of the rate of input of oxygen that can be achieved under given operating conditions. It is defined as the mass of oxygen absorbed in unit time at a specific temperature and pressure. Usual units are pounds per hour or milligrams per liter per hour. Aeration efficiency is the ratio of the quantity of oxygen absorbed to that blown

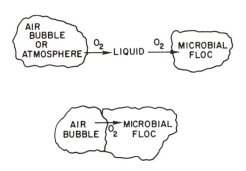

Fig. 7.2. Pathways of oxygen transfer in aerobic biological systems.

in for diffused aeration systems or the ratio of the oxygen absorbed into the mixed liquor per unit of energy of the mechanical equipment for mechanical and diffused aerators.

The oxygenation capacity of an aerator will govern the number of aerators that will be required for the treatment of a given flow of waste at a given detention period. The aeration efficiency will influence both capital and operational costs since it governs the size of the compressor or motor and the rate of consumption of energy required to achieve a given rate of oxygen absorption.

The oxygen capacity of aeration equipment usually is determined in water at zero dissolved oxygen and adjusted to 20°C. This is expressed as the standard oxygen transfer rate (SOTR) and provides a standard method of comparing the performance of different types of aeration equipment. The liquid is tap water to avoid interferences that may be in the wastes. A common temperature is used, and the dissolved oxygen of the water under testing conditions is reduced to zero by chemical means such as the addition of sodium sulfite. The SOTR is measured in terms of mass units, usually either pounds or kilograms per hour.

Under these conditions, Eq. (7.3) is modified to

$$\text{SOTR} = \frac{dC}{dt} \, V = (K_L a_{20}) V C_S \tag{7.5}$$

where V = volume of liquid and $K_L a_{20}$ = overall gas transfer coefficient at 20°C in tap water. A true measure of V is important. V should be the volume of the liquid that is under the direct mixing and pumping influence of the aeration systems. If the aeration unit is not completely mixed, the magnitude of V will be less than the liquid basin volume.

The oxygen demand of the microorganisms must be met by the oxygen transferred by the aeration equipment. In addition, it is desirable to have a measurable dissolved oxygen concentration (about 1–2 mg per liter) in the aerobic process to assure that aerobic conditions are maintained. Therefore, the oxygenation capacity of the aeration system should equal or exceed the oxygen demand of the microorganisms.

Under steady-state conditions in the aerobic biological unit, this means that the quantity of oxygen transferred by the aeration equipment (AOR) is equal to the oxygen demand (R_r) and a steady-state dissolved oxygen concentration (C_L) will result:

$$\text{AOR} = R_r = (K_L a)(C_S - C_L) \tag{7.6}$$

where AOR = actual oxygen transfer rate under the operating conditions; R_r = microbial oxygen demand rate in the aerobic unit under the operating conditions; and $K_L a$, C_S, C_L = oxygen transfer coefficient, dissolved oxygen saturation concentration, and actual liquid dissolved oxygen concentration, all under the operating conditions of the system.

Equation (7.6) is helpful in examining what happens if the AOR and R_r change as a result of different operating conditions. If AOR is greater than R_r, the aeration equipment will transfer more than enough oxygen to meet the microbial demand, and the liquid dissolved oxygen concentration C_L will increase until there is a new equilibrium and the system is again in balance. If AOR is considerably greater than R_r, C_L can approach the saturation concentration. Conversely, if AOR becomes less than R_r, C_L will decrease, and a dissolved oxygen concentration of zero may result in the mixed liquor.

Aeration equipment is rated on the basis of the oxygen transferred in tap water under standard temperature and barometric conditions with zero dissolved oxygen. This rating is rarely the quantity of oxygen transferred under process conditions, since the dissolved oxygen level normally is kept at a level of at least 1–2 mg per liter to assure aerobic conditions. The effect that the actual dissolved oxygen concentration (C_L) will have on the pounds transferred is illustrated in Table 7.1. As noted, the difference from standard conditions, i.e., zero dissolved oxygen, can be significant. It is important to recognize that aerator ratings reflect conditions that are unlikely to occur under process conditions.

Because standard conditions rarely exist in practice, modifications to Eq. (7.6) must be made to reflect the aerator oxygenation capacity under actual process conditions. These modifications include factors to adjust for different $K_L a$, C_S, temperature, and pressure conditions. Field oxygen transfer rates are calculated using alpha (α), beta (β), and theta (θ) correction factors:

$$K_L a_{ml} = \alpha(K_L a_W) \tag{7.7}$$

TABLE 7.1

Oxygen Transfer as Affected by Residual Dissolved Oxygen Concentration[a]

DO (mg/liter)	DO deficit ($C_S - C_L$) (mg/liter)	Oxygen transfer rate (mg/liter/hr)	Pounds of oxygen transferred per hour[b]
0	8.8	17.6	147
1	7.8	15.6	130
2	6.8	13.6	113
3	5.8	11.6	97
4	4.8	9.6	80

[a]$K_L a$ assumed as 2 mg/liter/hr/mg/liter. The uptake rate was calculated using Eq. (7.4) and assuming that the system was operating at equilibrium. Values for α and β assumed to be 1.0.
[b]Basin of 1 million gallon capacity assumed.

$$C_{S,ml} = \beta(C_{S,20}) \tag{7.8}$$

$$K_L a_T = (K_L a_{20})\theta^{(T-20)} \tag{7.9}$$

$$C_{S,P} = C_{S,760}\left(\frac{P - \rho}{760 - \rho}\right) \tag{7.10}$$

where α = the ratio of the $K_L a$ in the mixed liquor ($K_L a_{ml}$) to $K_L a$ in clean water ($K_L a_w$) at the same equivalent conditions of temperature, and pressure, geometry, mixing, etc.; β = the ratio of the oxygen saturation in a mixed liquor ($C_{S,ml}$) at temperature T and pressure P to the oxygen saturation ($C_{S,20}$) at 20°C and pressure P; θ = the temperature correction factor; P = atmospheric conditions at process conditions (millimeters of Hg); and ρ = saturated water vapor pressure (millimeters of Hg).

Generally, the standard and field oxygen transfer rates are compared in aeration tanks at the treatment site. If so, there are no pressure differences. If not, the field oxygen transfer rate will have to be corrected for pressure differences. $C_{S,P}$ also can be obtained from tables in standard references books that relate C_S to different temperatures and altitudes.

With these modifications, the actual field oxygen transfer rate (AOR) can be calculated by

$$AOR = \alpha(K_L a_{20})(\beta C_{S,P} - C_L)\theta^{(T-20)} \tag{7.11}$$

If the standard oxygen transfer rate (SOTR) is known from the manufacturer of the aeration equipment or from other testing, the AOR value can be calculated as:

$$AOR = (SOTR)\alpha\theta^{(T-20)}\left(\frac{\beta C_{S,P} - C_L}{C_{S,20}}\right) \tag{7.12}$$

At 20°C, $C_{S,20} = 9.17$ mg per liter.

Using the appropriate values of α, β, and θ is important. For example, when $C_L = 2$, $\alpha = 0.8$, $\beta = 0.9$, $\theta = 1.024$, and $T = 15$°C, AOR is only 50% of the standard transfer rate.

Aeration systems are designed before the aerobic biological system is in operation and before R_r is known. By establishing α, β, and θ or the actual $K_L a$ and applying them in Eq. (7.5), the oxygenation capacity of the system can be estimated and the aeration system designed. Since it is not practical to determine these parameters at full-scale conditions prior to construction of the treatment facilities, laboratory or pilot-plant-scale conditions resembling estimated process design conditions are used to estimate the preceding parameters that may occur in the full-scale aeration units.

A temperature correction value (θ) of 1.024 is commonly used for most aeration systems. Values of from 1.008 to 1.047 have been reported, with 1.024 fitting most experimental data except in very cold water (1). Based upon testing of surface aerators, a value of 1.012 has been suggested as being more correct where the water temperature is between 5 and 25°C (2).

Temperature affects both the oxygen saturation concentration and the oxygen transfer coefficient. As the temperature increases, C_S decreases and $K_L a$ increases. The overall oxygenation capacity is not as widely affected by temperature as are these two factors (Fig. 7.3) and varies about 3–5% over a temperature range of 10–30°C. The waste characteristics affect α and β directly and influence the oxygenation concentration of an aeration system to a greater degree than does temperature.

$K_L a$ can vary with the conditions of the test and the actual conditions in practice. The constituents in an aeration unit can affect $K_L a$ primarily by affecting the ability of the oxygen to be transferred from the gaseous to liquid phase. An increase in viscosity will decrease $K_L a$ (3). Therefore, constituents which increase the viscosity of the liquid can decrease $K_L a$ and α of the aerated mixture. Such constituents include total solids above 2% (3), and detergents (1, 4).

Alpha (α) is a gross term for the factors that affect $K_L a$ and usually is used to indicate factors other than temperature. Values for α vary with the waste type and strength. Values of α generally are less than 1.0, with most values in the range of 0.5 to 0.9. Occasionally, the characteristic of the waste may be such that α is greater than 1.0, but this is the exception rather than the rule. Alpha (α) will vary with temperature, basin geometry and size, intensity of mixing, and organics in the wastewater under aeration. In a specific aeration basin, α is primarily a function of the unmetabolized organic material in the system. The greater the organic material in the system, the lower the α value.

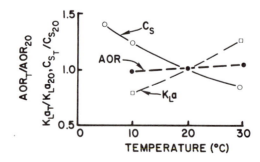

Fig. 7.3. Effect of temperature on $K_L a$, C_S, and the field oxygen transfer rate (AOR).

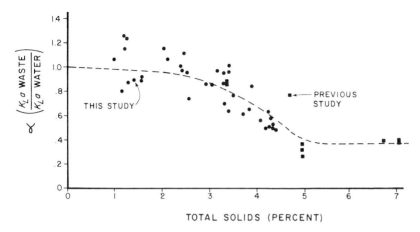

Fig. 7.4. Plot of α as a function of solids concentration in an oxidation ditch (data not corrected for temperature) (3).

Figure 7.4 indicates how the total solids concentration of an oxidation ditch mixed liquor affects the oxygen transfer characteristics of the aeration unit. The total solids had little effect until the solids concentration was about 2.0%. Beyond that concentration, oxygen transfer rates decreased as the solids concentration increased. When the solids concentration reached about 4.5%, the rotor was capable of transferring less than half of its maximum capability of oxygen. The relationship shown in Fig. 7.4 emphasizes the inadequacy of assuming a value for α, without exploring the factors that affect α in a particular aeration unit.

Beta (β) values also are usually less than 1.0, as many of the characteristics of wastewaters decrease the oxygen saturation capacity of a mixed liquor. Values of β generally are in the range of 0.7 to 0.9. The saturation of oxygen in a wastewater or mixed liquor sample can be obtained by permitting the sample to over-aerate until it has reached a saturation dissolved oxygen level. The ratio of C_S in the sample to C_S in tap water at standard conditions is defined as β. Standard conditions for tap water have been defined as sea level barometric pressure and zero chlorides.

Determination of the α, β, and $K_L a$ values should be based on conditions that exist or are expected in the biological unit and not in the raw waste, since it is the conditions that exist in the aeration unit that will influence the performance of the aeration equipment. Where wastes are not available to use in laboratory or pilot plant experiments to establish design parameters, it may be possible to estimate the parameters from ongoing systems using similar wastes or from existing published values. The best method is to use the specific waste in question.

Dissolved Oxygen Relationships

While aerators may be rated at zero dissolved oxygen to establish a common base, aeration systems are designed to maintain a specific residual in the biological unit. The residual should be above the critical dissolved oxygen concentration that will limit the respiration of the microorganisms. The rate of consumption of dissolved oxygen in mixed liquor is substantially independent of the dissolved oxygen concentration above about 0.5 mg per liter.

The dissolved oxygen concentration in the aeration unit need not be maintained higher than the minimum critical dissolved oxygen concentration, since the oxygenation capacity and the size of the aeration system is a function of the dissolved oxygen deficit that is maintained [Eqs. (7.6) and (7.11)]. The larger the deficit, the more efficient will be the aeration equipment in transferring oxygen.

In practice, waste loads, characteristics, and temperature vary. The aeration equipment should be able to produce an acceptable effluent under maximum conditions. To accommodate these variations, aeration units usually are designed to maintain a minimum dissolved oxygen concentration in the mixed liquor (C_L) of about 2 mg per liter.

Energy Relationships

Energy is required to transfer the required oxygen and to mix the aeration unit. In general, the power required to mix an aerobic unit varies directly with the volume under aeration. In contrast, the power required to meet the oxygen demand can be represented as a first-order reaction when related to the volume under aeration (Fig. 6.12). The volume and flow rate being treated are related since for a given detention time, the volume increases directly with the flow rate.

The power and total operation and maintenance (O & M) costs for several types of aeration systems used for agricultural wastes are compared in Table 7.2. The power costs are related directly with the flow rate of the system. As the plant size and flow rate increases, the power costs are a greater percentage of the total annual O & M costs.

Oxygen transfer rates frequently are stated in terms of the power required to transfer the oxygen, i.e., pounds of oxygen transferred per hour per horsepower. A description of the source of the horsepower should be provided. The power can refer to delivered horsepower, brake horsepower, or wire horsepower. The magnitude of these values can be different due to the efficiencies of the units that are involved (Fig. 7.5). The oxygen transfer can be stated as either the SOTR divided by the delivered horsepower or the SOTR divided by the supplied wire

TABLE 7.2

**Comparative Annual Operation and Maintenance Costs
for Aerobic Treatment Processes**

Aerobic process	Design flow (mgd)[a]	Power costs (dollars per year)	Total O & M costs (dollars per year)[b]	Power costs (% of total costs)
Activated sludge, con-ventional, diffused aeration	1.0	4800	18,000	27%
	10	48,000	82,000	59%
Activated sludge, con-ventional, mechan-ical aeration	1.0	4200	15,000	28%
	10	42,000	70,000	60%
Aerated lagoon	1.0	6000	16,000	37%
	10	60,000	72,000	83%
Rotating biological con-tactors, RBC units	1.0	6000	17,000	35%
	10	60,000	90,000	67%

[a]Million gallons per day.
[b]Estimated from figures in (5); data from the mid-1970s.

horsepower. Most coarse diffusion equipment, jet aerators, static aerators, and mechanical surface aerators transfer between 2 and 4 lb of oxygen per hour per wire horsepower.

Determination of Aeration Parameters

The engineer must understand the wastes and treatment system with which he will work. He must specify the operating conditions for which the system is to be designed, e.g., specify the dissolved oxygen concentration to be maintained under aeration for a given oxygen uptake rate, at a fixed temperature, and in a tank of definite dimensions. The manufacturer will need an idea of α to select the

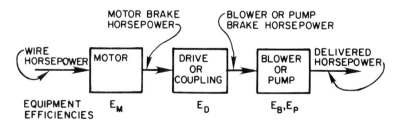

Fig. 7.5. Schematic of the power relationships involved in equipment used to transfer oxygen.

proper equipment. The equipment manufacturer should be responsible for providing accurate information on the $K_L a$ value of the aeration equipment in tap water since he is most familiar with his equipment under standard conditions and in most basins. Changing wastewater characteristics may preclude an accurate prediction of α and β. It should be the responsibility of the design engineer to determine the above critical factors that are valid for the expected aeration conditions since he is most familiar with the facility, its wastewater characteristics, its operation, and expected future changes.

One of the most complex aspects of designing an aerobic biological waste treatment unit is the selection of suitable aeration equipment. The only sure way to determine that aeration equipment will perform satisfactorily is to test it after installation. While estimates of oxygen transfer coefficients and oxygen uptake rates can be obtained using small-scale laboratory studies, more exact and useful data are obtained in full-scale aeration basins especially with the wastes to be treated. When full-scale units are not available, small-scale studies offer a way to develop useful estimates of parameters for design of actual aeration systems and equipment. Turbulence is the main difference between full-scale and small-scale aeration studies of a specific waste. The basic oxygen transfer processes are small-scale and are influenced by the smallest turbulence which produces diffusion of oxygen to the microorganisms. Even with similar power inputs per unit volume, different turbulence can result in units with different geometry, aerator location, and type of aerator.

Even with large-scale equipment for testing, shop testing of aerators may not adequately predict the performance of aerators after they have been installed. For surface aerators, at a given impeller submergence, the power consumption in the field may vary as much as 50% from the power consumption measured during shop tests (6). The difference in power consumption was not related to impeller design, volumetric energy level in the aeration basin, or the difference in volumes between the field and shop test conditions. Field tests represent the aeration equipment performance as installed and should be relied on as much as possible.

In lieu of relationships of energy input to turbulence regimes, empirical relationships based upon power input per unit volume have been developed from full-scale aeration systems to provide reasonable design parameters for mixing and solids suspension in aerobic biological treatment systems. Power levels of 0.015–0.02 hp per 1000 gal of basin volume have been suggested to obtain uniform dissolved oxygen concentrations but not necessarily all solids in suspension in aerated lagoons (7). Data from two full-scale aerated lagoons indicated that from 0.027 to 0.044 hp per 1000 gal was required if adequate mixing was maintained (8). European data indicated that in activated sludge basins the power input was 0.04–0.07 hp per 1000 gal to keep solids in suspension whereas in oxidation ditches, 0.015–0.023 hp per 1000 gal was sufficient (9).

Techniques commonly used to measure the oxygen transfer coefficient of aeration systems are (a) the non-steady-state method incorporating sulfite oxidation, (b) the steady-state method with microbial solids, and (c) the unsteady state method with microbial solids. An assumption inherent in the use of these techniques is that the test liquid is completely mixed and of uniform composition at any instant of time. A significant deviation from this assumption will result in erratic or erroneous data.

The non-steady-state method incorporating sulfite oxidation is useful in determining the oxygen transfer relationships in tap water. In the presence of metallic catalysts such as copper and cobalt, sulfite ion and oxygen react rapidly and irreversibly. The procedure is to add an excess of sulfite ion and a small amount of catalyst to the water. Approximately 0.8 lb of dry sodium sulfite per 1000 gal will be adequate per test and will provide an excess to compensate for oxidation during initial mixing. The aeration equipment is started, and the dissolved oxygen concentration is monitored either chemically or with a dissolved oxygen analyzer. The latter is preferred because of the minimum response time and avoidance of analytical errors.

Equation (7.13) identifies the reaction that takes place as the dissolved oxygen concentration is reduced to zero:

$$2Na_2SO_3 + O_2 \rightarrow 2Na_2SO_4 \qquad (7.13)$$

The theoretical sulfite requirement is 7.80 mg per liter of sulfite per 1.0 mg per liter of dissolved oxygen removal. The reaction of sodium sulfite with the dissolved oxygen is relatively slow unless there is an active catalyst. Cobaltous ion, added to the water as cobalt chloride ($CoCl_2 \cdot 6H_2O$) is the catalyst commonly used in the deoxygenation step. When repeated tests are run in the same tank, addition of catalyst with each run is not needed. If the catalyst is continually added, the catalyst concentration will increase to the point where it may affect the results.

The cobalt concentration should be kept low. Available test data indicate that a cobalt ion concentration in the range of 0.5 to 1.5 mg per liter as Co has no effect on the test data. A cobaltous ion concentration of greater than 0.05 mg per liter has been reported to have caused interference in the dissolved oxygen concentrations measured by the Winkler test (10). To avoid interferences with the dissolved oxygen test, the aeration test should be conducted at pH values of 6.9 or less.

Once the excess sulfite ion is oxidized, the dissolved oxygen concentration will increase. A plot of values of ($C_S - C_t$) against t on semilog paper should produce a straight line, the slope of which is $K_L a$ (Fig. 7.6).

The performance of the aeration equipment under process conditions is of considerable interest, and the sulfite oxidation method is not suitable for these

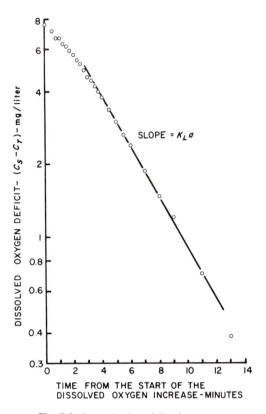

Fig. 7.6. Determination of $K_L a$ in tap water.

conditions. The value of $K_L a$ obtained under process conditions will depend upon the composition of the liquid that is aerated as well as the geometry of the unit and other process variables. In practice both the steady-state and non-steady-state methods are used with microbial solids since a convenient means of eliminating the oxygen demand of the mixed unit without altering its physical properties or without making difficult the estimation of dissolved oxygen has not been found.

Equation (7.4) is used with the steady-state method. At equilibrium, dC/dt is zero and the equation becomes

$$R_r = K_L a (C_S - C_t) \qquad (7.14)$$

C_t is the dissolved oxygen concentration that exists in the aeration unit under the test conditions and should be above the critical dissolved oxygen concentration.

In large units that may not be completely mixed, the tests should be run at a number of points in the unit. C_S is determined by saturating a sample of the mixed liquor with oxygen and measuring the dissolved oxygen at a given temperature. In Eq. (7.14), C_S is equal to $\beta C_{S,20}$.

The oxygen uptake rate R_r is determined by aerating a sample of the mixed liquor, sealing it from the atmosphere, providing continuous mixing, and observing the decrease in dissolved oxygen with time. A plot of dissolved oxygen with time should result in a straight line throughout most of the range. The slope of the straight line is R_r. Variation from a straight line may occur in the beginning of this test when equilibrium conditions are established and at the end of the test when the dissolved oxygen concentration becomes limiting. With all of the terms in the equation obtained from direct measurement, $K_L a$ can be determined by calculation.

When using the non-steady-state-method with microbial solids, dC/dt is not zero. Equation (7.4) is used by expressing it as

$$\frac{dC}{dt} = (K_L a C_S - R_r) - (K_L a C_L) \qquad (7.15)$$

Equation (7.15) has the form of a straight line. By plotting dC/dt against C_L a straight line with a slope equal to $K_L a$ will result.

This method involves turning off the aeration equipment and noting the decrease in dissolved oxygen that occurs with time. Since little oxygen is entering the aeration unit, the rate of dissolved oxygen decrease will be caused by microbial utilization of the oxygen and will be the microbial oxygen uptake rate R_r (Fig. 7.7a). In most cases this rate will deplete the oxygen within a short time after the aerator is turned off, and little solids settling will occur in this period. The dissolved oxygen probe should be agitated in the mixed liquor so that a suitable velocity, generally 1 ft per second, occurs in the area of the probe.

The aeration equipment remains off until a zero or low dissolved oxygen level is obtained. The aeration rate is then returned to normal and the dissolved oxygen concentration determined at frequent intervals until steady-state conditions are approached. The dissolved oxygen concentration (C_L) is plotted with time. Tangents to this curve are plotted at various values of C_L to establish values of dC/dt (Fig. 7.7a). These values are plotted against C_L as determined above (Fig. 7.7b) to determine $K_L a$ of the system under actual process conditions.

An advantage of this method is that $K_L a$ is determined independently of R_r or C_S and therefore is not influenced by any errors in their measurement. Any errors in measuring R_r will cause C_S to be in error. This method also permits information on $K_L a$, C_S, and R_r under actual process conditions rather than

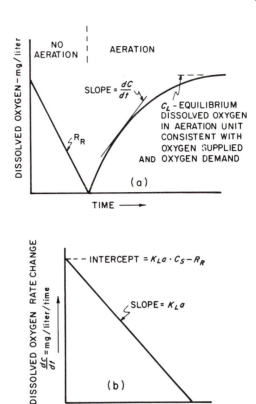

Fig. 7.7. Determination of $K_L a$ and C_S using non-steady-state methods.

taking aeration liquor samples and analyzing them for R_r or C_S under laboratory or other conditions that may not represent process conditions. Figure 7.8 illustrates the use of this approach with oxidation ditch mixed liquor treating poultry wastes.

If process conditions are unstable, the analysis cannot be applied. If a straight line is obtained from a plot of dC/dt versus C_L, the system is stable, and the parameters R_r and $K_L a$ are constant. An unstable system will give a curve when dC/dt is plotted against C_L and indicates that R_r and $K_L a$ are variable. Several factors contribute to the instability of an aeration system, the most important of which is lack of dissolved oxygen. The lack of dissolved oxygen at points in the system can create instability throughout the entire system, even though sufficient dissolved oxygen is present in specific parts of the system. If the equilibrium dissolved oxygen concentration under operating conditions falls below 1.0 mg per liter the evaluation of $K_L a$ by any analysis is questionable.

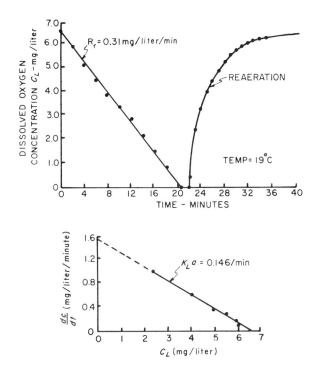

Fig. 7.8. Oxygen transfer and uptake characteristics—mixed liquor of oxidation ditch treating poultry wastes.

Aeration Equipment

Three equipment systems are presently available for the aeration of waste-waters in biological reactors: (a) diffused air systems, in which compressed air is forced into the liquid through a porous ceramic material or perforated pipe at a liquid depth of 10–15 ft or less; (b) the sparged air turbine system, in which compressed air is forced through a circular diffuser placed directly beneath a turbine-type mixer, which is placed at a significant depth in the liquid; and (c) a surface entrainment aerator, which mechanically creates turbulence at the liquid surface with a series of circular blades mounted so as to be partially submerged. The choice of an aeration system is based on such factors as capital and operating costs, flow and strength characteristics of the wastewater, maintenance and replacement experience, and designer's preference.

Diffused aeration, in which compressed air is put into a liquid through porous diffusers, has been a widely used aeration method. These units are installed such that the rising bubbles create turbulence and mixing. Figure 7.9

Fig. 7.9. An activated sludge unit using a diffused aeration system.

illustrates the aeration and mixing patterns in a large activated sludge unit using diffused aeration to treat combined municipal and meat-packing wastes. The rate of oxygen transfer by these systems is dependent upon the air–water interfacial area produced and the degree of mixing and turbulence generated. Because small bubble size provides a greater air–water interfacial area, early diffusers had small openings. Problems of clogging of these openings by chemical precipitates and biological growths prompted the development of diffusers with large openings and less maintenance. For standard conditions, the oxygen absorption efficiency of various types of diffuser systems will range from 3 to 10%.

In a submerged turbine system, a diffuser (sparger) ring delivers compressed air beneath the turbine through relatively large holes (Fig. 7.10). The turbine pumps a large quantity of water with the entrained air bubbles, and as this mixture passes through the turbine the bubbles are sheared, creating a large air–water interfacial area. The turbine generates considerable turbulence and mixing to aid in the transfer of the oxygen from the air bubbles. The diffuser ring supplies the oxygen, and the turbine supplies the mixing. In this manner, the two

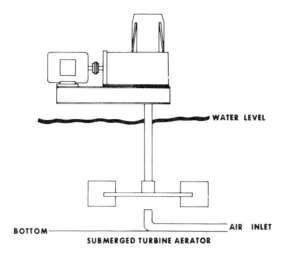

Fig. 7.10. Schematic of a submerged turbine aerator. [Courtesy of Mixing Equipment Co., Inc.]

functions of an aeration system, to meet the oxygen needs and to keep the solids in suspension, can be met by separate pieces of equipment. When the oxygen demand is low, the turbine can keep the solids in suspension while the air supply is decreased. Although oxygen absorption efficiencies of up to 50% may be obtained, high turbine speeds are required. From the practical standpoint an average efficiency of about 15–20% is accomplished with most wastewaters. These efficiency values are for standard conditions and must be corrected for the waste characteristics, temperature of the liquid, pressure, and desired dissolved oxygen in the liquid.

Submerged turbine aerators can satisfy high oxygen uptake rates that are beyond the capacity of slow-speed surface aerators. Submerged turbine aerators are independent of normal liquid level variations and icing problems in cold climates.

Surface aerators are generally slow speed with a high pumping capacity to move large quantities of liquid efficiently (Fig. 7.11). At least four types of surface aerators are manufactured, the plate type and updraft, downdraft, and brush types. The plate type is a rotating plate with vertical blades that create surface agitation to accomplish a high degree of oxygen transfer. The updraft type depends upon pumping large quantities of water at the surface with relatively low pumping energy. The downdraft type is an aerator where oxygen is supplied by air brought in by a negative head produced by a rotating blade or propeller. The brush type consists of a horizontal revolving shaft with blades attached to the shaft and extending below the water surface. The brush type is used extensively in oxidation ditches (Fig. 7.12).

Fig. 7.11. Fixed level surface mechanical aerator in an aerated lagoon.

Surface aerators achieve gas–liquid contact by means of an impeller or turbine located in the liquid surface of the water to be aerated. All types of surface aerators described previously promote gas–liquid contact in this manner. The rate of oxygen transfer is influenced by the total air–water interfacial area produced and the mixing and turbulence generated. Oxygen transfer occurs by (a) spraying the liquid in the air, (b) incorporation or dispersion of air into the liquid, and (c) generation of a high level of surface turbulence.

These actions are noted in Fig. 7.13. In the aeration zone, the dissolved oxygen concentration is increased from DO_1 to DO_2 where DO_1 and DO_2 are the average dissolved oxygen concentrations of the liquid entering and leaving the aeration zone, respectively. The oxygen transferred is related to the quantity of water moving through the aeration zone and the dissolved oxygen difference between discharged and inlet concentrations.

In a surface aerator, K_La is a function of the characteristics of the aerator, e.g., shape and height of blades, revolutions per minute, and geometry of the basin. The dissolved oxygen concentrations in the aeration basin will be determined by the dissolved oxygen concentrations leaving the aerator, the pumping rate of the aerator, the oxygen uptake rate, the configuration of the basin, and the location of the aerators in the basins. Surface aerators are essentially pumping units installed at the liquid surface and must pump a relatively large quantity of liquid to distribute the oxygen throughout the liquid and keep solids in suspension.

Fig. 7.12. Brush-type surface aerators in an oxidation ditch. (Courtesy of Lakeside Engineering Corp.)

Surface aerators mix effectively in their own zone of influence but not effectively throughout an aeration basin, even if there is more than one in the basin. Hydraulic translation of liquid from the zone of influence of one aerator to that of another aerator is slow. The transition between two zones of influence acts as a partition wall slowing intermixing in the basin.

Control of oxygen transfer capacity of surface aerators is related to aerator

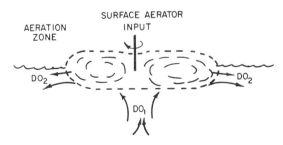

Fig. 7.13. Change in dissolved oxygen concentration through the aeration zone of a surface aerator.

design. Submerged aerators that disperse compressed air at the suction of the aerator can control oxygen transfer by varying air flow. Entrainment-type surface aerators (Fig. 7.10) control oxygen transfer by changing the speed or the submergence of the aerator. These latter measures adjust the horsepower of the aerator and thus the oxygen transfer capacity.

With a mechanical aerator, maximum oxygen transfer is at the aerator rather than through the surface of the aeration unit. When the surface of an aerated lagoon was covered to evaluate oxygen transfer at the aerator versus through the surface, the cover reduced the oxygen transfer rate by only about 10% (11). Other investigators have estimated the oxygen transfer through the surface to be as low as 2–5% during diffused aeration (4).

Surface aerators rarely can be selected without a good knowledge of the process operating conditions, basin geometry, and aerator spacing as well as the need for the aerator both to mix and to aerate. Where it is expected that these factors as well as capital and operating costs are critical, pilot and full-scale studies are valuable to select proper aeration equipment.

References

1. Stenstrom, M. K., and Gilbert, R. G., Effects of alpha, beta, and theta factor upon the design, specifications, and operation of aeration systems. *Water Res.* **15,** 643–654 (1981).
2. Landberg, G. G., Graulich, B. P., and Kipple, W. H., Experimental problems associated with the testing of surface aeration equipment. *Water Res.* **3,** 445–456 (1969).
3. Baker, D. R., Loehr, R. C., and Anthonisen, A. C., Oxygen transfer at high solids concentration *J. Environ. Eng. Div. Am. Soc. Civ. Eng.* **101,** 759–774 (1975).
4. Downing, A. L., and Boon, A. G., Oxygen transfer in the activated sludge process. *Int. J. Air Water Pollut.* **5,** 131–148 (1963).
5. U.S. Environmental Protection Agency, "Innovative and Alternative Technology Assessment Manual," Report 430/9-78-009. Office of Water Program Operations, Washington, D.C., (1978).
6. Stukenberg, J. R., and Wahbeh, V. N., Surface aeration equipment: field testing versus shop testing. *J. Water Pollut. Cont. Fed.* **50,** 2677–2686 (1978).
7. Eckenfelder, W. W., and Wanielista, M. P., "Theoretical Aspects of Aerated Lagoon Design," Presented at the Sanit. Eng. Div., Spec. Conf. Am. Soc. Civ. Eng., 1965.
8. McKinney, R. E., and Benjes, H. H., Evaluation of two aerated lagoons. *J. Sanit. Eng. Div. Am. Soc. Civ. Eng.* **91,** SA6, 43–56 (1965).
9. Zeper, J., and de Man, A., New developments in the design of activated sludge tanks with low BOD loadings. *Adv. Water Pollut. Res. Proc. Int. Conf.,* 112–123 (1970).
10. Naimie, H., and Burns, D., Cobalt interference in the 'Non-steady state clean water test' for the evaluation of aerator equipment—I. Causes and mechanisms. *Water Res.* **11,** 659–666 (1977).
11. McKinney, R. E., and Edde, H., Aerated lagoon for suburban sewage disposal. *J. Water Pollut. Contr. Fed.* **33,** 1277–1285 (1961).

8

Aerobic Treatment

Introduction

Aerobic treatment systems are utilized to avoid odor problems during waste storage, to meet specified effluent limitations, and to stabilize the wastes prior to land application. Animal waste slurries as well as wastes from food processing, grain milling, leather tanning, and other agricultural industries can be treated or stabilized using such systems. There are many types of aerobic systems that can be utilized, and a number of the more important ones are discussed in this chapter and in Chapter 6.

Activated Sludge Processes

General

Activated sludge processes are aerobic biological processes that can be used to treat many wastes. The processes are versatile and flexible, and an effluent of any desired quality can be produced by varying the process parameters. High BOD removal efficiencies can be achieved routinely. These processes require less land but more competent operation than do simpler processes such as oxidation ponds and aerated lagoons. A number of activated sludge processes exist; however, only those that have potential for agricultural wastes are discussed.

The term, "activated sludge" is applied both to the process and to the biological solids in the treatment units. Activated sludge is the complex biological mass that results when organic wastes are aerobically treated. The sludge will contain a variety of heterotrophic microorganisms, including bacteria, protozoa, and higher forms of life. The predominance of a particular microbial species will depend upon the waste that is treated and the way in which the process is operated.

In practice, most activated sludge plants operate in the stationary phase of microbial growth and are continuous flow systems. The biological growths will aggregate into flocculant masses which can be kept in suspension while the activated sludge unit is mixed but which will settle when the mixing is stopped. These properties are important since they permit the microbial solids to be distributed throughout the biological unit but to separate readily in a separation unit. A well-treated effluent containing few biological solids will result. Depending upon the activated sludge modification, the settled solids are recycled back to the biological unit or wasted to a solids handling and disposal system, or both.

For optimum treatment, the raw waste must be balanced nutritionally. If the microbial solids undergo extensive endogenous respiration in the same unit in which the influent BOD is metabolized, a portion of the microbial nutrients will be released to the mixed liquor. A tradeoff exists between the addition of nutrients and the length of aeration period for nutrient-deficient wastes. Activated sludge produced from nutrient-deficient wastes tends to have poor settling and filtering characteristics.

After endogenous respiration occurs, further aeration results in no greater removal of raw waste BOD but does reduce the quantity of sludge solids for ultimate disposal. Since nearly all the BOD entering a system is converted into cell mass, the effluent BOD will be caused almost entirely by the endogenous respiration of the organisms contained in the effluent suspended solids.

The settling characteristics of activated sludge floc have been correlated with the SRT of a system (1). Waste activated sludge solids from a unit with a SRT greater than 4 days were found to have better solids separation in a final clarifier. The settling characteristics of the floc were not correlated with either the soluble COD in the unit or the food-to-mass (F/M) ratio.

The Basic Process

The basic process was developed in England with E. Arden and G. J. Lockett generally given the credit for the experimentation that led to the process as it is now known. Their work was done in 1914. However, experiments in the late 1800s demonstrated that the process was possible (2).

A flow diagram for the basic process includes two units: an aeration unit, in which the aerobic treatment occurs, and a sedimentation tank, in which the solids are separated from the liquid. These two important functions of the system are separated so that each process can be controlled individually (Fig. 8.1).

The biological solids in the aeration tank are known as the aeration tank solids or mixed liquor suspended solids (MLSS). The volatile fraction of the mixed liquor (MLVSS) frequently is used as a measure of the active organisms in the tank. However, it is only a gross estimate of the active mass.

The return of a portion of the settled solids permits a higher MLSS and, hence, active mass to be kept in the aeration tank. The larger concentration of organisms purifies the wastes in a shorter time, allows a smaller aeration tank, and permits the solids retention time to be different from the hydraulic retention time. Without sludge return, the activated sludge process would be approximated by an aerated lagoon with its larger size and smaller microbial concentration. The general design and operational parameters that are relevant to the basic activated sludge process are identified in Table 8.1.

The basic system functions as a plug flow system in that the untreated waste and return sludge enter the aeration tank at one end of the tank and move as a unit to the other end of the tank. The incoming wastes and solids are not mixed uniformly with all of the contents of the aeration tank. To reduce the oxygen demand in the aeration tank, settleable solids in the raw wastes are removed in a primary sedimentation unit. With municipal wastes, the BOD of the raw wastes is reduced about 30–35% in the primary sedimentation unit while the suspended solids are reduced about 50–60%. Other wastes will have different removal relationships, since the percent removals in the primary units are related to the type of solids in the raw wastes. The dissolved oxygen concentration in the aeration unit should be kept above 0.5–1.0 mg per liter.

Because of the plug flow effect, the activated sludge MLSS in the basic process are on an alternating feed–nonfeed cycle, which can affect the ability of the microorganisms to handle variable hydraulic and organic loads. The microbial concentration is considerably less than the MLVSS concentration. The viable aerobic, heterotrophic microorganism content of conventional and low-rate activated sludge units is between 10 and 20% of the MLVSS.

When understood and properly operated, conventional activated sludge plants can produce high (90+%) BOD removals. The key to the quality of the effluent lies in a positive solids removal system. Compared to other aerobic biological treatment processes, the advantages of the activated sludge process include small land area, positive control over the degree of treatment, and flexibility of treatment to meet varying waste loads. Disadvantages include higher capital costs and the need for competent and constant control. The basic process can be sensitive to variations in BOD and hydraulic loads, varying air requirements throughout the aeration tank, a feed–nonfeed cycle for the microorganisms since the wastes are introduced only at one point in the tank, and inability to control nitrification.

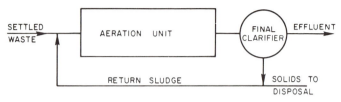

CONVENTIONAL ACTIVATED SLUDGE

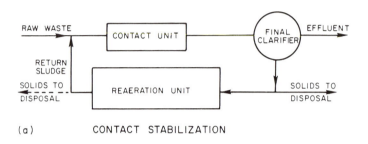

(a)   CONTACT STABILIZATION

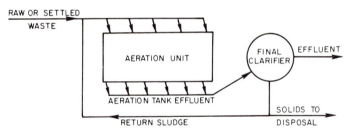

COMPLETE MIXING ACTIVATED SLUDGE

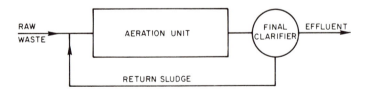

(b)   EXTENDED AERATION

Fig. 8.1. Flow diagram of activated sludge modifications.

TABLE 8.1

Design and Operational Parameters for the Conventional Activated Sludge Process

Organic loading rate	0.2–0.5 kg of BOD_5 per kg of MLSS per day
Volumetric loading rate	0.3–0.8 kg of BOD_5 per m³ per day (20–50 lb of BOD_5 per 1000 ft³ per day)
MLSS concentration	1,500–4,000 mg/liter
Hydraulic retention time	6–8 hr
Solids retention time	3+ days
Return sludge rate	25–50% of the influent flow
Oxygen requirements	1.2–1.5 kg of oxygen per kg of BOD_5 applied
Sludge production rate	0.5–0.7 kg per kg of BOD_5 removed
Aeration	Diffused or mechanical

On the average, the effluent suspended solids concentration and its variation are greater than corresponding effluent BOD values. A lognormal distribution generally can be used to describe effluent BOD and suspended solids data. Solids separation facilities, such as final clarifiers, are a limiting factor in the performance of activated sludge treatment systems.

Energy

Energy requirements are important when considering various waste treatment or management processes. Activated sludge processes have high energy requirements due to the aeration equipment needed to supply the oxygen and mix the aeration tanks. Table 8.2 provides comparative information on the energy required for various components of small systems such as might be used for food

TABLE 8.2

Energy Requirements for Components of a Conventional Activated Sludge Treatment System (kWh/year)[a]

	Capacity of wastewater treatment facility (million gallons per day)			
	0.1	0.5	1.0	5.0
Preliminary treatment[b]	3,125	5,200	6,410	11,180
Primary sedimentation	3,190	5,420	6,820	11,990
Mechanical aeration	16,000	80,000	160,000	800,000
Secondary sedimentation	5,010	10,390	16,400	54,870

[a]Adapted from (3).
[b]Bar screen, comminutor, nonaerated grit removal.

processing and related wastes. At an operating treatment plant, additional energy will be required to pump the wastewater to the treatment plant, for disinfection of the effluent if needed, for sludge treatment, and for miscellaneous purposes.

The energy required for the aeration equipment (Table 8.2) is much greater than that needed for any other component of an activated sludge treatment system. Other aerobic biological treatment processes can have smaller energy requirements (Table 8.3). Facultative lagoons essentially have no energy requirements, because photosynthetic action provides the oxygen, and mixing is not desired. Both trickling filters and rotating biological contactors have lower energy requirements because of the different methods of oxygen transfer and mixing that are used. Due to the long detention times and the exertion of the nitrogenous oxygen demand in extended aeration units, the aeration and hence energy requirements are greater than those in an activated sludge unit.

While energy requirements are important, other factors such as size, complexity, and maintenance also must be considered in determining the most appropriate treatment process. For instance, facultative lagoons may have little energy requirements but require a much greater land area than an activated sludge unit.

Application to Agricultural Wastes

Conventional activated sludge systems rarely are economic for livestock wastes, because of the high oxygen demand of such wastes, the high aeration energy requirements, and the skill and attention that are required. In addition, they are rarely used for food processing wastes. There are, however, three

TABLE 8.3

Energy Requirements for the Operation of Aerobic Biological Treatment Processes[a] (kWh/year)[b]

	Capacity of wastewater treatment facility (million gallons per day)			
	0.1	0.5	1.0	5.0
Activated sludge (mechanical aeration)	16,000	80,000	160,000	800,000
Extended aeration	35,000	175,000	350,000	1,750,000
Facultative lagoons	0	0	0	0
Rotating biological contractor	7300	36,500	73,000	365,000
Trickling filter (with recirculation)	7200	31,950	61,300	278,300

[a]Only the energy associated with the operation of the specific process is noted; the energy required for needed ancillary processes such as raw sewage pumping, preliminary treatment, primary and secondary sedimentation, and sludge treatment is not included.

[b]Adapted from (3).

modifications of the basic activated sludge process that have been used with the high and variable wastes that may originate from food processing operations. These are contact stabilization, complete mixing, and extended aeration. The application of these processes to agricultural wastes is noted in the following sections.

Contact Stabilization

In contact stabilization or biosorption (Fig. 8.1) the raw wastes are mixed and aerated with activated sludge for a short time and then settled in a sedimentation tank. The contact unit is used to adsorb the particulate and colloidal organic matter to the activated sludge floc. The detention time in the contact unit will depend upon the time necessary for adsorption and can be as low as 20–30 min.

Excess sludge may be wasted from the process either from the sedimentation tank or from the effluent of the reaeration tank. Solids wasted from the sedimentation tank will have a higher oxygen demand than those wasted from the reaeration unit since the latter have been oxidized in the reaeration unit where endogenous respiration is predominant.

The essential difference between the contact stabilization and conventional activated sludge treatment process is that the adsorption phase is separated from the biological oxidation phase. The sludge in the reaeration tank normally will range between 6000 and 8000 mg per liter. The solids in the contact tank can range from 3000 to 5000 mg per liter.

The detention time in the reaeration unit is from 4 to 6 hr. The reaeration unit volume is less than the volume in a conventional activated sludge plant since the tank is designed on the basis of the flow containing the settled solids and not the entire raw waste flow. Aerating concentrated solids in these units rather than more dilute solids, as in the conventional activated sludge process, reduces the aeration volume while providing high BOD reduction. Oxygen requirements for the system are about 40% for the contact tank and 60% for the reaeration tank. A higher oxygen demand rate will occur in the contact tank as the waste and microorganisms are mixed.

Generally accepted design criteria for contact stabilization systems indicate that as long as the average daily BOD_5 load applied does not exceed approximately 50 lb per 100 lb of mixed liquor suspended solids in the system, a reduction of approximately 90% or greater of the applied BOD_5 can be expected. The SRT of the system will be comparable to that in a basic activated sludge system.

The process is less sensitive to fluctuating or toxic loads, since it has a high concentration of microbial solids contacting the raw wastes. Additional concentrations of microbial solids are separated from the raw wastes and unaffected by transitory toxic conditions. The smaller contact tank permits better mixing to occur and a smaller plug flow effect.

Primary sedimentation rarely is a component of this process, since contact stabilization can handle the particulate material in the raw wastes. One of the main disadvantages for the process is that soluble wastes may not be adequately treated, due to the short detention time of the wastes and the microbial solids in the contact tank.

Conventional activated sludge systems can be converted readily to contact stabilization when increased treatment capacity is required. Prefabricated package plants for small waste flows are available from a number of manufacturers (Fig. 8.2).

Contact stabilization has been used with wastes from the killing and packing of broilers (4). At an aeration tank loading of 30–33 lb of BOD per 1000 ft³ per day the BOD was reduced from 1220–1720 mg per liter in the influent to 14–22 mg per liter in the effluent. Suspended solids removal was 94%.

In a comparative study of systems for the treatment of fruit processing wastes, both conventional activated sludge systems and contact stabilization provided greater than 90% removal of the organic load when the systems were loaded at low rates, less than 0.4 mg of COD removed per day per milligram MLVSS (5). The resulting solids were removed satisfactorily in clarifiers. Completely mixed aeration basins provided effective buffering capacity to avoid pH

Fig. 8.2. A contact stabilization plant treating combined municipal and cheese plant wastes. The sedimentation section is in the middle with the contact and reaeration basins on the periphery of the circular unit.

fluctuations in the aeration systems caused by abrupt changes in the pH of the waste flow.

Complete Mixing

The ideal activated sludge treatment system would include the following: ability to handle shock organic and hydraulic loads, uniform waste addition, uniform air utilization, complete waste mixing, uniform organic level in the aeration system, and simplicity of operation. Complete mixing activated sludge contains these attributes. Complete mixing is a process in which the incoming wastes are completely mixed with the entire contents of the aeration tank within a few minutes after the wastes enter the tank (Fig. 8.1). In small tanks the wastes and return sludge can be introduced at a single point, but in large tanks multiple inlets are required.

The aeration tank acts as a surge tank to level out any fluctuations in organic or hydraulic load. The microbial solids will respond to changes in the waste characteristics and adjust themselves accordingly to reestablish equilibrium conditions. The use of the entire mass of activated sludge results in a more uniform oxygen demand and uniform loading. The process has been utilized at a number of municipal and industrial treatment facilities.

A complete mix, activated sludge process satisfactorily treated potato processing wastewater, accomplishing over 90% BOD removal. BOD removals of 70–80% were obtained when the solids in the effluent from the aeration basin were not removed in a final clarifier (6). The high pH values, up to 11.5 at times, of the influent wastewater were buffered in the aeration basins and were not detrimental to the treatment efficiencies.

Citrus wastes required nitrogen and phosphorus additions to accomplish satisfactory complete mix, activated sludge treatment. The recommended design loading rate was 0.3 lb of BOD per lb of MLSS (7).

Extended Aeration

Where wastes can be held for long detention periods, extended aeration offers the opportunity to provide a high degree of treatment and to oxidize the entering and synthesized solids. This process includes long-term aeration, at least 24 hr for municipal wastes and longer for stronger wastes (Fig. 8.1). The solids are settled in the sedimentation tank and returned directly to the mixing tank, generally with separate wasting, producing a long SRT. Any excess sludge in the effluent has been aerobically digested in the aeration tank and should have little residual BOD. The fundamental research on extended aeration was initiated on studies of the treatment of dairy wastes (8).

TABLE 8.4

Design and Operational Parameters for the Extended Aeration Process

Organic loading rate	0.05–0.2 kg of BOD_5 per kg of MLSS per day
Volumetric loading rate	0.15–0.4 kg of BOD_5 per m^3 per day (10–25 lb of BOD_5 per 1000 ft^3 per day)
MLSS concentration	3,000–6,000 mg/liter
Hydraulic retention time	24+ hr
Solids retention time	10–20+ days
Return sludge rate	50–200% of influent flow
Oxygen requirements	2.0–2.4 kg of oxygen per kg of BOD_5 applied
Sludge production rate	0.2–0.4 kg per kg of BOD_5 removed
Aeration	Diffused or mechanical

General design and operational parameters that are relevant to the extended aeration process are presented in Table 8.4.

As measured by effluent quality, overall performance of extended aeration plants depend on the ability of the system to retain suspended solids. The discharge of excess suspended solids frequently results from hydraulic overload of the sedimentation section of the plants. Although extended aeration units are unsophisticated, they do require operational and maintenance attention to achieve the high BOD removals (90+%) that they can produce. Because of the long detention times, a high degree of nitrification may occur.

An extended aeration system for poultry processing wastes achieved 90% removal. The loading of the process was about 15 lb of BOD per day per 100 lb of MLSS, and the MLSS were 89% volatile. The average hydraulic detention time was 24 hr. BOD loadings approaching 40 lb of BOD per day per 100 of lb MLSS over a 12-hr period were successfully handled on an individual day basis. The air supplied was 1500 ft^3 per lb of BOD removed (9).

Citrus wastes were subjected to an extended aeration process. BOD removals ranged from 90 to 98%, with an average of 94% (10). Mixed liquor suspended concentrations of 3000–5000 mg per liter were held in the system when the aforementioned removals were achieved. The system was designed for a loading of about 7 lb of BOD per day per 100 lb of MLSS.

Extended aeration has been considered to represent best practicable technology for the treatment of dairy products wastewaters in Canada (11). Loadings that ranged from 0.08 to 0.14 kg of BOD_5 per kg of MLSS per day resulted in BOD reductions of 99% and suspended solids reductions of 95–99%. Hydraulic surges due to washing and cleaning in the dairy products plants adversely affected the performance of the treatment systems.

Completely mixed activated sludge systems and aerated lagoons can operate as extended aeration units. The key is the long detention time and the opportunity to oxidize the microbial solids. Total oxidation of the microbial solids will not occur, since portions of the microbial solids resist degradation and will be in the effluent. Extended aeration solids can be dewatered on sand beds and disposed of on the land.

Aerobic Digestion

Introduction

Waste treatment processes essentially are separation and concentration processes. Biological treatment processes use microorganisms to metabolize the organic matter and concentrate it in the form of synthesized microbial solids. Settling units can be used to separate the inert and microbial solids from the waste flow before discharging the effluent to the environment. Each separation and concentration process produces waste solids which require further handling and disposal.

A treatment process that can dispose of the organic solids and not produce any residue is desirable but has not yet been developed. Aerobic digestion, however, is a process which, if operated correctly, can produce a minimum amount of biodegradable solids. The residue from this process is a stable sludge and low oxygen demand, good settling characteristics, and no offensive odor. Aerobic digestion is an aerobic biological process designed to have a long solids retention time. The gross oxidation in the system includes the direct oxidation of any biodegradable matter by the active mass of organisms and endogenous respiration, the oxidation of microbial cellular material.

When aerobic digestion is used with waste solids that have had previous treatment, the microorganisms in these solids have removed the organic matter in the influent waste flow and have metabolized any stored and adsorbed food, and microbial synthesis is minimal. Therefore, the net decrease in sludge accumulation that occurs is caused by endogenous respiration.

The decrease in solids that occurs in an aerobic digester will be caused only by a decrease in the active mass in the digester, and that decrease is an exponential one. The inorganic and nonbiodegradable organic solids will not decrease. Although percent volatile solids reduction is convenient to measure aerobic digestion process efficiency, it is not the fundamental relationship. The reduction in active mass would be a better parameter to describe the performance of an aerobic digester.

Little change in the cellulosic material occurs during aerobic digestion. The nitrogenous matter is the most susceptible to degradation. As aerobic oxidation proceeds, cellular nitrogen is broken down and released to solution in the form of ammonia, and the ammonia is further oxidized. Nitrates in excess of 900 mg per liter were obtained in one study (12). Parameters related to nitrification can be used to note the progress of aerobic digestion since the oxygen supply is in excess of the demand and primarily is used for mixing. These parameters are a pH and alkalinity decrease and a nitrate increase.

Drying of aerobically digested sludge solids on sand beds can be an efficient method of disposal if the treatment facility is operated properly and the sludge drying beds designed adequately. Well-conditioned aerobically digested sludges have been shown to drain more rapidly than anaerobically digested sludges (13).

Aerobic digestion offers advantages, when compared to anaerobic digestion, which include the lack of any need for insulation, added heat, or special covers, the ability to handle low waste sludge concentrations (eliminating the need for sludge thickening devices), and the production of solid residue and liquid effluent with a low oxygen demand. The disadvantages include the cost of supplying air and the lack of any usable end product such as methane, which is produced in anaerobic digestion.

Figure 8.3 indicates the use of surface aerators in an aerobic digestion lagoon used for the excess solids from the aerobic biological treatment of food processing waste.

Fig. 8.3. Aerobic digestion of excess biological solids from an aerobic system treating food processing wastes.

The Oxidation Ditch

Introduction

Oxidation ditches are feasible aerobic biological treatment systems for agricultural wastes, whether they are liquid, such as food processing wastes, or slurries, such as livestock wastes. The largest number of ditches used for agricultural wastes have been installed in confined animal production facilities. The majority of the ditches have been utilized for swine operations although they have been incorporated in dairy and beef cattle as well as in poultry facilities. The reasons for the use of oxidation ditches in livestock confinement operations include (a) odor prevention and control, (b) saving labor, (c) waste treatment, and (d) ease of incorporation in confinement housing.

Both in-house and external oxidation ditches are possible. Oxidation ditches are shallow and can be incorporated in most confined livestock facilities. External oxidation ditches are similar to aerated lagoons. The choice between an aerated lagoon and an external oxidation ditch will be related to the shape and quantity of available land and to the oxygenation and power requirements of the surface aerators.

With in-house oxidation ditches, labor will be saved by avoiding moving the animal wastes to a treatment system and in the ability to use pumps to transport the liquid wastes from the ditch. Because of the evaporation by the rotor, the total waste volume will be less than if the wastes were not so aerated.

Because the oxidation ditch is an aerobic biological treatment system, the odors associated with anaerobic conditions will not occur. The system is not odor-free, but the odors that exist are rarely considered objectionable. Ammonia emission will occur if the oxygen supply is inadequate or if nitrification does not exist.

Operational problems occur, but with proper maintenance and an understanding of the microbial processes occurring in the ditch, they can be held to a minimum. The resulting waste management scheme is low in odors, requires little manual labor, and is convenient for the operator. Effluent from a continuous overflow oxidation ditch must be kept in aerated holding units or disposed of in a short time. Residual organic content and microbial activity in this effluent can cause anaerobic conditions and odors if the effluent is not aerated while being held prior to disposal.

The oxidation ditch was developed by the Institute of Public Health Engineering in the Netherlands. The basic process is known by a variety of names, such as the aeration ditch or the Pasveer ditch, after the name of an early advocate of the process. The system is an aerobic biological waste treatment system with a long liquid detention time and adequate mixing. The basic system is economical for small communities or industries and requires a minimum of supervision.

The oxidation ditch is similar to the aerated lagoon in that a surface aerator is used to supply the necessary oxygen. The oxidation ditch has a long, narrow shape with a center partition and can be modified to fit into available space. The key components of the system are a continuous open channel and a surface aeration rotor to mix the ditch contents as well as to supply the oxygen (Fig. 8.4). Depending on specific wastes and modes of operation, other components may be added, and many alternatives exist for further treatment or ultimate disposal (Fig. 8.5). The components will vary, depending upon the objectives of a particular system.

Attempts to have construction costs as low as possible have caused some oxidation ditches to be designed both as an aeration unit and as a sludge separation unit. This is accomplished by cyclic operation of the rotor. While the rotor is off, the mixed liquor solids can settle, and a portion of the clarified liquid is removed. This procedure can be effective with municipal wastes or other high-volume, low-strength wastes. It also is effective with animal wastes except when the mixed liquor in these ditches reaches high levels, from 1 to 5+% total solids, and solids separation by sedimentation does not occur.

Untreated wastes can be added directly to the ditch. Prior treatment such as solids separation is not needed. Depending upon the quality of the untreated wastes, the effluent may be of suitable quality for discharge to surface waters. Such may be the case for municipal wastes or for some food processing wastes, especially if the solids in the effluent are removed before the effluent is discharged. The effluent from oxidation ditches treating livestock wastes is not suitable for discharge to surface waters because of the color, mineral salts, and residual BOD in the effluent. The effluent is suitable for land disposal. Odors will be a minimum with the process if sufficient oxygen is supplied by the rotor.

Experience with municipal as well as animal wastes has shown that BOD

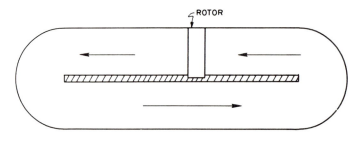

BASIC OXIDATION DITCH

Fig. 8.4. Diagram of the basic oxidation ditch.

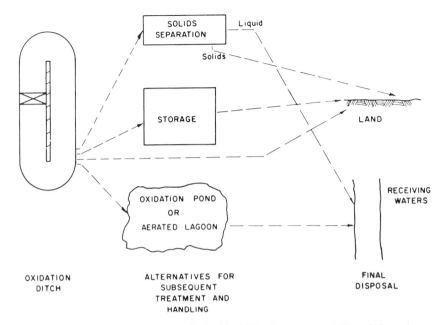

Fig. 8.5. Integration of an oxidation ditch with additional treatment and disposal alternatives.

reductions of 80–90% can be obtained. The advantages of the process include no odors, a treated effluent, and oxidation of waste solids. The large volume of the ditch, as compared to the influent flow, and the long detention time make the oxidation ditch less sensitive to changes in the quality and quantity of the untreated waste.

Process Design

Two types of oxidation ditches can be used: the in-house ditch commonly found in livestock confinement facilities and the external ditch that is separate from the waste production site. The in-house ditch is located beneath self-cleaning slotted floors of swine and cattle confinement buildings or below cages in poultry buildings. The in-house ditch takes advantage of the continuous and uniform waste-loading to the unit, the controlled temperature in the confinement building, and the continuous mixing and aeration to produce a near ideal biological waste treatment process. The in-house ditch also eliminates the equipment that otherwise would be required to transfer the livestock wastes from the source to the ditch, since the animals continuously push the waste through the slots to the ditch.

The external ditch is the preferred method for municipal and food processing wastes. The external ditch is uncovered and exposed to ambient temperatures.

In an oxidation ditch, the volume of the ditch per animal or case of food packed is of less importance than the oxygenation capacity of the rotor. The rotor must supply the required oxygen to meet the oxygen demand of the wastes entering the ditch. If adequate oxygen is not supplied, the ditch becomes oxygen-limited; poorer process efficiencies result, and odors can occur.

For in-house oxidation ditches, the size of the ditch is controlled to a large extent by its ability to fit into a confinement building. Ditch volume is important since it must be large enough to permit adequate detention time for treatment, stabilization, and the storage that is desired. The degree of solids stabilization and waste treatment purification that occurs is a function of the loading rate of the ditch. With low-volume wastes such as animal wastes, long detention times are the rule.

Design recommendations for in-house oxidation ditches for a variety of livestock wastes are noted in Table 8.5 (14). Additional investigations are necessary to verify the recommendation for dairy and beef cattle, sheep, and poultry since the majority of work has been done with swine wastes. The data in Table 8.5 provide a reasonable estimate of equipment and power requirements for these units. The rotor and ditch volume requirements per animal are based on the type of rotors currently in use. Future improvements in rotors will alter these relationships should better oxygen transfer efficiencies result. The data in Table 8.5 have been corrected to 80% of the oxygenation capacity of the rotor in tap water, i.e., an alpha (α) value of 0.8. Such values may or may not represent the actual conditions in an oxidation ditch. The oxygenation capacity of a rotor in an actual system is most important, and an estimate of the transfer relationships, i.e., α and β, in a specific waste slurry must be known to correctly size a rotor (Chapter 7). The general design and operational parameters for oxidation ditches treating dilute wastes, such as municipal and food processing wastewater, are shown in Table 8.6.

The detention times in oxidation ditches range from one or more days for continuous systems treating dilute wastes to months for what are essentially storage systems treating animal wastes. Because of its large capacity relative to its influent flow, the oxidation ditch acts as a surge tank to level out peaks of loading.

Oxygenation Relationships

Aeration is accomplished by a surface rotor rotating about a horizontal shaft. Except for a portion of the rotor blade immersed in the liquid, the entire mechanism is above the liquid. Usual immersion depths are from 2 to 8 in. Common rotors consist of iron plates or angles attached perpendicular to a

TABLE 8.5

Design Recommendations for In-the-Building Oxidation Ditches[a]

Animal	Weight of animal (lb per unit)	Daily BOD₅ produced (lb per unit)[b]	Daily required oxygenation capacity (lb per unit)[c]	Number of animals per foot of rotor (units per foot)[d]	Ditch volume (ft³ per unit)[e]	Daily power requirement (kWh per unit)[f]	Daily cost (cents per unit)[g]
Swine							
Sow with litter	375	0.79	1.58	16	23.7	0.83	1.66
Growing pig	65	0.14	0.28	91	4.2	0.15	0.30
Finishing hog	150	0.32	0.62	41	9.6	0.33	0.66
Dairy cattle							
Dairy cow	1300	2.21	4.42	6	66	2.33	4.66
Beef cattle							
Beef feeder	900	1.35	2.70	10	40	1.42	2.84
Sheep							
Sheep feeder	75	0.053	0.11	230	1.6	0.06	0.12
Poultry							
Laying hen	4.5	0.0198	0.0396	650	0.6	0.021	0.042

[a]From Jones et al. (14).
[b]Use specific production data when known.
[c]Twice the daily BOD₅.
[d]Based on 25.5 lb of O_2 per ft of rotor per day.
[e]Based on 30 ft³ per lb of daily BOD₅.
[f]Based on 1.9 lb O_2 per kWh.
[g]Based on electricity at 2 cents per kWh.

TABLE 8.6

General Design and Operational Parameters for the Oxidation Ditch Process

Organic loading rate	0.05–0.2 kg of BOD_5 per kg MLSS per day
Volumetric loading rate	0.15–0.5 kg of BOD_5 per m^3 per day (10–30 lb of BOD_5 per 1,000 ft^3 per day)
MLSS concentration	3000–6000+ mg/liter
Hydraulic retention time	24+ hr
Solids retention time	20+ days
Return sludge rate	50–100% of the influent flow
Oxygen requirements	2.0–2.4 kg of oxygen per kg of BOD_5 applied
Sludge production rate	0.2–0.5 kg per kg of BOD_5 removed
Aeration	Mechanical
Horizontal velocity in the ditch	At least 1 ft per second (30 cm per sec)

rotating shaft or a circular cage with rectangular pieces of metal attached to the periphery. The amount of oxygen required depends upon the microbial oxygen demand. Design relationships for rotors require an estimate of the effectiveness of transferring oxygen to a slurry over a range of rotor speeds, diameters, configurations, liquid waste properties, and desired dissolved oxygen levels in the wastes. The oxygen transfer coefficient $K_L a$ and the pounds of oxygen transferred to the ditch contents increase as the rotor immersion increases (Fig.

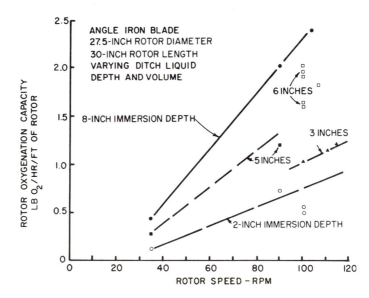

Fig. 8.6. Effect of rotor speed and immersion depth on oxygenation capacity (15).

8.6) (15), while the oxygen transferred per kilowatt hour decreases as the immersion increases. The power consumption for a given rotor increases as immersion of the rotor blades increases (Fig. 8.7). Similarly the pounds of oxygen transferred increases as the rotor speed increases although the oxygen transferred per kilowatt hour decreases with an increase in speed. Multiple rotors can be installed in ditches where oxygen demand is high.

Because of the low loading and long liquid detention time in the ditch, nitrification can take place if dissolved oxygen concentrations above 2.0 mg per liter are maintained. The oxygen demand of nitrification increases the total oxygen demand of the system. The rotor must be of sufficient size to supply both the carbonaceous and nitrogenous demand.

The rotor is designed to provide adequate mixing as well as to aerate the ditch contents (Fig. 8.8). As a general rule, adequate velocity and oxygenation will occur when the immersion of the rotor is approximately one-fourth to one-third the liquid depth in the ditch. Oxidation ditches are designed as shallow units having depths of from 15 to 30 in. When rotor immersion is less than that suggested previously, lower velocities will result, and solids sedimentation can occur. Velocities of 1.5–1.8 ft per sec should be maintained in the ditch to minimize settling of the solids. The aeration process takes place primarily within the immediate vicinity of the rotor. Additional aeration takes place during the movement around the ditch, but this aeration is minor compared to that occurring at the rotor.

The dissolved oxygen of the mixed liquor will decrease from a maximum value immediately after the rotor (Fig. 8.9). Depending upon the oxygen demand of the mixed liquor and the oxygen supplied by the rotor, the dissolved oxygen concentration can be reduced to zero prior to the liquid again passing through the rotor. The low or zero dissolved oxygen concentration in part of the ditch

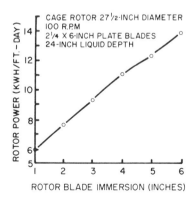

Fig. 8.7. Variation of power requirements of a cage rotor as a function of rotor immersion depth (14).

Fig. 8.8. Aeration and turbulence caused by rotors in an oxidation ditch. [Courtesy of Lakeside Equipment Corp.]

enhances nitrogen removal by denitrification. Nitrification can occur in the portion of the ditch where there is adequate oxygen, and the nitrates can be reduced where anoxic conditions occur. Nitrogen losses in oxidation ditches have ranged from 30 to as much as 90% (16, 17).

In lightly loaded systems, such as the oxidation ditch, the power requirements needed to keep the solids in suspension may exceed the power needed for oxygen transfer. This leads to a high power use per BOD removed.

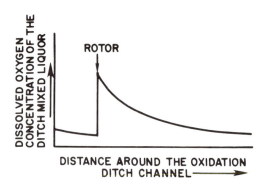

Fig. 8.9. Dissolved oxygen concentration variations throughout an oxidation ditch.

Operational Characteristics

The performance of oxidation ditches has been found to be better in terms of both effluent quality and reliability of meeting specific effluent standards (Table 8.7) than other aerobic secondary treatment processes. Although this comparison is for plants treating municipal wastes, the same relative relationship should hold when the processes are used to treat food processing wastewaters.

An oxidation ditch is capable of 95–99% nitrification without design modifications, due to the long operating solids retention times. Nitrogen removal by single-stage combined nitrification and denitrification also has been achieved at properly designed and operated oxidation ditch plants. Such removal is achieved by controlling the oxygen input to produce both aerobic and anoxic zones in the channel.

Reports from Europe and Canada have indicated that the oxidation ditch will operate successfully in cold weather. The rotor and surrounding area should be enclosed and heated where necessary. The ditch may have an ice cover in parts, and a decrease in efficiency will occur. However, the long detention time in the ditch can partially compensate for the lower efficiency per organism in the colder weather.

There are certain problems associated with oxidation ditches. These include foaming, humidity, odors, and rotor maintenance. Foaming can be a nuisance and in some cases objectionable. Even in a properly operating ditch, some foam is present because of the agitation by the rotor and surface active agents resulting

TABLE 8.7

Performance of Oxidation Ditches and Other Aerobic Biological Processes Treating Municipal Wastes[a]

Biological process	Average effluent concentration (mg/liter)		Average removal (%)		Average reliability (%) in meeting a 30 mg/liter effluent standard for	
	BOD	TSS	BOD	TSS	BOD	TSS
Oxidation ditch	12	11	93	94	99	99
Activated sludge (1.0 mgd)	26	31	84	81	85	90
Activated sludge (package plants)	18	28	—	—	80	50
Trickling filters	42	26	79	82	15	—
Rotating biological contactor	25	23	78	79	90	70

[a]Adapted from (18).

from degradation of the organic wastes. Antifoam agents such as kerosene or proprietary chemicals can relieve excess foam problems. Excess foam usually occurs during start-up conditions and when the ditch is overloaded. When the ditch is designed and operated properly and when there is an adequate microbial population in the ditch, foaming problems are a minimum. The total depth of the ditch should include freeboard of at least a foot to allow for some foam during start-up conditions.

Problems with bearings, improperly designed rotors, motors, and belts have caused many problems with earlier rotors and drive units. The experience gained from these problems has been incorporated into current equipment. Equipment breakdown is less frequent. It is important that the maintenance instructions provided by the equipment manufacturer be followed for maximum equipment life.

Experimental observations on the survival of potentially pathogenic bacteria in an oxidation ditch indicated that 90% of the naturally occurring population of *Staphylococcus aureus* was eliminated in 7 days. Ninety percent of an inoculated *S. aureus* population died in 6 days, and 90% of an inoculated population of *Salmonella typhimurium* died in 12 hr, indicating that the oxidation ditch liquor is a hostile environment for these organisms (19).

Application to Agricultural Wastes

The oxidation ditch has been applied successfully to many liquid agricultural wastes. Currently, however, oxidation ditches are rarely used for the stabilization of livestock wastes, because of high energy (aeration) costs and because there are more appropriate alternatives for livestock waste handling and disposal. The following examples illustrate the use of oxidation ditches with both food processing and animal wastes.

An oxidation ditch treating milk processing wastewater was capable of removing 99% and 98% of the influent BOD_5 and suspended solids, respectively (20). The ditch operated at an average loading of 0.8 kg of BOD_5 per kg of MLSS per day and achieved an effluent BOD_5 and suspended solids concentration of 7 and 28 mg per liter, respectively.

An oxidation ditch was used to treat the wastewaters from a large meat-packing plant (21). The process included an aerated grit chamber, a settling and flotation unit to remove grease and solids prior to the oxidation ditch, and secondary clarifiers. The overall treatment plant removal was 96% for BOD_5, 85% for suspended solids, and 96% for grease.

Swine wastes have been treated in a below-house oxidation ditch, and the approach has been considered suitable for either partial or complete treatment. A ditch with a single aeration channel provided odor control and a 90% reduction in BOD_5 when operated on the extended aeration principle (22). The ditch capacity was 0.28 m^3 per pig, the oxygen required was 0.30 kg per day per pig, and the power requirement of the rotor was 0.20 kWh per day per pig.

An oxidation ditch can be a technically feasible dairy cattle waste management method capable of odor control, abatement of fly problems, and waste stabilization. The process parameters that can be used for the design of such a ditch are aeration requirements for odor control—1.6 kg of oxygen per day per 500 kg liveweight of animal, aeration requirements for odor control plus nitrification—2.4 kg of oxygen per day per 500 kg liveweight, and minimum volume—1600 liters per 500 kg liveweight (23). An economic comparison indicated that an oxidation ditch was not competitive with alternative dairy waste management approaches.

Hog waste treatment has been accomplished in a full-scale oxidation ditch with an oxygenation capacity-to-load ratio of 1.8 (24). Foaming was excessive at the start. Effluent BOD and COD values after settling were about 15 and 600 mg per liter, respectively, although the raw pig wastes had BOD and COD concentrations of about 30,000 and 80,000 mg per liter, respectively. Sludge accumulation was about 18% of the total BOD loading per day.

When an oxidation ditch was operated as a nonoverflow unit with beef cattle wastes, the COD increased at a linear rate, reaching about 50,000 mg per liter after 160 days. The BOD remained low and relatively constant, ranging from 900 to 2500 mg per liter during this period. A high cellulose ration was fed to the animals, and the COD increase was caused by the nonbiodegradable fraction in the waste. In one experiment the total solids reached 11.8%, and in another, 7.3% before the solids had to be removed. The solids entering the ditch averaged 5.1 lb per day per head, and 34% solids reduction was obtained in 143 days of treatment (25).

If a batch- or storage-type oxidation ditch is used, the COD and solids will increase almost linearly until solids are removed. BOD values and volatile solids levels will remain low and constant or slightly increase since the organic demand and solids are being oxidized. In an oxidation ditch with an overflow, the COD and solids levels will build up to equilibrium levels consistent with the volume and dilution of the ditch.

Carrousel

Introduction

The Carrousel reactor is a modified oxidation ditch process which permits deeper tanks to be used and therefore requires smaller land areas. It has the potential for lower energy inputs for aeration and mixing (26). In the Carrousel reactor, there is a separation between a specific aeration zone and an aeration circuit (Fig. 8.10). The aeration zone is designed in accordance with the type and size of the surface aerator, and the aeration circuit functions as a mixing and detention basin to achieve the required solids detention time. This separation permits lower initial costs of aeration per kilogram of oxygen transfer, higher rates of aeration efficiency, and the use of deep units to achieve adequate flow rates as compared to the traditional shallow oxidation ditch with horizontally mounted cage rotors.

In the Carrousel reactor, vertically mounted mechanical surface aerators are used (Fig. 8.11) to provide the necessary oxygen and provide the required horizontal velocity in the aeration circuit. The aerator commonly used has an impeller which can be raised or lowered to adjust oxygenation input to BOD loads and oxygen demand.

The aerators provide the velocity necessary to keep the mixed liquor solids in suspension in the deep parallel channels that form an endless circuit for the mixed liquor movement. Complete mixing occurs in the aeration zones, with plug flow occurring in the channels. The dimensions of a specific system will vary. Common configurations have rounded ends and from 2 to 10 aerators. Each aerator is capable of supplying sufficient oxygen to treat up to 3000 lb BOD per day.

This type of unit was introduced in 1968 in Holland and subsequently has been constructed in many locations in Europe for municipal, industrial, and agricultural wastes. The low F/M ratio and the long detention time of the process results in high removal efficiencies: BOD removals greater than 95% and COD removals greater than 90% being reported, and the excess sludge is stabilized. Secondary settling tanks can be used with the settled sludge returned to the aeration tanks and the excess sludge periodically wasted. Sand drying beds are a common method of handling the wasted excess sludge.

The aeration pattern allows for nitrogen control by nitrification and de-nitrification. The high sludge age that can be achieved is conducive to nitrification. When oxygen control is used to produce a zone of low or zero dissolved

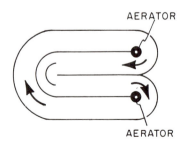

Fig. 8.10. Schematic of a Carrousel treatment system.

Fig. 8.11. Vertical mechanical surface aerator used in Carrousel treatment systems. (Courtesy of Envirobic Systems, Inc.)

oxygen in the aeration channel, denitrification will occur. Nitrogen removals of 40–80% have been reported in full-scale units.

Application to Agricultural Wastes

The major applications have been with liquid wastes that require a high degree of treatment prior to discharge. The process has been used for dairy (Fig. 8.12), meat-packing, brewery, slaughterhouse, and other wastes in Europe.

Autothermal Aerobic Digestion

This process, which also has been called liquid composting, is an aerobic treatment process for concentrated organic wastes in which the energy from microbial oxidation of the waste results in elevated operating temperatures. The process can be used for waste management objectives ranging from odor control to complete stabilization. The process was developed in the late 1960s in Europe. The greatest application of the process has been in Europe with animal wastes, although it is capable of being utilized with other concentrated organic wastes including sewage sludge.

When 1 gm of carbon is oxidized to carbon dioxide, about 0.14 gm of hydrogen are oxidized to water. The oxidation of this carbon and hydrogen releases about 12.7 kcal of energy. Part of this energy is utilized by micro-organisms for growth, and the rest can increase the temperature of the mixed

Fig. 8.12. Carrousel system in Holland used to treat wastewater from a dairy. (Courtesy of Envirobic Systems, Inc.)

liquor. An increase in temperature in conventional aerobic biological systems, such as an activated sludge unit, rarely is noticed, because of the low concentration of organic matter in the systems, the quantity of water that must be heated, and the fact that the tanks are uninsulated. In the oxidation of concentrated organic matter, such as in composting (Chapter 10), considerable heat can be generated.

The increased temperatures increase the microbial reaction rates and help produce rapid stabilization of the organic matter and destruction of pathogens. The elevated temperatures result in a reduction of the viscosity of the high total solids under aeration (27). This reduction reduces mixing requirements and enhances oxygen transfer. The result is to make it possible to aerate concentrated wastes, such as up to 10% total solids, and to reduce reactor volume.

The key to the success of the liquid composting process is an aerator capable of transferring sufficient oxygen to meet the oxygen demand of the concentrated mixed liquor. Aeration systems used for conventional processes, i.e., aerated lagoons or activated sludge, may not have the required oxygen transfer capacity. Other mechanical aerators have been used (27).

Intense foam is generated, requiring mechanical foam cutters for control. The continuous foam layer provides some surface insulation for the heat produced by the microbial mass.

With reactor total solids concentrations of 5–10%, temperatures in the range of 40–54°C can be maintained. With lower total solids concentration (2–7%), the process operates in the mesophilic range (27–32°C). The mesophilic or "warm" process functions with larger volumes per aerator and has the objectives of odor control and making the wastes more suitable for pumping. Total solids, COD, and BOD reduction are not as great as in the thermophilic process. The thermophilic process results in a greater degree of waste stabilization and in disinfection of pathogens such as *Salmonella*.

Liquid composting has been used with dairy, beef, and swine manure (27–29), and autothermal aerobic digestion has been used for sludge stabilization (30–32). The process can achieve thermophilic temperatures with sludges that are 50% biodegradable or greater. The temperatures can be in excess of 45°C and can reach 60°C. Maximum temperatures and organic removal rates were obtained at loading rates of between 8 and 10 kg of biodegradable COD per m^3 per day (32). Pathogen destruction efficiency in the process was excellent, with almost none of the samples containing any virus.

Rotating Biological Contactor

Introduction

The rotating biological contactor (RBC) is analogous to a rotating trickling filter. The disks, connected to a common shaft and spaced a short distance from each other, rotate in a semicircular tank through which the liquid waste flows. The large diameter, lightweight plastic disks are submerged about 35–45% in the tank. A biological film develops on the disk surface in a manner similar to that developed on the surface of a trickling filter. When submerged in the wastewater, the microbial film absorbs organic matter. In rotation the discs carry a layer of wastewater into the atmosphere where it trickles down the disk surfaces, and the microbial film absorbs oxygen. Organisms on the disk surface use the oxygen and organic matter for growth, reducing the oxygen demand of the wastewater. A schematic cross-section of a typical unit is illustrated in Fig. 8.13, and a large-scale unit is shown in Fig. 8.14. Disk speed can be varied and generally is in the range of 2–5 rpm.

The rotating disks serve to provide support of a fixed biological growth, contact of the growths with the wastewater, and aeration of the wastewater. The growths continuously shear off the disks. The mixing action of the system keeps these solids in suspension until the treated wastewater carries them out of the unit. The aerated mixed liquor in the unit also serves to treat the wastewater. The discharged wastewater contains metabolic products of degradation and the excess solids synthesized in the unit. The solids settle well in a final clarifier. The process can be used independently or in conjunction with other treatment processes.

Performance criteria for rotating disc treatment systems generally have been based on removal efficiency, i.e., percentage of reduction of some wastewater component between influent and clarified effluent, and on the hydraulic loading rate of the unit. With constant-strength wastes, such parameters can be useful. However, it is difficult to use such criteria to relate wastes of different strengths. Removals should be related to a parameter incorporating influent flow rate, waste strength, and disc surface area, such as the amount of waste characteristic applied per disc surface area per unit time, e.g., pounds of BOD per 1000 ft^2 per day (gm/m^2/day). The general design and operating characteristics for RBC units are noted in Table 8.8.

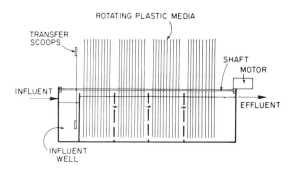

Fig. 8.13. Schematic cross section of a rotating biological contractor.

Fig. 8.14. Full-scale rotating biological contactor module. (Courtesy of Envirex. Inc., a Rexnord company.)

Application

In cold climates, freezing of the disk surfaces is a possibility. Contact with outside air can be limited by a simple covering with closed side walls. Insulation is not required to avoid freezing, because the warm wastewater should keep the ambient air around the unit from freezing if it is covered. As expected, the temperature of the liquid in the unit will affect the efficiency of the micro-

TABLE 8.8

General Design and Operating Parameters for RBC Units

Organic loading rate	5–40 gm of BOD_5 per m^2 per day (1–8 lb of BOD_5 per 1000 ft^2 per day)
Hydraulic loading rate	20–40 liters/m^2/day (0.5–1.0 gal/ft^2/day)
Number of stages in series	2–6
Hydraulic retention time	4–24 hr
Solids retention time	Greater than HRT; not controlled
Aeration	Mechanical rotation and oxygen diffusion through exposed microbial film

organisms and be reflected in the treatment efficiency. If the unit is enclosed, adequate ventilation through the enclosure should be assured to provide adequate oxygen for the microorganisms.

The operating power requirements in these units are for rotation of the disks. Because of the low density of the plastic media, the bouyancy of the disks helps offset the weight of the attached biomass, and power requirements are low. The comparative energy requirements or RBC units are noted in Table 8.3.

Rotating biological contactor units have been used to treat municipal wastes, recreational camp wastewaters, and industrial wastewaters such as those from meat packing, swine production, wineries and distilleries, milk and cheese plants, glue factories, vegetable processing, poultry production, and paper production.

The RBC units can be feasible for the treatment of wastewaters. For this reason, they have been used for the treatment of food processing wastes but not for livestock wastes. The following summarizes results that have been obtained with RBC units.

The effluent from an anaerobic lagoon has been treated successfully (33) by an RBC unit. The influent BOD was between 125 and 150 mg per liter, and an overall BOD reduction of 83% was achieved when the hydraulic loading rate was 4 gal/ft^2/day.

A full-scale, three-stage RBC plant, consisting of a primary clarifier, RBC unit, and secondary clarifier, has produced effluent concentrations of less than 20 mg per liter when treating the variable wastes from turkey and meat-packing plants (34). The design BOD was 628 mg per liter, the design flow was 1.16 million gallons per day (mgd), and the overall design hydraulic loading was 0.7 gal/ft^2/day.

A two-stage RBC system provided satisfactory treatment of combined cheese manufacturing and municipal wastewater (35). The effluent BOD was less than 11 mg per liter, with organic loadings of 3–4 lb of BOD per 1000 ft^2 per day.

In an experimental situation, a small-scale RBC unit was able to remove 88–71% of the COD from cheese processing wastewater (36). The hydraulic loading rates ranged from 0.19 to 0.75 gal/ft^2/day.

Trickling Filters

General

Trickling filters are designed to treat dilute liquid wastes aerobically. Wastes with a high concentration of particulate matter, such as liquid animal manures, are inappropriate wastes to be treated by trickling filters. High concentrations of organic or inorganic solids will result in clogging, reduced efficiencies, and increased maintenance problems if wastes containing such solids are applied to trickling filters.

Trickling filters are not filters but are aerobic oxidation units which absorb and oxidize the organic matter in the wastes passing over the filter media. The media used in trickling filters generally is crushed stone or rock of large size, generally 2–4 in. in size, or plastic media of various configurations. Horizontal redwood slats and other inert material have been used as trickling filter support media. The properties of common filter media are noted in Table 8.9 (37).

The waste to be treated is distributed either intermittently or continuously over the top of the media and flows through the media. The rotary distributor is the most common type of liquid distribution system used with trickling filters today (Fig. 8.15). Fixed nozzle distribution systems, in which the distribution nozzles were permanently positioned in the filter media, were used in the early part of this century but are no longer used, because of clogging and poor liquid distribution problems. The wastewater is discharged above the trickling filter by the rotary distributor, permitting some aeration of the liquid to take place before contact with the media. Further aeration takes place as the liquid flows over the media.

The filter ranges in height from 4–7 ft for stone or rock to 10–40 ft for plastic media. The surface of the media contained in the filter supports the microorganisms that metabolize the organic matter in the waste. The filter should have media as small as possible to increase the surface area in the filter and the number of active organisms that will be in the filter volume. However, the media must be large enough to have enough void space to permit liquid and air flow and remain unclogged by the microbial growth. Large-size media such as 2- to 4-in. rock and plastic media function satisfactorily.

The microbial slime layer and water flowing over the media will consider-

TABLE 8.9

Physical Properties of Common Filter Media[a]

Media	Nominal size (in.)	Weight (lb/ft^3)	Surface area (ft^2/ft^3)	Void space (%)
Granite	1–3	90	19	46
Granite	4	90	13	60
Slag	3	68	20	49
Plastic	21 × 38	6	27	94

[a]From (37). [Reprinted with permission from *Int. J. Air Water Pollut.* **5**, by E. H. Bryan and D. H. Moeller, Aerobic biological oxidation using Dowpac. Copyright 1963, Pergamon Press, Ltd.]

Fig. 8.15. Circular trickling filter indicating size of typical rock media, rotary distributor, and liquid distribution pattern.

ably increase the weight of the material in the filter. For plastic media, the combined weight of the slime, water, and media may be 4–5 times more than the weight of the media alone. The support structure for the filter should be designed to hold such loads.

The filter media sit on an underdrain system which collects the liquid from the filter and transports it to a final sedimentation basin. The underdrain systems assist in keeping the filter aerobic by facilitating air movement through the filter. Solids are removed from the system continuously from the final sedimentation unit. With common trickling filter systems, settled solids are not returned to the trickling filter.

Heterotrophic facultative bacteria are the largest population of microorganisms in a trickling filter. Protozoa and higher forms of animal life can be present in a filter. Algae will grow on the surface of filters that are not overloaded but will not grow below the surface, since sunlight cannot penetrate. A gray-to-brown microbial growth will exist within the filter, even though the surface may be green due to algal growth. The algae and higher forms of animal life contribute little to the efficiency of the filter.

The organic matter in the wastewater stimulates biological growth on the surface of the media. The growth is established first in areas where the flow does not wash it from the media and will spread throughout the media. Even in warm weather when microbial growth is most rapid, it takes a considerable period of time for the microbial growth to become established in a filter and for equilibrium performance conditions to occur. The period can be from 4 to 6 weeks. It is important that the microbial growth not be killed by toxic conditions in a waste, since the filter will not function at design efficiency until the growth is reestablished, which can be another lengthy period. Where seasonal wastes are to be treated, such as at some canning plants, trickling filters may not be a satisfactory solution to the waste treatment problem, because of the length of time required to build up the necessary microbial population and to produce the desired effluent.

After the microbial layer on the media is established, the liquid flows over the layer rather than through it. The liquid waste flows down the media as a wave, creating turbulence between the waste and liquid film in the microbial surface (Fig. 8.16). The organic matter in the waste is transferred to the liquid film, and waste products of metabolism are transferred from the liquid film to the waste. This transfer is continuous throughout the filter depth.

The outer layers of microorganisms are exposed to the fixed liquid layer and metabolize most of the waste. Aerobic metabolism is maintained by continuous transfer of oxygen from the void spaces in the filter to the fixed liquid layer. The rate of oxygen transfer is related to the oxygen differential between the air and

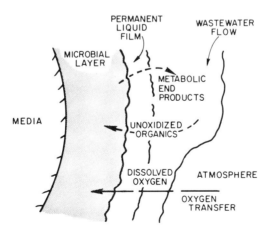

Fig. 8.16. Schematic of processes in a trickling filter.

fixed liquid layer. For a given size of filter and type of filter media, the available surface for oxygen transfer is established, and there is a fixed quantity of oxygen that can be transferred into the filter per unit of time. This fixed quantity will limit the maximum organic loading that can be applied to the filter and still maintain aerobic conditions.

Most filter loading rates have been established such that the rate of metabolism is not limited by the oxygen transfer. In trickling filters the organic matter limits the rate of microbial metabolism. Trickling filters are food-limited systems operating in the declining growth phase.

The thickness of the aerobic microbial layer has been estimated at about 0.005 cm, while the actual thickness of the entire microbial layer is much greater. Because only the surface of the microbial layer obtains most of the food and oxygen, the microorganisms attaching the microbial growth to the media die and are carried from the media by the wastewater flow. Microbial growth becomes reestablished in areas where older growths have been removed. This cycle is continuous in a trickling filter.

The effluent from the filter contains unmetabolized organic matter in the applied wastewater, biological solids separated from the media, and metabolic end products. The biological solids must be separated from the trickling filter effluent before the liquid is discharged from the treatment plant if a high quality effluent is desired. This is done in a final sedimentation tank, which is an integral part of a trickling filter system. If any liquid is used for recirculation in a trickling filter, it is generally a portion of the effluent from the sedimentation tank.

Filter Types and Loadings

Loadings to trickling filters are expressed in terms of the amount of oxidizable matter or hydraulic flow applied to the surface of the media. General characteristics of trickling filters are noted in Table 8.10. Process flow patterns for various types of trickling filters are illustrated in Fig. 8.17.

In standard rate filters, the wastewater is applied to the media and passes through the media to the final sedimentation tank (Fig. 8.17). Effluent recirculation may be done during low-flow periods to keep the media moist and to provide some organic matter for the microorganisms, but it is rarely done continuously.

A high-rate filter generally uses continuous recirculation (Fig. 8.17) to produce an acceptable effluent since the applied loading rates are higher. Recirculation equalizes variable waste loads and reduces waste loads.

The light weight of the plastic media (Table 8.10) reduces the need for heavy supporting structures and permits tall units, reducing land requirements. The large void space permits higher application rates without clogging and free passage of air to supply the necessary oxygen. A honeycomb-type structure commonly is used with plastic media to increase surface area and porosity. When

TABLE 8.10

General Characteristics of Trickling Filters

Characteristic	Standard rate	High rate	Synthetic media
Organic loading			
gram $BOD_5/m^3/day$	100–200	300–1000	1,000+
lb $BOD_5/1000\ ft^3/day$	Less than 12	20–60	60+
Hydraulic loading			
$m^3/m^2/day$	1–3	4–8	8+
gallons/ft^2/day	20–60	100–200	200+
Depth			
meter	2–3	2–3	6–9
feet	6–10	6–10	20–30
Media	Rock, slag	Rock, slag	Plastic, redwood slats

placed in the trickling filter, the layers of the media are placed at angles to each other, so that the possibility of wastewater falling through the media without contacting the media is very small (Fig. 8.18). The larger surface area per unit volume increases BOD removals and can permit larger hydraulic loads.

Performance

Typical treatment efficiencies for trickling filters treating municipal wastes range from 85 to 95% BOD removal for standard-rate filters and from 75 to 85% BOD removal efficiencies for high-rate filters. Lightly loaded trickling filters can

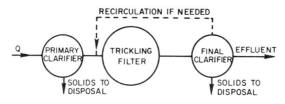

SINGLE STAGE, STANDARD RATE

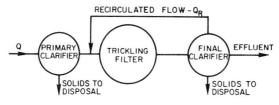

SINGLE STAGE, HIGH RATE

Fig. 8.17. Flow diagrams of trickling filter treatment processes.

Fig. 8.18. Plastic media trickling filter and rotary distributor. (Courtesy of American Surface Corp.)

produce a high degree of nitrification. The relative energy requirements of trickling filters are noted in Table 8.3.

As with all biological treatment units, the performance of trickling filters is influenced by the temperature of the liquid. The liquid temperature determines the rate of microbial metabolism, whereas the air temperature differential between the air in the media and that outside the media determines the air flow through the filter. Since currently designed trickling filters are food- rather than oxygen-limited systems, induced aeration through the filters by fans is not effective.

The movement of air through trickling filters is essential to the maintenance of aerobic conditions and will be upward or downward, depending upon the relative temperature of the liquid and ambient air. When the liquid is colder than the air, air movement will be downward and outward through the underdrain system or ventilation ports. The draft causing a movement of air in a filter is a function of the depth of the filter and the average air densities inside and outside thc filtcr and should be directly related to the difference between the temperature of the air and liquid.

The temperature in the filter will decrease during the winter and will decrease the rate of microbial metabolism and the efficiency of the filter.

Pretreatment

The biological growth on a filter removes the organic material in a waste-water by adsorption and metabolism, and the filter is intended to be aerobic. The wastewater applied to the filter should consist primarily of dissolved and colloidal organic matter. Removal of large particles that may clog a filter is necessary to avoid closing of the voids in the media which would prevent air movement and oxygen transfer to the microorganisms. Clogging can also result from excessive microbial growth filling or bridging the void space.

When clogging causes a filter to become anaerobic and/or if the removal efficiency decreases, the problem may be handled by improved removal of the particulate matter and by reducing the organic concentration applied to the filter by incorporating of increasing effluent recycle.

Application to Agricultural Wastes

Trickling filters have been used more with food processing wastes than with any other type of agricultural waste. Depending upon the type of waste and the effluent quality required, a trickling filter treatment system can include primary sedimentation to remove settleable solids in the raw waste, recirculation pumps, and lines and final sedimentation units. When the fundamental relationships for design and use of trickling filters and not understood, difficulties and poor performance can result.

Trickling filters have been explored for the treatment of potato starch processing wastes. Standard rate rock filters, loaded up to 30 lb of BOD per 1000 ft^3 per day obtained BOD reductions of about 90%. The effluent pH was generally between 7 and 8, although the wastes had an acid pH when they were applied to the filters. Loadings up to 70 lb of BOD per 1000 ft^3 per day in a high-rate filter, with a recirculation ratio of 10, also obtained above 90% BOD removals. At higher BOD loadings, the filter, which contained media of about 0.75 in. in diameter, became clogged with the biological solids sloughed from the media (38). Media with larger openings may have permitted higher loadings without clogging problems.

Plastic media did not exhibit clogging problems when loaded up to rates of 130 lb of BOD per 1000 ft^3 per day of potato processing wastes. BOD reductions were about 70% prior to the final sedimentation unit and about 85% after the sedimentation unit (39). The recirculation ratio was 6.

Settled peach and pear canning wastes were applied to plastic media trickling filters at loading rates from about 300 to 1800 lb of BOD per 1000 ft^3 per day (40). The raw waste and effluent pH values were 10.5 and 6.2, respectively. Recycle ratios were 1 and 2. At 316 lb of BOD per 1000 ft^3 per day, 86% BOD removal occurred. When the loading was 1030 and 1760 lb of BOD per 1000 ft^3

per day, BOD removals were 33 and 36%, respectively. Addition of nitrogen and phosphorus to these wastes resulted in a BOD removal of 45% when the loading rate was 1510 lb of BOD per 1000 ft^3 per day.

The wastes from the canning of peaches and fruit cocktail have been treated by a pilot plant trickling filter using plastic media (40). A heavy slime growth, composed primarily of fungi, became established when the raw wastes were treated. This growth grew so thick that anaerobic conditions occurred. An evaluation of the wastes indicated that they were nitrogen deficient. The addition of nitrogen in the raw wastes to obtain a BOD: N ratio of 20:1 changed the characteristics of the filter growth. A thin growth, comprised mainly of bacteria, became established. This growth sloughed continuously and settled rapidly. The filter acted as a roughing filter.

A combined system of anaerobic lagoons followed by trickling filters has been used to treat meat-packing wastes (41). The anaerobic lagoon served to remove the gross solids in wastes. The plastic media trickling filters were loaded at about 73 lb of BOD per 1000 ft^3 per day at a hydraulic load of 0.8 gal per minute (gpm) per square foot of surface area. The BOD and COD removal averaged 74 and 73%, respectively. The wastes received adequate aeration in the filter, with the dissolved oxygen in the filter effluent averaging about 4 mg per liter. Grease removal was about 69%. No phosphate removal occurred, although about 20% nitrogen removal resulted, chiefly as a decrease in ammonia concentration. Some nitrification occurred in the filter although denitrification took place in the final clarifier. The aforementioned removals refer to changes that took place in the combined trickling filter–final clarifier system.

References

1. Bisogni, J. J., and Lawrence, A. W., Relationships between biological solids retention time and settling characteristics of activated sludge. *Water Res.* **5,** 753–763 (1971).
2. Martin, A. J., "The Activated Sludge Process." MacDonald & Evans, London, 1927.
3. Middlebrooks, E. J., and Middlebrooks, C. H., "Energy Requirements for Small Flow Wastewater Treatment Systems," Special Report 79–7. U.S. Army Corps of Engineers, Cold Regions Research and Engineering Laboratory, Hanover, New Hampshire, 1979.
4. Dart, M. C., The treatment of meat trade effluents. *Effluent Water Treat. J.* **7,** 20–33 (1967).
5. Esvelt, L. A., and Hart, H. H., Treatment of food processing wastes by aeration. *J. Water Pollut. Contr. Fed.* **42,** 1305–1326 (1970).
6. R. T. French Co., "Aerobic Secondary Treatment of Potato Processing Waste," Final Rep. Proj. WPRD 15-01-68. Water Quality Office, U.S. Environmental Protection Agency, 1970.
7. Winter Garden Citrus Products Cooperative, "Complete Mix Activated Sludge Treatment of Citrus Process Wastes," Final Rep., Proj. 12060 EZY. U.S. Environmental Protection Agency, Washington, D.C., 1971.
8. Hoover, S. R., Jasewicz, L., Pepinsky, J. B., and Porges, N., Dairy waste assimilation by activated sludge. *Sewage Ind. Wastes* **23,** 167–178 (1951).
9. Teletzke, G. H., Chickens for the barbecue—Wastes for aerobic digestion. *Wastes Eng.* **8,** 134–138 (1961).

10. Coca-Cola Co., "Treatment of Citrus Processing Wastes," Final Rep., Proj. WPRD 38-01-67. Water Quality Office, U.S. Environmental Protection Agency, Washington, D.C., 1970.

11. Guo, P. H. M., Fowlie, P. J., and Jank, B. E., Activated sludge treatment of wastewaters from the dairy products industry. *In* "Proceedings of the Ninth National Symposium on Food Processing Wastes," Report EPA 600/2-78-188, pp. 178–202. Industrial Environmental Research Laboratory, Office of Research and Development, U.S. Environmental Protection Agency, Cincinnati, Ohio, 1978.

12. Jaworski, N., Lawton, G. W., and Rohlich, G. A., Aerobic sludge digestion. *Int. J. Air Water Pollut.* **5,** 93–102 (1963).

13. Randall, C. W., and Koch, C. T., Dewatering characteristics of aerobically digested sludge. *J. Water Pollut. Contr. Fed.* **41,** R215–R238 (1969).

14. Jones, D. D., Day, D. L., and Dale, A. C., "Aerobic Treatment of Livestock Wastes," Bull. No. 737. University of Illinois, Urbana Agricultural Experiment Station in cooperation with Purdue University (revised bulletin, April 1971).

15. Day, D. L., Jones, D. D., and Converse, J. C., "Livestock Waste Management Studies— Termination Report," Agric. Eng. Res. Rep., Proj. 31-15-10-375. University of Illinois, Urbana, 1970.

16. Prakasam, T. B. S., Srinath, E. G., Anthonisen, A. C., Martin, J. H., and Loehr, R. C., Approaches for the control of nitrogen with an oxidation ditch. *In* "Proceedings of the Agricultural Waste Management Conference" (R. C. Loehr, ed.), pp. 421–435. Cornell University, Ithaca, New York, 1974.

17. Dunn, G. G., and Robinson, J. B., "Nitrogen losses through denitrification and other changes in continuously aerated poultry manure. *In* "Proceedings of the Agricultural Waste Management Conference" (R. C. Loehr, ed.), pp. 545–552, Cornell University, Ithaca, New York, 1972.

18. Ettlich, W. F., "A Comparison of Oxidation Ditch Plants to Competing Processes for Secondary and Advanced Treatment of Municipal Wastes," EPA 600/2-78-051. Municipal Environmental Research Laboratory, Office of Research and Development, U.S. Environmental Protection Agency, Cincinnati, Ohio, 1978.

19. Sotiracopoulos, S., and Dondero, N. C., Airborne microorganisms in high density poultry management systems. *In* "Proceedings of the Agricultural Waste Management Conference," pp. 159–174. Cornell University, Ithaca, New York, 1974.

20. Guo, P. H. M., Fowlie, P. J., Cairns, V. W., and Jank, B. E. "Performance Evaluation of an Oxidation Ditch Treating Dairy Wastewater," EPS 4-WP-79-7. Environmental Protection Service, Environment Canada, Ottawa, 1979.

21. Paulson, W. L., and Lively, L. D. "Oxidation Ditch Treatment of Meatpacking Wastes," EPA-600/2-79-030. Industrial Environmental Research Laboratory, Office of Research and Development, U.S. Environmental Protection Agency, Cincinnati, Ohio, 1979.

22. Scottish Farm Buildings Investigation Unit, "The Treatment of Piggery Wastes," North of Scotland College of Agriculture, Aberdeen, Scotland, 1974.

23. Martin, J. H., Loehr, R. C., and Cummings, R. J., The oxidation ditch as a dairy cattle waste management alterative. *In* "Livestock Wastes: A Renewable Resource," pp. 346–350. Am. Soc. Agric. Eng., St. Joseph, Michigan, 1980.

24. Scheltinga, H. M. J., Farm wastes. *J. Inst. Water Pollut. Contr. (G.B.)* **68,** 403–413 (1969).

25. Moore, J. A., Larson, R. E., Hegg, R. O., and Allred, E. R., Beef confinement systems— oxidation ditch. *Annu. Meet. Am. Soc. Agric. Eng.* Paper 70-418 (1970).

26. Koot, A. C. J., and Zeper, J., Carrousel, A new type of aeration system with low organic load. *Water Res.* **6,** 401–406 (1972).

27. Terwilleger, A. R., and Crauer, L. S., Liquid composting applied to agricultural wastes. *In* "Managing Livestock Wastes," pp. 501–505. Am. Soc. Agric. Eng., St. Joseph, Michigan, 1975.

28. Grabbe, K., Thaer, R., and Ahlers, R., Investigations on the procedure and the turn-over of organic matter by hot fermentation of liquid cattle manure. *In* "Managing Livestock Wastes," pp. 506–509. Am. Soc. Agric. Eng., St. Joseph, Michigan, 1975.
29. Riemann, U., Aerobic stabilization and land disposal of liquid swine manure. *In* "Proceedings of the Agricultural Waste Management Conference" (R. C. Loehr, ed.), pp. 455–463, Cornell University, Ithaca, New York, 1974.
30. Popel, F., and Ohnmacht, C., Thermophilic bacterial oxidation of highly concentrated substrates. *Water Res.* **6,** 807–815 (1972).
31. Gould, M. S., and Drnevich, R. F., Autothermal thermophilic aerobic digestion. *J. Environ. Eng. Div. Am. Soc. Civ. Eng.* **104,** 259–270 (1978).
32. Jewell, W. J., and Kabrick, R. M., Autoheated aerobic thermophilic digestion with aeration. *J. Water Pollut. Contr. Fed.* **52,** 512–523 (1980).
33. Chittenden, J. A., and Wells, W. J., Rotating biological contactors following anaerobic lagoons. *J. Water Pollut. Contr. Fed.* **43,** 746–754 (1971).
34. Chou, C. C., and Hynek, R. J., Field performance of three RBC aeration modes treating industrial wastes. *Proc. Purdue Ind. Waste Conf.* **35,** 855–865 (1981).
35. Hammer, M. J., Rotating biological contactors treating combined domestic and cheese-processing wastewaters. *Proc. Purdue Ind. Waste Conference* **37,** 29–38 (1983).
36. Mikula, W. J., Reynolds, J. H., Middlebrooks, E. J., and George, D., Performance characteristics of a rotating biological contactor (RBC) treating a cheese processing wastewater. *Proc. Purdue Ind. Waste Conference* **34,** 335–353 (1980).
37. Bryan, E. H., and Moeller, D. H., Aerobic biological oxidation using Dowpac. *Int. J. Air Water Pollut.* **5,** 341–347 (1963).
38. Buzzell, J. C., Caron, A. J., Ryckman, S. J., and Sproul, O. J., Biological treatment of protein water from the manufacture of starch. *Water Sewage Works* **111,** 327–330 and 360–365 (1964).
39. Guttormsen, K., and Carlson, D. A., "Potato Processing Waste Treatment—Current Practice," Water Pollut. Contr. Res. Ser., DAST-14. Federal Water Pollution Control Administration, U.S. Dept. of the Interior, Washington, D. C., 1969.
40. National Canners Association, "Waste Reduction in Food Canning Operations," Rep. Grant WPRD 151-01-68. Federal Water Quality Administration, U.S. Dept. of the Interior, Washington, D.C., 1970.
41. Baker, D. A., and White, J., Treatment of meat packing wastes using PVC trickling filters. *In* "Proceedings of the Second National Symposium on Food Processing Wastes," pp. 237–312. Pacific Northwest Water Lab., U.S. Environmental Protection Agency, Denver, Colorado, 1971.

9

Anaerobic Treatment

General

Anaerobic treatment is a natural and widely used process for the stabilization of organic matter. Other terms that have been used for this technology are anaerobic digestion and fermentation. In the process, decomposition of organic matter occurs in the absence of molecular oxygen. Almost all natural and many synthetic organic compounds can be stabilized by anaerobic processes. The processes can be applied to animal manures, crop residues, food processing wastes, human wastes, and sewage sludges. If the process is carried to completion, the end products are a stabilized sludge and gases such as methane and carbon dioxide. Both end products have beneficial aspects (Fig. 9.1). The objectives of anaerobic treatment are to stabilize the organic compounds in the untreated waste or residue and to produce methane, which can be used as a source of energy. Typical advantages and potential concerns of anaerobic treatment are noted in Table 9.1.

Factors Affecting Anaerobic Treatment

Process Microbiology

Anaerobic digestion is a biological process in which organic compounds are metabolized and stabilized by facultative and anaerobic microorganisms that

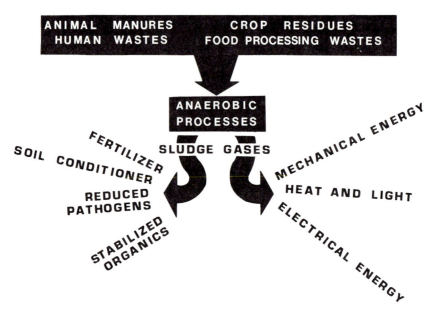

Fig. 9.1. Potential uses of end products from the anaerobic stabilization of human wastes and agricultural residues.

function in the absence of molecular oxygen. The manner by which digestion occurs can be considered in a simplified way as a multistage process (Fig. 9.2).

The predominant groups of bacteria involved in anaerobic metabolism are fermentative bacteria, hydrogen-producing acetogenic bacteria, and methanogenic bacteria (1). The fermentative bacteria hydrolyze and convert polysaccharides to organic acids, alcohols, hydrogen, and carbon dioxide. In addition, proteins and fats are converted to simpler compounds. The hydrogen-producing acetogenic bacteria convert the alcohols and organic acids produced by the fermentative bacteria to acetate, carbon dioxide, and hydrogen. The by-products of both groups of bacteria are converted by the methanogenic bacteria to methane and carbon dioxide.

The methanogenic bacteria appear to have two important roles in an anaerobic environment. One is to convert organic acids to methane, and the other is to function as an electron sink by producing methane by carbon dioxide reduction. By doing so, methanogens maintain a low partial pressure of fermenter hydrogen (H_2). Only under such low partial pressures may the electron carrier nicotinamide adenine dinucleotide (NAD) relinquish its hydrogen atoms to form hydrogen. If hydrogen formation is not energetically favorable (such as when there is a process imbalance), organic acids of nearly equivalent energy content to that of the original substrate will accumulate (1).

TABLE 9.1

Advantages and Concerns Associated with Anaerobic Treatment Processes

Advantages

The organic content of the waste is reduced and stabilized so that the final product represents a lesser pollution hazard

The fertilizer constituents of the input material are conserved, the carbon-to-nitrogen ratio is in balance, and the resultant sludge or solids are useful as a soil conditioner and supplemental fertilizer

Combustible gases are produced, which can produce energy for use by the farmer, food processor, or other generator of the organic matter if sufficiently high rates of digestion are maintained

Weed seeds and some pathogens may be inactivated during the process

Rodents and flies are not attracted to the resultant sludge

Potential Concerns

Suitable management is needed for continuous or daily feeding of the anaerobic unit and for proper maintenance and supervision

Temperatures, pH values, and volatile acid concentrations must be maintained within proper boundaries for successful operation

Toxic compounds, such as metals, organics, and oxygen, that may inhibit the anaerobic process must be avoided

A suitable and economic use for the methane and digested residues must be available

The volume of the digested material may be more than that of the raw material if dilution is required for operation of the digester

The digested residues contain high concentrations of organic matter and nutrients and can pollute surface and groundwaters if not handled and utilized properly

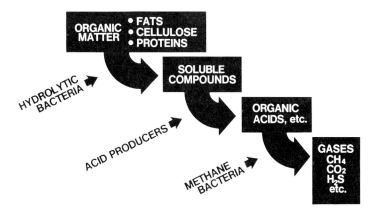

Fig. 9.2. Microbial reactions that occur in anaerobic processes.

The acid formation stage is required before the organic matter can be utilized by the methane-forming bacteria. Stabilization of organic compounds occurs due to fermentation of the organic acids by methane bacteria to methane and carbon dioxide. The methane bacteria are strictly anaerobic. Several different groups of methane bacteria, each fermenting a limited number of organic compounds, stabilize the complex mixtures of organic materials in wastes. Some of the methane bacteria grow rapidly, requiring only a short solids retention time. The slowest growing and most important methane bacteria utilize acetic and propionic acids. These bacteria require longer solids retention times and are the growth-limiting bacteria in the design and operation of anaerobic treatment systems.

Acetic acid is an important intermediate acid and is a source of most of the methane that is produced. Reduction of carbon dioxide results in most of the remaining methane that is generated. Carbon dioxide acts as a hydrogen acceptor and is reduced to methane gas.

Only a small amount of microbial cell growth occurs in anaerobic processes, due to the small amount of energy available to the bacteria under anaerobic conditions (Chapter 5). In contrast to aerobic processes, anaerobic processes convert a smaller fraction of the added organic wastes or residue to cell mass. Most of the biodegradable organic material is converted to methane.

Because of the microbial process steps that are involved in anaerobic processes (Fig. 9.2), it is essential that the facultative acid-forming bacteria and the methane bacteria be in proper balance. If liquefaction and acid formation occur faster than the acids can be converted to methane, the acids increase, the pH decreases, the methane bacteria are inhibited, and the anaerobic process no longer functions. Operationally, the rate at which the methane bacteria convert the organic acids to methane is the rate-limiting step in the anaerobic process. Due to the relatively low microbial yield and therefore low microbial concentration in anaerobic processes, it can be difficult for such processes to recover from upsets and inhibitions.

The concentration of volatile acids in the anaerobic mixture and the percentage of carbon dioxide or methane in the gas are parameters that are good indicators of routine process performance or impending problems. If the volatile acid concentration and percentage of methane or carbon dioxide stay relatively constant in a satisfactorily performing anaerobic unit, the unit is in equilibrium. If there is a continuing increase in volatile acids concentration or carbon dioxide percentage or a decrease in the methane percentage, this is an indication that adverse conditions are beginning to occur. The reasons for the adverse conditions should be determined, and corrective action should be taken.

The anaerobic bacteria can be affected by factors such as temperature, rapid variations in loading, inhibitors, solids retention time, and buffer capacity. Engineering process control and an understanding of the factors affecting facultative and anaerobic bacteria are needed for a successful anaerobic process.

pH–Alkalinity Relationships

If volatile acids are produced at a faster rate than they are utilized, adverse conditions will not occur as long as the buffer capacity of the system can neutralize the excess acids. The buffer capacity can be expressed as the alkalinity of the system. With a fixed detention time, the alkalinity in an anaerobic unit will be in proportion to the organic loading to the unit.

Maintaining the pH of an anaerobic process between about 6.8 and 8.0 will result in satisfactory performance. In this range the pH is controlled primarily by the natural carbonate buffer equilibrium. The decarboxylation of organic acids produces carbon dioxide, which reacts with water to form carbonic acid. Ionization of the carbonic acid produces bicarbonate and hydrogen ions [Eq. (9.1)].

$$CO_2 + H_2O \leftrightarrows H_2CO_3 \leftrightarrows HCO_3^- + H^+ \tag{9.1}$$

Should the concentration of hydrogen ions tend to increase and the pH tend to decrease, the reactions noted in Eq. (9.1) would tend to shift to the left and maintain a stable equilibrium and pH. Conversely, should the pH tend to increase, the reactions would tend to shift to the right; additional hydrogen ions would be in solution, and again a stable pH would be maintained.

Another reaction that occurs in anaerobic processes is the deamination of proteins. This reaction tends to increase the alkalinity and the pH of the process. Ammonium hydroxide is formed by the hydrolysis of ammonia which has been released from the proteins. The ammonium hydroxide reacts with carbonic acid to form water and a buffer in the form of ammonium bicarbonate.

When the volatile acids begin to increase, bicarbonate alkalinity is neutralized, and volatile acid alkalinity results. Under these conditions, the bicarbonate alkalinity can be approximated by Eq. (9.2).

$$BA = TA - 0.71TVA \tag{9.2}$$

where BA = bicarbonate alkalinity (milligrams per liter as $CaCO_3$), TA = total alkalinity (milligrams per liter as $CaCO_3$), and TVA = total volatile acid concentration (milligrams per liter as acetic acid).

When the bicarbonate alkalinity is in the range of about 2500–5000 mg per liter or more, an increase in the volatile acids in an anaerobic unit will be neutralized by the buffer capacity. If the bicarbonate alkalinity becomes too low, the buffering capacity may not be adequate, the pH decreases, and the process is inhibited.

Maintenance of adequate alkalinity and pH can be obtained by the addition of alkaline substances such as lime and sodium bicarbonate. An example of the addition of lime to anaerobic lagoons to maintain anaerobic activity was presented in Chapter 6. The addition of such chemicals requires careful control and

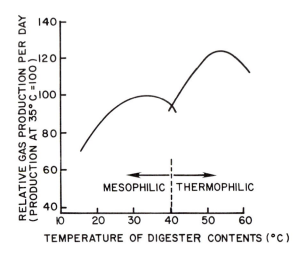

Fig. 9.3. Temperature will affect the activity of the microorganisms in anaerobic processes.

adequate mixing. The addition of lime and other basic compounds should be considered only as a temporary measure until the cause of the increased volatile acids and low alkalinity is identified and corrected.

Temperature

Temperature will affect the performance of anaerobic systems since it affects the activity of the microorganisms. The optimum temperature of mesophilic anaerobic treatment is 30–37°C (Fig. 9.3). Although it takes less time to obtain a desired degree of treatment with thermophilic rather than mesophilic anaerobic conditions, thermophilic conditions are not widely used and have not yet been demonstrated as economical. Most organic matter digests well at mesophilic temperatures. In addition, mesophilic conditions require less heat to bring the wastes and residues to the appropriate temperature and to offset heat losses to the environment.

However, it does not follow that anaerobic treatment must occur at optimum mesophilic temperatures. Satisfactory treatment can occur at lower temperature if an adequate mass of active microorganisms and a sufficiently long solids retention time (SRT) are provided. At reduced temperatures microbial growth will be less, and there will be less active organisms in the system. At low temperatures, the SRT must increase if comparable degrees of anaerobic treatment are to occur. This can be accomplished by solids recycle. Relationships between temperature and minimum SRT values have not been well established for anaerobic processes. The minimum SRT values at 25 and 15°C may be about 1.2–1.5 and 2.0–3.0, respectively, times the values at 35°C.

The rate of digestion is not the only factor to be evaluated in selecting an operating temperature. Other factors include the amount of heat required to raise the influent to the operating temperature and the characteristics, especially the settleability, of the anaerobic sludges produced at different temperatures. Reasonably constant temperature is important to the process. When anaerobic treatment occurs at suboptimal temperatures, the decrease in rate constants should be compensated for by an increase in the active mass and by an operating SRT greater than the minimum SRT at the lower temperatures.

Retention Time

Adequate time is required for the microorganisms to decompose the organic matter and generate the methane. The time that the organic matter must remain in the anaerobic unit determines the size of the unit. In a continuous-flow, completely mixed anaerobic unit, the hydraulic retention time and the SRT are equal. In units that are not completely mixed or employ recycle, the SRT is greater than the hydraulic retention time. The relative importance of these retention times in biological systems was discussed in Chapter 5.

If the SRT of the system is less than the minimum microbial reproduction time, the microorganisms are removed from the system faster than they can reproduce. In anaerobic systems, the minimum SRT values are much longer than those for aerobic systems. Minimum SRT values for anaerobic systems are in the range of 2–6 days, and operating SRT values generally are in excess of 10 days, since by that time most of the methane has been generated (Fig. 9.4).

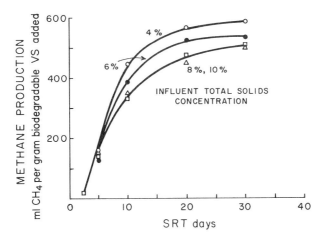

Fig. 9.4. Methane production resulting from the digestion of dairy cattle manure as a function of SRT (2).

Nutritional Requirements

The nutritional requirements are related to the net growth of the microbial cells (Chapter 5). In anaerobic systems treating nutritionally balanced organic material such as manures and most food processing wastes, a deficiency of any of the necessary elements is unlikely. However, when anaerobic systems are designed to stabilize residues or wastes that may not be nutritionally balanced, the possibility of nutrient limitation should be investigated.

Inhibition

Excessive volatile acids are not the only items that can be detrimental to anaerobic units. Certain wastes may contribute inhibitory concentrations of inorganic and organic materials. Alkaline metal salts, such as those of sodium, potassium, calcium, or magnesium, which may be in high concentrations in certain wastes, can be the cause of the inefficiency and/or failure of anaerobic systems. The inhibition has been shown to be associated with the cation portion of the salt. Examples of the concentrations that may cause process inhibition are presented in Table 9.2.

In anaerobic systems, ammonia will be in the form of either the ammonium ion or the dissolved ammonia gas (free ammonia). As the pH of the system increases, the free ammonia concentration in solution increases. The ammonium ions result primarily from the degradation of nitrogenous compounds, such as proteins and urea. Additional information on ammonia inhibition in biological systems is presented in Chapter 12. Free ammonia concentrations can be toxic to the microorganisms in an anaerobic unit.

Inhibition by ammonia can be a problem with anaerobic processes, particularly when wastes high in nitrogen, such as swine and poultry manures, are digested. Generally, at ammonia concentrations above 3 gm/liter, inhibition occurs, regardless of the pH. However, in one study, although ammonia inhibition did occur, complete stopping of methane production did not occur, even at ammonia nitrogen concentrations of 7.0 gm per liter (3).

TABLE 9.2

Concentrations of Cations That Can Cause Inhibition in Anaerobic Processes

Cation	Inhibitory concentration (mg/liter)
Calcium	3000–8000
Magnesium	1000–3000
Potassium	2500–12,000
Sodium	3500–8000

The impact of nitrogen can be related to the carbon-to-nitrogen (C/N) ratio of the wastes. Optimum methane production occurred with swine manure when the C/N ratio was in the range of 10:1 to 15:1. Digesters with ratios less than 10:1 exhibited ammonia inhibition, while digesters with ratios greater than 18:1 had low bicarbonate alkalinity and hence greater instability when the volatile acid concentrations increased (4).

High concentrations of sulfides can be toxic and should be prevented. Sulfates are the major precursors of sulfides in anaerobic units. Sulfides also result from the anaerobic degradation of sulfur-containing organic and inorganic compounds and from sulfides in the raw wastes. Soluble sulfide concentrations up to 200 mg per liter as sulfur may exert no significant effect on anaerobic treatment. Insoluble sulfide complexes have had no significant effect on anaerobic treatment up to at least 400 mg per liter of sulfide. Excessive concentrations of sulfides can be precipitated by iron salts to reduce the sulfide toxicity in an anaerobic unit.

While oxygen is important in aerobic biological units, the addition of oxygen to anaerobic units is detrimental and should be avoided. The advantages of anaerobic treatment occur because the degradation of organic matter takes place in the absence of oxygen. As noted in Chapter 5, microorganisms will use metabolic pathways that yield the greatest amount of energy. Satisfactory anaerobic digestion will occur when the microorganisms obtain a minimum amount of energy from the metabolic reactions and when the end products of the reaction are those (such as methane) containing considerable usable energy.

Oxygen and oxidized material such as nitrates should not be added to anaerobic units, since they will be used as hydrogen acceptors in preference to oxidized organic matter and carbon dioxide. When nitrates have been added to anaerobic units as a source of nitrogen, volatile acids have increased and gas production has decreased.

Mixing

The maximum rate of reaction occurs when a biological reactor is well mixed and the organisms are in continuous contact with degradable organic material. Mixing also serves other functions, such as (a) maintenance of uniform temperatures throughout the unit, (b) dispersion of potential metabolic inhibitors, such as volatile acids, and (c) disintegration of coarser organic particles to obtain a greater net surface for degradation.

In an unmixed digester, if the viscosity of the mixture is low, solids will settle and a scum will form. Both the scum and the settled solids can cause problems in the operation of anaerobic digesters. Mixtures of manure and urine are viscous enough so that separation is not a problem if the influent solids concentration is above about 10%.

Mixing can occur naturally due to the gas generation in the anaerobic unit or can be mechanically induced with modern gas and mechanical mixing equipment. The effectiveness and economic value of mixing are related to the type of anaerobic process and the type and concentration of material being treated or digested and must be considered separately for each process and waste or residue.

Feedstock Material

A wide range of plant and animal matter can be digested anaerobically. The gas yield and degree of stabilization that result from anaerobic digestion are a function of the fraction of the total waste that is available to the anaerobic bacteria, i.e., the biodegradable fraction. Lignin is unaffected by the bacteria in an anaerobic unit, and cellulose is decomposed very slowly.

The biodegradable fraction of a waste is a function of how the wastes were generated and handled prior to digestion. For effective anaerobic treatment, the addition of inert material such as sand and dirt in wastes should be minimized, and fresh wastes should be utilized. The longer the wastes are exposed to natural environmental degradation, the smaller will be the biodegradable fraction that remains for methane production. Fresh manure will have a greater biodegradable fraction than manure scrapped from a feedlot.

Biodegradability

Natural organic matter, such as manures, crop residues, and food processing wastes, contains a refractory or nonbiodegradable organic fraction that will not be digested and converted to methane. The volatile solids added to an anaerobic unit consist of biodegradable volatile solids and refractory volatile solids. Refractory volatile solids resist microbial decomposition over long periods of time and remain after the rate of volatile solids degradation decreases to a very low level. The refractory fraction (R) of a waste or residue is the ratio of the refractory volatile solids to the initial or influent volatile solids. Several methods of determining R are available (5–7).

Information on the R value of manures and residues has been developed in the past decade. Table 9.3 summarizes available biodegradable and refractory fraction data. Several aspects are apparent from the data in Table 9.3. One is that within the temperature range of 25–55°C, temperature does not seem to have an obvious effect on the R value of various substrates. Another aspect is that there appears to be a range of R values for a specific type of manure or residue. For instance, R values for dairy cattle manure range from 0.38 to 0.63, with most of the values in the 0.52–0.59 range. This variation probably is due to the type of

TABLE 9.3

**Refractory and Biodegradable Fractions of Manures, Crop Residues,
and Other Organic Matter**

Source	Temperature (°C)[a]	Refractory value (R)[b]	Biodegradable fraction[b]	Reference
Manures				
Dairy cattle manure	25	0.55	0.45	8
Dairy cattle manure plus bedding	25	0.55	0.45	8
Dairy cattle manure	32.5	0.58	0.42	9
Dairy cattle manure	35	0.56	0.44	8
Dairy cattle manure plus bedding	35	0.52	0.48	8
Dairy cattle manure	35	0.60	0.40	2
Dairy cattle manure	35	0.43	0.57	2
Dairy cattle manure	35	0.38	0.62	8
Dairy cattle manure	36	0.63	0.37	2
Dairy cattle manure	35	0.41–0.47	0.53–0.59	10
Pig manure	35	0.27	0.73	10
Chicken manure	35	0.24	0.76	10
Elephant manure	35	0.47	0.53	10
Crop residues				
Wheat straw	25	0.38	0.62	11
Wheat straw	35	0.33	0.67	11
Wheat straw	55	0.44–0.49	0.51–0.56	11
Wheat straw	35	0.45	0.55	10
Corn stalks	35	0.23	0.77	10
Corn leaves	35	0.28	0.72	10
Corn stover	35	0.26	0.74	11
Corn stover	55	0.32	0.68	11
Old grass	35	0.27	0.73	11
Old grass	55	0.4–0.48	0.52–0.6	11
Other				
Cattails	35	0.41	0.59	10
Water hyacinth	35	0.41	0.59	10
Corn meal	35	0.15	0.85	10
Newsprint	35	0.72	0.28	10

[a]Temperature of the anaerobic study from which the results were obtained.
[b]Fraction of initial volatile solids.

ration fed the animals and the degree of manure stabilization that may have occurred, such as in the barn or barnyard, before the manures were used in the methane generation studies.

With a few exceptions, manures and residues that result from forages (cattle manure and elephant manure) have R values that are in the same range (0.5–0.6). Material that is prepared to be more readily digested (e.g., corn meal) or that contains more concentrates and less roughage (e.g., pig and chicken manure)

TABLE 9.4

**Biodegradability Constants for Different
Animal Manures**[a]

Manure	Biodegradability fraction[b]
Swine	0.90
Beef (confinement)	0.65
Beef (dirt)	0.56
Dairy	0.36
Poultry	
Broiler	0.70
Layer	0.87

[a]Adapted from (12).
[b]Grams of VS destroyed per gram of VS added to
an anaerobic digester.

have lower R values (0.13–0.27) than do common crop residues used as forage
(e.g., wheat straw, corn stalks and stover, and grasses). It is understandable that,
after the digestible components are extracted from the forages by the animals, the
manures have a higher R value than do the forages. Newsprint has a very high R
value (0.72).

Additional data on the biodegradable fraction of various manures have been
developed (Table 9.4) from a computer simulation of results from several studies. The data in Table 9.4 are similar to those in Table 9.3, but there are
differences in the value of the biodegradable fraction. Such differences may be
due to the rations the animals were fed and the freshness and other characteristics
of the manures.

Knowledge of the biodegradable fraction of a waste or residue is fundamental to the feasibility of anaerobic treatment, especially when the objective is
methane generation and energy production. The greater the biodegradable fraction, the greater will be the quantity of methane generated from the material and
the fewer the nondegradable (refractory) solids that must be removed and transported following anaerobic treatment. A manure with a biodegradable fraction of
0.45 ($R = 0.55$) will produce less methane and biogas and will have greater inert
solids in the resulting sludge, at the same loading and operating conditions, than
will a manure that has a biodegradable fraction of 0.75. Table 9.5 compares the
methane and biogas production and the nonbiodegradable fraction that results
from anaerobic digestion of material having different fractions of biodegradable
material. In determinations of the feasibility of an anaerobic treatment process,
the biodegradable fraction of the material to be treated or digested should be
known or determined since it will have an impact on the methane yield and the
solids remaining for disposal (Table 9.5).

TABLE 9.5

Comparative Methane and Biogas Yield and Nondegradable Solids from an Anaerobic Unit[a]

Biodegradable fraction	Refractory fraction	Methane yield (m^3)	Biogas yield (m^3)	Nondegradable solids in resulting sludge (kg)[b]
0.35	0.65	0.18	0.30	1.07
0.45	0.55	0.23	0.39	0.94
0.55	0.45	0.29	0.48	0.81
0.65	0.35	0.34	0.56	0.68
0.75	0.25	0.39	0.65	0.55
0.85	0.15	0.44	0.74	0.42

[a]Assumptions: 10 kg of manure with a dry matter (total solids) content of 15%; ash is 15% dry matter; methane yield is 0.4 m^3 per kilogram of biodegradable VS destroyed; methane is 60% of total biogas yield; all biodegradable VS are converted to methane and carbon dioxide.

[b]Inorganics (ash) plus the nondegradable VS (refractory VS) in the manure added to the digester.

Lignin can limit the extent of organic matter degradation by trapping nutrients and inhibiting cellulolytic enzymes or by forming lignin–carbohydrate linkages that resist enzymatic separation. Table 9.6 presents the reported lignin content of various manures.

A relationship between the lignin content determined by the sulfuric acid method (lignin$_s$) of the total volatile solids of a substrate and volatile solids

TABLE 9.6

Reported Lignin Content of Manures[a]

Manure	Lignin (% of dry matter)
Dried poultry waste	1.4
Broiler litter	9.7 (9.4–10.4)[b]
Dairy cow manure	14.4 (7.2–27.1)
Beef cattle manure	8.6 (5.0–15.0)
Beef cattle oxidation ditch settled solids	6.3
Steer manure	3.1
Dairy cow manure screenings	11.1 (10.0–12.0)

[a]From (13).

[b]Values in the parentheses indicate the range of reported data; the single value represents the average.

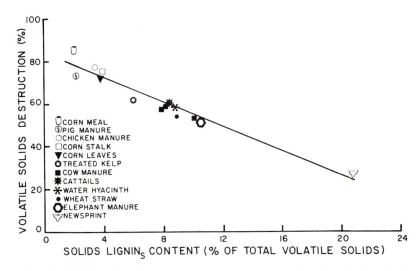

Fig. 9.5. Volatile solids destruction during anaerobic treatment can be related to the lignin content of the wastes being digested. [Adapted from (10).]

destruction during anaerobic digestion has been shown (Fig. 9.5). This relationship was developed from a broad range of substrates. The line of best fit for the data in Fig. 9.5 is

$$Y = -2.82X + 83 \qquad (9.3)$$

where Y is the percentage of volatile solids destruction and X is the lignin$_s$ content of the material as a percentage of total volatile solids.

The biodegradable fraction of the material to be digested was estimated from the lignin$_s$ content of the total volatile solids (9):

$$B = -0.028X + 0.83 \qquad (9.4)$$

where B is the biodegradable fraction of the total volatile solids and X is the lignin$_s$ content in terms of percentage of the total volatile solids. Equation (9.4) has been used successfully to identify the volatile solids biodegradable fraction of other substrates such as grass and corn stover (11). Relationships such as in Fig. 9.5 and Eq. (9.4) help determine the biodegradability of specific wastes and residues and the feasibility of anaerobic treatment.

The relationship between the volatile solids destruction and lignin content [Fig. 9.5 and Eq. (9.3)] was compared with independent data obtained from the digestion of beef cattle manure (14). While there was good correlation between the lignin content and methane yield [cubic meters of methane per kilogram of volatile solids (VS) fed] for the results obtained from the beef cattle study, it was not the same correlation noted with other manures. Other factors besides the lignin content may affect methane yield and destruction. For a given type of manure or residue, it may be possible to predict the methane yield and VS destruction as a function of the lignin content.

Carbon-to-Nitrogen Ratio

Adequate nitrogen should be available for the microbial processes. Too little nitrogen (a high C/N ratio) will slow the digestion process. Carbon-to-nitrogen ratios of greater than 25:1 are considered high. Data in the literature suggest that C/N ratios of between 8 and 25 could be satisfactory for anaerobic systems fermenting mixtures of crop residues and manures (15). Ratios between 16 and 19 have been recommended (16). Manures and sludges have suitable C/N ratios, although some crop residues may not. Wastes having a low C/N ratio can be added to those with a high ratio to achieve satisfactory conditions for digestion. Table 9.7 identifies C/N values of manures and agricultural residues.

Availability

Adequate raw material should be available on a consistent basis and in an amount capable of meeting the methane generation requirements if energy production is the major objective. The material also should be in a form suitable for

TABLE 9.7

Carbon/Nitrogen Ratios of Manures and Crop Residues

Item	C/N ratio
Manures	
Hog[a]	14
Cow	20–25
Chicken[a]	10
Carabao[a]	23
Crop residues[a] (air-dried material)	
Corn stalk	57
Corn cobs	50
Rice straw	51
Peanut hulls	31
Other	
Grass clippings	12–16

[a]Adapted from (17).

the operation of the anaerobic unit. Wide variations in waste characteristics and loading are not conducive to satisfactory performance.

Gas Production and Use

Production

The gas produced by the anaerobic treatment of organic waste (biogas) is colorless and burnable and contains about 50–65% methane, 30–45% carbon dioxide, and small percentages of hydrogen sulfide, hydrogen, carbon monoxide, and other gases. The exact composition of the gas is related to the constituents of the waste or residue and the operating conditions of the anaerobic unit.

Biogas burns well with a blue flame, is nontoxic, possesses a slight smell, and has a calorific value of about 600–650 Btu per cubic foot. The fuel value of biogas is due mainly to the methane, which has a heating value of about 995 Btu per cubic foot. The other combustible gases are hydrogen and carbon monoxide.

The quantity of methane produced per mass of substrate added to an anaerobic unit depends on the methane production per mass substrate destroyed and the efficiency of VS destruction. Information on the potential VS destruction (Fig. 9.5) or biodegradability [Eq. (9.4)] will not provide a reasonable estimate of methane generated unless the methane produced per unit of VS destroyed or per unit of VS added to an anaerobic unit is known.

Different substrates produce different amounts of methane per unit of VS added or destroyed. Table 9.8 shows the methane production of various substrates. Although cornmeal had the largest VS reduction (85%), the methane generated per gram of VS reduced was less than that obtained from several other substrates. Of the substrates evaluated, pig manure produced the largest quantity of methane per unit of VS destroyed.

Other data also identify the differences in methane production in terms of volume per quantity of VS added to or destroyed in a digester that can occur with various agricultural residues. The differences are due to the fact that each residue can be digested to a different extent. As noted earlier, only that portion of the VS that is biodegradable serves as substrate for the acid-forming bacteria (Tables 9.3 and 9.4).

The ultimate methane generation that should result from various types of beef cattle manures and manure mixtures is identified in Table 9.9. Manure from cattle that were fed higher silage rations (91.5 versus 7%) produced less methane in terms of quantity both per unit of VS added and per unit of VS destroyed. The difference was more obvious in terms of quantity per unit of VS added. Older wastes contaminated with dirt (from a dirt feedlot) also had a lower methane yield. Manures that are the freshest, that have the least amount of inert material,

TABLE 9.8

Methane Generation and Volatile Solids (VS) Reduction Resulting from the Anaerobic Digestion of Organic Substrates[a]

Substrate	Methane generation		VS reduction (%)
	liter/gm VS added[b]	liter/gm VS reduced	
Pig manure	0.43	0.56	73
Elephant manure	0.30	0.54	53
Cow manure	0.28	0.47	56
Treated kelp	0.27	0.42	62
Chicken manure	0.32	0.39	76
Wheat straw	0.22	0.38	55
Corn stalks	0.30	0.36	77
Corn leaves	0.27	0.35	72
Corn meal	0.32	0.35	85
Water hyacinth	0.25	0.35	59
Cattails	0.22	0.33	59
Newsprint	0.06	0.18	28

[a]Adapted from (10).
[b]Liter per gram = cubic meter per kilogram.

TABLE 9.9

Ultimate Methane Generation from Animal Manures[a]

Type of manure	Ultimate methane generation		Reference
	m³/kg VS added[b]	m³/kg VS destroyed	
Beef cattle	0.31–0.34	—	18
Beef cattle fed			
91.5% corn silage	0.17	0.36	14
40% corn silage	0.23	0.40	14
7% corn silage	0.29	0.42	14
Beef cattle, 6–8 weeks old from dirt feedlot	0.21	—	14
Beef cattle—manure:molasses			
100% manure	0.32	—	15
75:25	0.33	—	15
50:50	0.36	—	15

[a]Estimated from long-term studies or by calculation.
[b]m³/kg = liter/gram; m³/kg = 15.7 ft³/lb.

TABLE 9.10

Methane Generation from Operating Digesters[a]

Type of manure	Actual methane generation		Methane content of biogas (%)	Temperature of digester (°C)	Reference
	m³/kg VS added[b]	m³/kg VS destroyed			
Beef cattle					
Young steers	0.22–0.26	0.45–0.50	55–56	55	19
Beef cattle	0.22–0.31	0.50–0.58	50–55	55	20
Manure 1–10 days old	0.17–0.29	0.30–0.60	54–59	50	20
Beef cattle—manure:molasses					
100% manure	0.27–0.30	—	50–54	55	15
75:25	0.28–0.31	—	48–51	55	15
50:50	0.30–0.34	—	48–51	55	15
Dairy cattle	0.25–0.40	0.80–0.93	53–57	35	21
Mostly feces and urine	0.22–0.27	0.66–0.79	49–54	60	21
Swine	0.31–0.43	0.57–0.74	61–62	55	22
	0.26–0.38	0.60–0.76	59–64	35	22

[a]Generally from laboratory or small-scale units.
[b]m³/kg = liter/gm; m³/kg = 15.7 ft³/lb.

and that contain the greatest biodegradable content have the greatest potential for methane production.

The actual methane produced from various manures is shown in Table 9.10. The methane content of the biogas was in the range of 50–60% for cattle manure and in the range of 60–65% for swine manure. The actual methane production reported in the noted studies was less than the ultimate, since the studies were used to identify practical operating conditions rather than total methane production. The biodegradability of the manures is related to the type of feed fed to the animals. Swine manure generally is more biodegradable than cattle manure, because swine are fed more readily digestible, high-energy rations. Data from several studies indicate (22) that the ultimate methane yield (cubic meters of CH_4 per kilogram of VS added) is in the range of 0.44–0.52.

Additional information on the biogas and methane generated from agricultural residues is presented in subsequent sections of this chapter.

Use

The successful utilization of the biogas requires a sufficient demand, the ability to store the gas, and the ability to use the gas or convert it to another form of energy. The combustible content and the fuel value of biogas can be increased by removal of the carbon dioxide. This can be done by passing the biogas through an alkaline solution, which will absorb the carbon dioxide. However, since carbon dioxide does not interfere in most uses of the biogas, removal is not required.

The most common use of biogas is for lighting, cooking, drying, and other direct heating purposes. With a properly designed burner, as much as 60% of the fuel value of the biogas can be used for heating. Figure 9.6 shows a biogas stove and burner used for cooking in a commune in the People's Republic of China. Simpler burners also can be used, although the efficiency will be lower. However, simple burners are easy to construct and are less expensive.

Appliances designed for other gaseous fuels can be converted to use biogas. To do so it may be necessary to adjust air-to-gas ratios and injector or flame orifice openings.

Internal combustion engines can be converted to use biogas. For gasoline engines, biogas can replace the gasoline. The carburetor should be changed to a biogas carburetor. For diesel engines, biogas cannot completely replace the fuel, and some diesel fuel is needed for ignition. Again a biogas carburetor is needed.

Fig. 9.6. A stove and burner using biogas generated by a mixture of manure and grasses.

Fig. 9.7. A biogas cogenerator used for the production of electricity and hot water.

Figure 9.7 shows a methane fired cogenerator used for the production of electricity and hot water.

The hydrogen sulfide in biogas can cause problems in the use of the gas. In the presence of water vapor, the hydrogen sulfide will cause corrosion of pipes, valves, burners, and compressors. Manufacturers of engines and compressors have specific recommendations concerning the allowable concentration of hydrogen sulfide in the gas used in their equipment. These recommendations should be followed and, where necessary, hydrogen sulfide should be removed from the biogas before use. Chemical removal, primarily using iron, is the common method used for hydrogen sulfide removal.

Sludge Characteristics and Use

Characteristics

The residual slurry (sludge) from an anaerobic unit digesting human wastes, crop residues, or manures will contain lignin, lipids, synthesized microbial cells, volatile acids and other soluble compounds, inert material in the original waste, and water. Anaerobic treatment conserves nutrients for the production of crops. Practically all of the nitrogen present in the material entering an anaerobic unit is conserved. To minimize subsequent ammonia nitrogen losses, the sludge should be stored in lagoons or tanks which present a minimum of surface area for ammonia volatilization. Except for carbon, other chemical elements contained in the material added to the anaerobic unit will be conserved in the sludge.

Use

The most common use of the anaerobically treated material is as a fertilizer and soil conditioner. The sludge is an organic fertilizer that adds major and trace nutrients to the soil and improves the physical condition of the soil by the addition of organic matter. The sludge should be applied to agricultural land in accordance with sound guidelines and appropriate loading rates, such as those used for the application of manures.

With the exception of the "dry fermentation" process, anaerobic units will have a liquid or slurry discharge that can be handled by most agricultural manure pumps. Storage facilities normally are needed so that a desired land application schedule can be followed. The liquid or slurry can be stored in an open or covered lagoon or tank. The size of the storage unit is related to the difficulty in applying the material to the land and the weather and crop production patterns.

HEAT PROCESSING UNIT

Fig. 9.8. Heat processing of anaerobic digester slurry for use as part of feed for swine.

The sludge also can be used as part of animal feeds. As discussed in Chapter 10, animal feeds rarely will contain more than 10% of such material. Figure 9.8 shows a unit used to process the slurry from an anaerobic digester for hog feed in the Philippines. The slurry was cooked at high temperatures with biogas. The cooking reduced pathogens and parasites before the material was added to the rations to the animals.

Where appropriate, the digested material can be added to fish ponds to support the growth of plankton on which the fish feed. The air-dried solids resulting from the anaerobic digestion of animal manures have been used both for mushroom and for earthworm production.

Management Systems

Anaerobic treatment processes are part of an overall waste or resource management system. Such a system may be (a) one that has the objective of treatment and disposal of a waste or residue and possibly the generation of methane as a potential fuel or (b) an integrated system in which the anaerobic unit is a key part of an agricultural operation. Figure 9.9 shows how an anaerobic unit can be integrated into an agricultural operation. Manure, crop residues, and human wastes can be added to a digester, with the biogas used in the agricultural operation for lighting, fuel, or power and the digester residues being returned to agricultural land for crop production. Such a closed cycle is being utilized in many parts of the world such as the People's Republic of China, the Philippines, and India to increase the self-sufficiency of farmers and to sustain agricultural production.

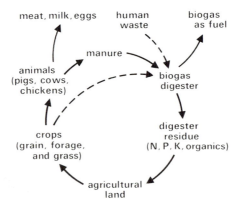

Fig. 9.9. Integration of crop residue, manure, and other wastes in a biomass energy and utilization cycle.

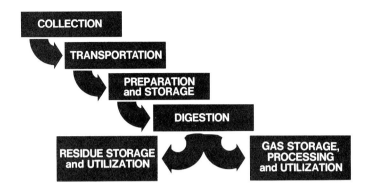

Fig. 9.10. The collection, transportation, preparation, storage, and utilization components must be considered when the feasibility of an anaerobic treatment system is determined.

In other countries, the various wastes or residues are kept separate and are handled in different ways. At large cattle or poultry operations, only the manure may be used as the substrate for anaerobic digestion. However, no matter what the waste or residue, anaerobic treatment must be considered as part of a larger system. This system will include (Fig. 9.10) equipment for collection, transportation, preparation, and storage of the material to be treated; the anaerobic unit; equipment to store and utilize the residue or slurry that results; and equipment for the handling, storage, processing and utilization of the methane and biogas that results.

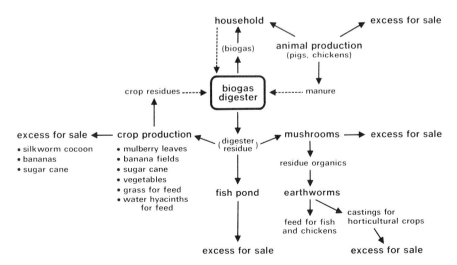

Fig. 9.11. An integrated agricultural production system in which anaerobic treatment is a key component.

Overall technical and economic feasibility of anaerobic processes requires that the total system and its components be identified and evaluated. A narrow focus only on the anaerobic process will not provide the proper perspective.

In some parts of the world, a very closely integrated agricultural production and biomass energy system exists. The advantages of such integrated systems include increased resource utilization, enhanced self-sufficiency, increased yields, more diversified products, and protection of public health and the environment. Figure 9.11 schematically shows a highly integrated agricultural system in the People's Republic of China. The anaerobic digester is a key component since it produces needed energy, uses resources (residues) that otherwise would be wasted or cause public health or environmental problems, and conserves nutrients for use in subsequent operations. Variations of the integrated operation shown in Fig. 9.11 exist in many countries (23).

Anaerobic Processes

There are several types of anaerobic processes that can be used for food processing wastes, manures, crop residues, wastewaters, and other wastes with similar liquid or solid characteristics. The terminology used to describe anaerobic processes is noted in Table 9.11. The selection of an anaerobic process is dependent on the substrate that will be treated or digested and the objectives of the total waste or energy management system. The processes described in the following paragraphs are those which have been used in practice or have shown promise.

Single Stage

The single-stage anaerobic process (Fig. 9.12) is one of the oldest processes used for waste management. Both biological stabilization and liquid–solid separation occur in the same unit. The heavier solids will settle, and the lighter solids will float to form a scum when dilute wastes are used in a single-stage unit. Such

TABLE 9.11

Terms Used to Describe Anaerobic Processes

Single stage	Batch
Two stage	Continuous
Plug flow	Random mix
Attached film	"Dry" fermentation

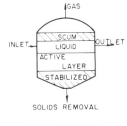

SINGLE STAGE

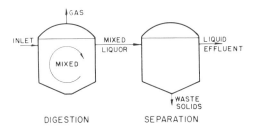

TWO STAGE

Fig. 9.12. Single- and two-stage anaerobic processes.

separation usually does not occur with substrates such as manure if the influent solids concentration is above 10%.

Conditions optimum for both biological activity and solids separation are difficult to achieve in a single-stage unit. Adequate mixing will inhibit solids separation by sedimentation, which is the usual method of solid–liquid separation in anaerobic systems. Quiescent conditions for satisfactory gravitational solids separation will not allow adequate food and microorganism contact for optimum biological action. Hydraulic retention times are long, on the order of 20+ days. Solids retention times are even longer; the active mass is small because of the low loading and poor contact with the wastes, and the system is more sensitive to load fluctuations than are other anaerobic processes.

Single-stage units can be operated as either batch or continuous processes. Anaerobic lagoons and animal waste storage tanks are examples of the practical use of a single-stage anaerobic unit for agricultural wastes. Single-stage units have been used widely in rural areas of many countries. Figure 9.13 is a schematic of a typical single-stage unit common to many parts of the world. Single-stage digesters have many shapes, such as a concrete box, a vertical cylinder, a horizontal cylinder, a flattened sphere, or a hemisphere. The shape frequently depends upon the construction materials available, the skills of those doing the construction, and the success of similar digesters in the country or region.

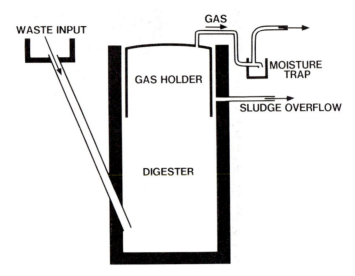

Fig. 9.13. Schematic of a single-stage anaerobic digester used in many parts of the world.

Figure 9.14 illustrates the aboveground portion of several single-stage units used at a farm in Asia. Both floating and fixed-cover gas collectors have been used with single-stage units, and both can be seen in Fig. 9.14. Figure 9.15 shows a rubberized bag being used as a floating gas collector. An advantage of flexible collectors is that they are less expensive than steel collectors.

Fig. 9.14. The aboveground portion of anaerobic units at a small farm in Asia, illustrating the gas collection devices and tubes transporting the biogas to various points of use on the farm.

Fig. 9.15. Rubberized bag used as a floating biogas collector.

The contents of a single-stage unit are removed at long intervals, such as a year, and applied to available agricultural land. The length of the interval between digester content removal is related to the buildup of solids in the digester and to whether the digester is operated as a batch digester. Batch digesters generally are cleaned out whenever gas production decreases significantly and new organic residues are to be added.

Two Stage

Realization that optimum conditions for both solids separation and biological action are incompatible in a single anaerobic unit led to the development of the two-stage system (Fig. 9.12). The first stage is a continuously fed, well-mixed, temperature-controlled biological unit. The second stage is a solids concentrator and separator. Two advantages accrue from a continuously and completely mixed, first-stage anaerobic unit: (a) the entire volume of the first stage may be used for biological stabilization of the organic waste, and (b) such a unit can be designed for optimum biological action.

The development of two-stage systems has permitted higher loadings. The

SRT and HRT in the first stage are the same. The capacity of the first stage is controlled by the HRT, which must be more than the minimum SRT. No flexibility to alter the SRT exists after the system is built.

As noted in Chapter 5, greater flexibility in the operation of a biological process can occur if the process contains solids recycle. With sludge recycle, a portion of the solids from the separation unit is recycled to the first-stage unit. A system with solids recycle would have a small HRT and a large SRT. Such an anaerobic process is analogous to the activated sludge process and has been called the "anaerobic contact process."

The advantages of sludge recycle in biological systems have been apparent for many years, although sludge recycle has not occurred in anaerobic systems to the extent that it has in aerobic systems. Laboratory and pilot plant studies on the anaerobic digestion of meat-packing wastes (24) indicated feasibility of high-rate, completely mixed, anaerobic systems with sludge return for these wastes. A full-scale system was designed and built, based on the pilot plant studies (25). The detention time in the first stage, based on influent flow, was approximately 12 hr, and that based on total flow was approximately 3.5 hr. The sludge retention time was greater, a matter of 10–15 days.

Attached Film

Attached-film units are columns or beds which contain solid media to which anaerobic bacteria become attached. The passing of the substrate through the media in an upflow manner allows substrate–microbe contact. Because bacteria are retained on the medium and not washed out with the effluent, extremely long SRT values can be obtained, whereas HRT values may be as short as a few hours. Attached-film units have also been known as an anaerobic filter.

Attached-film units can be either fixed bed or expanded bed. In a fixed-bed unit, the medium does not move. Fixed-bed reactor units have worked well with soluble wastes but can become clogged when used with wastes containing many particulates. In an expanded-bed unit, the upward velocity of the liquid passing through the media is great enough to provide a media expansion of 10–20%. With expanded-bed units, wastewaters containing particulate matter can be treated without clogging problems occurring. Because of the high degree of contact between the organic matter and the organisms attached to the media, anaerobic treatment can operate at ambient temperatures and high loading rates and with low power requirements, minimum solids production for further disposal, and high reaction rates. The anaerobic attached-film, expanded-bed process has been shown to achieve high (greater than 60%) organic removal percentages at low temperatures (10 and 20°C) when treating low-strength wastewaters (less than 600 mg per liter) at short HRT and at high organic loading rates (up to 8 kg of COD per cubic meter per day) (26). The process also has been shown to be successful at 55°C (27) with high-strength soluble wastes (5–16 gm of COD per liter).

Both anaerobic fixed-bed and expanded-bed units have been shown to be unaffected by large fluctuations in temperature, flow rate, organic concentrations, and organic loading rates (28–30). With hydraulic overloadings (30), anaerobic fixed-film units returned to steady-state conditions once the overloading ceased. The major reason for the minimum impact of the preceding variables is the high biomass concentrations and long SRT in the fixed-film units. In an expanded-bed unit, volatile solids concentrations on the order of 35 gm per liter and SRT values in excess of 1 year would not be unusual when the unit was treating a medium-strength wastewater (29).

The anaerobic rotating biological contactor is another anaerobic fixed-film process that can obtain high removals of organic carbon (31).

Upflow Anaerobic Sludge Blanket

The Upflow Anaerobic Sludge Blanket (UASB) has been investigated extensively in the Netherlands since 1971 for the direct treatment of wastewaters and has been considered as a process step for denitrification. Use of the UASB process in the United States has been with several food processing wastes.

The UASB is an anaerobic unit in which the influent enters at the bottom, travels through a blanket of anaerobic sludge, and is treated by the organisms in the sludge blanket. The treated effluent and the generated biogas exit from the top of the unit at different points. Mechanical mixing and sludge recirculation are kept at a minimum to avoid disruption of the sludge blanket. The anaerobic sludge has good settling characteristics. The process permits high organic and hydraulic loading rates since the sludge blanket has a high specific activity as well as good settleability. Examples of the wastes that have been treated and the results that have been obtained with the UASB process are noted in Table 9.12.

Dry Fermentation

Dry fermentation is the fermentation of organics at solids concentrations higher than that at which water will drain from the substrate (11). This type of process permits anaerobic treatment to be expanded to relatively dry material, such as crop residues. Until the development of the process, anaerobic treatment was limited to liquids and slurries.

Three substrates, corn stover, wheat straw, and old grass, have been used with the process (11). Due to the high volumetric density of the organics in dry fermentation, it has been used as a batch process. At normal biodegradable fractions, such as those noted in Table 9.3, most of the available carbon will be converted to biogas in a 75-day period. A dairy cow manure seed of 15–30% was

TABLE 9.12

Anaerobic Treatment of Wastes Using the UASB Process[a]

Waste	Temperature (°C)	Maximum loading rates		Removal efficiency (%)[c]
		kg COD/m³/day	m³/m³/day[b]	
Sugar beet	28–32	30–32	4–6	80–95
	30–34	14–16	3–4	87–95
Potato processing	19	3–5	1.2	95
	26	10–15	3.0	95
	30	15–18	4.0	95
	35	25–45	6–7	93
	30–35	7	0.6	91–97

[a]Adapted from (32).
[b]Cubic meter of waste applied per cubic meter of reactor volume per day.
[c]Based on the COD of centrifuged influent and effluent samples.

desirable to provide an inoculum of anaerobic bacteria and sufficient nitrogen and other nutrients. Dry fermentation units with initial solids contents of up to 40% were able to successfully produce methane over 6 months. A biodegradable VS destruction of 90% was achieved in 60–90 days at 55°C or in 120–200 days at 35°C (11). Dry fermentation is a promising anaerobic process that could be used where sufficient quantities of crop residues or similar wastes occur.

Batch and Continuous Units

A batch unit is one in which neither solids nor liquids enter or leave until the desired reactions or degradation have been achieved. A unit is filled, the reactions are allowed to go to completion, and the unit is emptied before it is again started. If a continuous supply of product, such as methane, is desired and if there is a continuing supply of raw material, a series of batch units may be used.

In a continuous unit, there is a constant input and discharge of material. Proper loading rates and detention times are used to achieve the degree of stabilization, solids reduction, or methane generation that is desired.

Plug Flow

An idealized plug flow unit is one in which individual portions of the input pass through the unit in the same sequence in which they entered. In order to achieve plug flow performance, there should be no intermixing of the unit; each particle should be retained in the unit for a time equal to the theoretical retention time; and microbial activity, substrate concentration, temperature, and any other characteristic are constant for all points in any cross section of the unit.

Fig. 9.16. A plug flow anaerobic digester stabilizing dairy cattle manure. The flexible cover is used to collect the biogas.

Neither idealized plug flow nor complete mixing is achieved in any biological unit. However, plug flow reactors are effective and satisfactory with dairy manures and similar residues. To avoid liquid–solid separation and related operational problems, high solids concentrations should be used. Figure 9.16 shows a plug flow anaerobic unit treating dairy manure.

Random Mix

A random mix unit is mixed intermittently, such as when waste is added to a unit. Such mixing can be useful for substrates which would undergo liquid–solid separation. This approach requires a low amount of energy, since mixing requirements are minimal. The needed degree of mixing is determined on a case-by-case basis.

Application to Agricultural Residues

Since the early 1970s, there has been expanding interest in the production and utilization of biogas and methane from agricultural residues. This interest has resulted from increases in the cost of gasoline and other fossil fuels and from a desire to identify alternative energy sources and to make maximum use of available resources. Many laboratory and pilot plant studies have been conducted to determine the technical and economic feasibility of methane production and to develop sound design and operating procedures for digestion systems stabilizing agricultural residues. A number of full-scale farm or feedlot systems also have been constructed and operated. The large-scale systems have provided valuable operating, design, cost, and gas utilization information that has further improved the engineering and scientific knowledge of anaerobic processes and systems.

The results from the laboratory, pilot-scale, and large-scale studies have pushed back the frontiers of knowledge concerning anaerobic treatment, anaerobic stabilization of agricultural residues, and the types of systems that can be used. Processes that are less expensive to construct and operate and that fit in with the agricultural operation producing the residue or manure have been developed. The breadth of residues that can be treated by anaerobic processes, ranging from very dilute wastewaters to manure slurries and crop residues, has expanded. New anaerobic processes such as attached-film, plug flow, expanded-bed, up-flow sludge blanket, and dry fermentation have been developed. The role of mixing has been delineated, and the relative biodegradability of various residues has been determined.

As a result of the activity in this field, the feasibility and viability of anaerobic treatment have increased considerably, and there are many alternatives and approaches that can be considered. The following sections (a) describe relevant information that has resulted from this activity and can be used to understand and design anaerobic processes and (b) provide information about the operating and performance characteristics of various anaerobic processes.

Manures

Of all agricultural residues, manures have been studied the most. As a result, there is a large amount of information describing processes that can be used, performance that can be obtained, design and operating relationships that can be considered, and factors that affect the processes. Data from many studies have been presented in previous sections of this chapter. Other useful information is presented in the following paragraphs.

General. A kinetic model of the following type has been proposed (3,33) to describe the steady-state production of methane from dairy cattle manure:

$$\gamma_v = [B_0 S_0 / HRT][1 - K/(HRT \cdot \mu_m - 1 + K)] \qquad (9.5)$$

where γ_v is the volumetric methane production rate (cubic meter of methane per cubic meter of fermenter per day); S_0 is the influent VS concentration (kilogram per cubic meter), B_0 is the ultimate methane yield (cubic meter of methane per kilogram of VS fed) as the HRT approaches infinity; HRT is the hydraulic retention time (days); μ_m is the maximum specific growth rate of micro-organisms (day^{-1}); and K is a dimensionless kinetic parameter. For a given S_0 and HRT, γ_v is a function of B_0, μ_m, and K.

The value of B_0 can be determined by (a) plotting the steady-state methane yield against the reciprocal of the HRT and extrapolating to an infinite HRT or (b) incubating a known amount of substrate until a negligible amount of CH_4 is produced. The latter is a long-term batch digestion. Both methods gave similar results when beef cattle manure was fermented at temperatures ranging from 30 to 65°C (33). The value of B_0 did not change with temperature and averaged 0.32 m^3 of methane per kilogram of VS fed for the steady-state method and 0.33 for the batch method. As discussed earlier, B_0 depends on the animal species; ration; manure age, collection, and storage methods; and the amount of biodegradable material in the manure.

Temperature affects the maximum specific growth rate of microorganisms (μ_m) in a digester. The effect of temperature on μ_m has been described by the following empirical relationship (3, 33):

$$\mu_m = 0.013T - 0.129 \tag{9.6}$$

where T is the digester temperature between 20 and 60°C. At temperatures above 60°C, the specific growth rate decreased. This relationship was obtained from results of the anaerobic digestion of sewage sludge, dairy cattle manure, and beef cattle manure.

The dimensionless parameter K was found to be represented by Eq. (9.7) when cattle manure was digested (33):

$$K = 0.8 + (0.0016)e^{0.06(S_0)} \tag{9.7}$$

This relationship was developed using data from several independent studies with influent VS concentrations of 40–100 kg of VS per cubic meter. There was no statistical difference in K when data from digesters operating at 35 and 55°C were analyzed.

A relationship having the form of

$$K = 0.6 + (0.0206)e^{0.051(S_0)} \tag{9.8}$$

was developed using data from several swine manure digestion studies (34).

As indicated earlier in the chapter, factors such as un-ionized ammonia and volatile acids can inhibit an anaerobic digester. In addition, chemicals that are added to animal feed to enhance feed efficiency, such as monensin, which is added to ruminant diets, or antibiotics, such as chlortetracycline, also may have an inhibitory effect. Monensin in beef cattle manure was found to delay the start of active fermentation of such manure (14) and in some cases reduced active methane production. Chlortetracycline had only a minimal effect.

The effect of mixing on methane production from beef cattle waste has been evaluated (19). The digesters were mixed continuously or intermittently every 1,

2, or 3 hr. The methane production rates from the continuously mixed digesters were only 8–11% greater than the rates from the intermittently mixed digesters. The results indicated that random mixing at long intervals was adequate and that there was little potential for increasing the fermentation rates of manures by increased mixing. These results also were confirmed in pilot-scale studies.

Swine manure has been digested successfully at suboptimal temperatures (22–25°C) (35). Digester loading rates ranging from 0.61 to 1.80 gm of VS per liter of digester volume per day were used without any indications of process failure. At the low temperatures, an SRT about twice as long as that needed at mesophilic temperatures was needed. Volatile solids reductions ranged from 22 to 41%, and the methane content of the biogas ranged from 63 to 68%.

Swine manure also has been digested at 35°C. At a loading rate of from 2.4 to 2.9 kg of VS per cubic meter of digester capacity, stable digestion occurred, gas production was about 1.0 m^3 per kilogram of VS destroyed, and the methane content of the biogas ranged from 60 to 62% (36).

A fixed-film, anaerobic unit treated pig waste at 35°C (37). A mixture of urine and feces was screened prior to use in the digester. The material added to the unit had an average suspended solids content of 18 gm per liter, a total COD of 14.6 gm per liter, and an ammonia nitrogen concentration of 2700 mg per liter. Based on an analysis of the microbial film on the media, about 1.1–1.4 gm of COD was removed per gram of microbial film per day. Liquefaction of the organic solids appeared to be the rate-limiting step in the process. The anaerobic unit had excellent stability and was able to adjust to drastic overloading or intermittent feeding.

Large-Scale Systems.   A number of large-scale systems have been built and operated in the United States. Descriptions and, where available, performance of some of these systems are presented in the following paragraphs. The operational and performance characteristics of several large-scale digesters are summarized in Table 9.13.

A 25-ton-per-day digestion facility has been constructed for beef cattle feedlot manure at Bartow, Florida (42). The digestion units were two epoxy-coated steel reactors, each equipped with a 3.7-kW mixer and insulated with urethane foam. Desired temperatures were maintained by steam injection. The solids in the effluent were separated in a solid bowl centrifuge and used as a cattle feed supplement.

In Pennsylvania, a plug flow digester designed for a 700-head cow dairy has been constructed and operated (38). The digester was made of concrete with a hypalon cover. Warm-water pipes ran the length of the digester. The biogas was burned in a diesel engine generator set to produce electricity. Waste heat from the engine–generator was used to maintain the temperature of the digester. Scum problems, possibly due to the wood chips used for bedding at the dairy, have been an intermittent problem.

TABLE 9.13

Operational and Performance Characteristics of Large-Scale Anaerobic Digesters

| Manure | Operating temperature (°C) | Biogas generation | | Methane content of biogas (%) | VS destruction (%) | VS loading rate (kg/m³/day)[a] | Reference |
		m³/kg VS added	m³/kg VS destroyed				
Dairy cattle	38	0.37	0.83	60	42	4.0	38
Beef cattle	35	—	1.75[b]	66	39	3.1	39
Poultry	35	0.44	—	62	—	1.95	40
Dairy cattle	35	0.23–0.36	0.77–0.99	55–60	30–40	—	8
Dairy cattle	35	0.20[c]	—	58	—	—	41

[a]One pound of VS per cubic foot of digester capacity = 15.7 kg VS/m³.
[b]Estimated results.
[c]Methane production.

A farm-scale digester to handle the manure from a 1000-head beef feedlot was constructed in Illinois (39). Raw manure was pumped from a holding pit to a 150,000-gal insulated, aboveground, bolted steel panel digester. The digester was stirred periodically. General maintenance procedures required about 4 hr per month. The biogas was scrubbed of hydrogen sulfide, compressed, stored, and burned to replace propane for space heating, crop drying, hot water, and other uses.

A 97-m^3 digester has been used for the manure from a 15,000-bird flock of poultry (layers) (40). The digester was constructed with monolithic concrete sidewalls and a floating steel silo cover coated with fiberglass and insulated with polyurethane foam. The digester was mixed periodically, especially when manure was added and before the digester contents were removed. The biogas was used for maintaining the digester temperature and for heating the building. The hydrogen sulfide content of the gas ranged from 2.0 to 8.7 mg of sulfide per liter of gas. The higher concentrations occurred when copper sulfate was included in the feed for the birds.

A full-scale plug flow reactor with a flexible polyurethane liner and cover was constructed and operated for use with dairy cattle manure (8). A closed-loop, hot-water recirculation system maintained digester temperatures. A full-scale, completely mixed unit was compared to the plug flow unit. The results from both units were similar.

Eleven biogas systems that utilize agricultural residues were in operation or under construction in California in 1982 (41). These include plug flow digesters at four dairies, a lined lagoon biogas system, and a pilot-scale dry fermentation system at one dairy. The plug flow digesters had a length-to-width ratio of about 4:1, were in or below the ground, had flexible or rigid liners, and did not include mixing. The digesters were fed a mixture of manure and milking parlor wash water to produce a pumpable slurry (greater than 12% solids) which did not separate into liquid and solids fractions in the digesters.

Construction of manure biogas units has occurred in many countries. As an example, about 105 units are reported to be operating on farms in Switzerland, with 36 built in 1981 (43). With a few exceptions, the units are continuous-flow reactors operating at mesophilic temperatures. In addition, a vast number have been reported to be operating on communes in the People's Republic of China and at small agricultural operations in India, Taiwan, and other countries in Asia.

Summary.   The available data make it very clear that anaerobic digestion of manures is a technically feasible process and can be an economically feasible one. The economic feasibility depends upon the biodegradability of the manure, the quantity of biogas or methane produced, and the substitution of the biogas for other fuels. Improved anaerobic processes, such as the plug flow and random mix processes, provide for low construction and operational costs and can be part of the other operations at a farm or feedlot. Whenever possible, fresh manures uncontaminated with dirt or other inert material should be used.

There now are adequate data from small- and large-scale manure digestion processes that can be used for the design and operation of digestion systems for farms or feedlots. It is not necessary or advisable to use data obtained from the digestion of sewage sludge for the design of anaerobic processes for manures. Animal manures have different characteristics, and the type and size of sewage sludge digesters are considerably different from those that should be used for manures.

Food Processing Wastes

Anaerobic processes also have been used with food processing residues. These processes have been used (a) as part of a treatment system in which the anaerobic process reduces the organic load to subsequent treatment processes or (b) to produce energy for use at the site. The following paragraphs illustrate the variety of anaerobic processes that have been used and the diversity of wastes that have been treated or stabilized by these processes.

The methane yield from food processing wastewaters has been estimated from laboratory digestibility studies (44). The results are summarized in Table 9.14. Many of the wastewaters were readily treated by anaerobic processes. Others, such as those from piggeries and from brewery, yeast, palm oil, and wool industries, required long detention times to achieve 50% biogasification.

TABLE 9.14

Methane Yield of Food and Fiber Processing Wastewaters[a]

Source of wastewater	Methane yield	
	m^3/kg VS destroyed[b]	m^3/kg COD removed
Flax retting	—	0.39
Enzyme production	—	0.38
Brewery	—	0.60
Yeast production	0.06–0.34	0.3–0.85
Palm oil production	0.86	0.50
Slaughterhouse	0.33–0.90	—
Wool processing	0.84	0.42
Margarine production	0.36	0.14
Piggery slurry	1.04	0.34

[a]Adapted from (44).
[b]m^3/kg = liter/gm.

TABLE 9.15

Anaerobic Treatment of Whey[a]

Anaerobic process	Temperature (°C)	Waste loading rate (kg/m^3/day)[b]	Removal (%)
Anaerobic filter	22–25	1.9 COD	97–98 (COD)
Conventional	35	2.1 BOD	99 (BOD)
Contact process	35	4.3 BOD	99 (BOD)
Fixed film	28–31	8.9–27 COD	7–93 (COD)
Expanded bed	35	8.2–22 COD	61–92 (COD)

[a]Adapted from (45).
[b]Kilogram of noted parameter per cubic meter of reactor volume per day.

Sulfate inhibited methane production at 20 gm per liter but not at 0.6 gm per liter.

Whey has been treated by anaerobic processes such as the conventional processes, the anaerobic contact process, the anaerobic filter, and the anaerobic expanded bed (45). Results that have been obtained are summarized in Table 9.15.

Tomato peeling and bean blanching wastes having total volatile solids concentrations from 5500 to 25,000 mg per liter have been treated by downflow, fixed-film reactors (46). About 0.4 m^3 of methane was produced per kilogram of VS degraded, and 60–75% VS removal was achieved.

Shellfish processing wastes also have been treated successfully by the anaerobic filter(47). Organic removals of 80% for COD and 88% for soluble BOD were obtained. Effluent solids concentrations did not exceed 20 mg per liter. The biogas produced contained about 80% methane. The anaerobic filter exhibited resiliency to periods of inoperation and to fluctuations in wastewater loadings.

Methane generation from brewery waste can be technically and economically feasible (48). Methane production ranged from 0.3 to 0.4 liters at standard temperature and pressure (STP) per gram of dry substrate added (0.3–0.4 m^3 per kilogram) and increased as the detention time increased. The methane content of the biogas was 60–65%.

Two-stage anaerobic treatment with solids return was used successfully with pear processing wastes over a loading rate range of 0.10 to 0.46 lb of VS per cubic foot per day (49). At HRTs of 0.5 and 1 day, treatment efficiency decreased because of difficulties in solids flocculation and separation. COD removal efficiencies of up to 95% were obtained.

Potato processing wastes have been treated in pilot-scale anaerobic filters. Average organic removals of 70% were obtained, and the average suspended solids removal was about 60%. The filter operated at a temperature of 25°C, and

TABLE 9.16
Anaerobic Stabilization of Agricultural Wastes by Anaerobic Filters[a]

Waste	Biogas production (m³/kg COD)[b]	Methane content of biogas (%)	COD removal (%)	Filter loading (kg COD/m³/day)
Pig slurry supernatant	0.34	80–85	80–90	19.6
Milk processing wastewater	0.36	82	82	4.9
Silage effluent	0.50	85	86	7.9

[a]Adapted from (51). [First published in *Symposium Papers Energy from Biomass and Wastes VI*, Institute of Gas Technology, Chicago, Illinois, June 1982, pp. 443–478.]
[b]Cubic meter of biogas (STP) per kilogram of COD added to the filter.

the entering waste had an average pH of 10.2. The gas had a methane content of 70% (50).

Anaerobic filters have stabilized many other agricultural wastes (51). Examples of the performance of such filters are summarized in Table 9.16. The filters were able to withstand shock loadings, without a significant decrease in digestion efficiency, and the units had tolerance to variations in pH, ammonia, and volatile acids.

Downflow stationary fixed-film reactors have been used with food processing wastes and have accomplished satisfactory methane production and COD reduction (52). Illustrative results are noted in Table 9.17. Such reactors were able to maintain high rates of methane production under adverse conditions such

TABLE 9.17
Methane Production and COD Reduction of Food Processing Wastes in a Stationary Fixed-Film Reactor[a]

Waste	Concentration of waste fed to reactor (gm COD/liter)	Methane production (m³/m³/day)[b]	COD removal (%)
Bean blanching	10	3.0	90
Pear peeling	110–140	1.2	58
Rum stillage	50–70	2.5	57
Sugar waste	10	3.0	91
Barley stillage	53	2.7	82
Whey	66	3.1	95

[a]Adapted from (52). [First published in *Symposium Papers Energy from Biomass and Wastes VI*, Institute of Gas Technology, Chicago, Illinois, June 1982, pp. 401–420.]
[b]Cubic meter of methane (STP) per cubic meter of reactor per day.

as low temperatures, hydraulic overloadings, sudden changes in waste composition, and lack of feeding.

The upflow anaerobic sludge blanket process has successfully treated sugar wastes and wastewater from the production of potato and corn products (53). In a pilot plant study, loadings exceeding 10 kg of COD per cubic meter of digester volume per day were maintained consistently. COD removals approached 90%, methane was about 70% of the biogas produced, and methane production was about 0.35 m³ per kilogram of COD removed.

Wastewaters from a vegetable cannery have been digested in a full-scale anaerobic unit in France (54). The digesters received about 12,000 kg of BOD per day, COD removals were about 90%, gas production was 0.4–0.45 m³ per kilogram of COD added, and the methane content of the gas was about 60%. The gas was used for steam generation at the cannery.

Other Residues

Anaerobic processes are applicable to any fermentable organic residues, as the following examples indicate.

The potential of woody biomass for the production of methane by biological gasification has been evaluated (55). Six woody species were evaluated for methane production in the presence and absence of added nutrients, and the anaerobic biodegradability of the species was determined. Methane production increased when adequate nutrients were present. Results that were obtained are summarized in Table 9.18. In each case the methane yields were below 1 stan-

TABLE 9.18

**Methane Production and COD Anaerobic Biodegradability
of Woody Biomass**[a]

Woody species	Total gas yield SCF/lb VS added[b]	Methane content (%)	VS reduction (%)
Black alder	0.51	72	3.2
Cottonwood	1.19	69	7.4
Eucalyptus	0.31	72	2.0
Hybrid poplar	0.99	67	6.1
Loblolly pine	0.27	77	1.6
Sycamore	1.35	63	8.5

[a]Adapted from (55). [First published in *Symposium Papers Energy from Biomass and Wastes VI,* Institute of Gas Technology, Chicago, Illinois, June 1982, pp. 341–367.]

[b]SCF/lb VS added = standard cubic feet of biogas generated per pound of VS added to the digester; 15.7 ft³/lb = 1.0 m³/kg.

dard cubic foot (SCF) per pound of VS added. The biodegradability of the woody substrate was the factor that limited the production of methane.

Methane also can be produced from seaweed (56). Washing the seaweed prior to digestion was not necessary, and nitrogen and phosphorus additions were not required. The methane production variations were related to the species of seaweed being digested. The highest methane production observed was 8.7 ft^3 of CH_4 per pound of VS degraded (0.55 m^3 per kilogram of VS). The biogas contained 64–72% methane.

Summary

The information provided in this chapter and that in the ever-increasing literature related to anaerobic treatment leaves no doubt that anaerobic processes can be an important component of the treatment, stabilization, and energy production systems used for agricultural residues. The technical success of these processes with a wide variety of organic wastes indicates that they should receive strong consideration.

When considering the role of anaerobic processes in the attainment of environmental or other goals, both the advantages and the disadvantages of such processes should be recalled so that the processes are used correctly. The advantages are discussed elsewhere in this chapter. The disadvantages include the fact that anaerobic processes rarely will produce an effluent acceptable for direct discharge to surface waters. In addition, a volume comparable to that of the material added to the process will remain, and the transport problem of the contents or effluent of the digester will be of the same order as that of the untreated waste. Also, if incomplete digestion occurs, an odor problem results.

Management is important to the success of anaerobic processes. With recognition of the relative advantages and disadvantages of such processes and with adequate management, anaerobic processes can be a very useful component of agricultural waste management systems.

References

1. Bryant, M. P., Microbial methane production—theoretical aspects. *J. Am. Sci.* **48,** 193–200 (1979).
2. Morris, G. R., Anaerobic fermentation of animal wastes: A kinetic and empirical design evaluation. M.S. thesis, Cornell University, Ithaca, New York, 1976.
3. Hashimoto, A. G., Chen, Y. R., and Varel, V. H., Theoretical aspects of methane production—state-of-the-art. *In* "Livestock Waste: A Renewable Resource," pp. 86–91. Am. Soc. Agric. Engrs., St. Joseph, Michigan, 1980.
4. Georgacakis, D., Sievers, D. M., and Iannotti, E. L., Buffer stability in manure digesters. *Agric. Wastes* **4,** 427–441 (1982).

5. Wood, J. L., and O'Callaghan, J. R., Mathematical Modeling of animal waste treatment. *J. Agric. Eng. Res.* **19**, 245–256 (1974).

6. Jewell, W. J., and McCarty, P. L., Anaerobic decomposition of algae. *Environ. Sci. Tech.* **5**, 1023–1031 (1971).

7. Miller, T. C., and Wolin, J. J., A serum bottle modification of the Hungate technique for cultivating obligate anaerobes,'' *Appl. Microbiol.* **27**, 985–987 (1974).

8. Jewell, W. J., Dell'Orto, S., Fanfoni, K. J., Hayes, T. D., Leuschner, A. P., and Sherman, D. F., "Anaerobic Fermentation of Agricultural Residue: Potential for Improvement and Implementation," Final Report, Project DE-AC02-76ET20051. U.S. Dept. of Energy, Washington, D.C., 1980.

9. Morris, G. R., Jewell, W. J., and Loehr, R. C., Anaerobic fermentation of animal wastes: Design and operational criteria. *Proc. Purdue Ind. Waste Conf.* **31**, 689–696 (1977).

10. Chandler, J. A., Predicting methane fermentation biodegradability. M.S. thesis, Cornell University, Ithaca, New York, 1980.

11. Jewell, W. J., Cummings, R. J., Dell'Orto, S., Fanfoni, K. J., Fast, S. J., Gottung, E. J., Jackson, D. A., and Kabrick, R. M., "Dry Fermentation of Agricultural Residues," Final Report, Project SERI/TR-9038–10. U.S. Dept. of Energy, Solar Energy Research Institute, Golden, Colorado, 1982.

12. Hill, D. T., Simplified Monod kinetics of methane fermentation of animal wastes. *Agric. Wastes* **5**, 1–16 (1983).

13. Martin, J. H., Loehr, R. C., and Pilbeam, T. E., Animal manures as feedstuffs: Nutrient characteristics. *Agric. Wastes* **6**, 131–166 (1983).

14. Hashimoto, A. G., Varel, V. H., and Chen, Y. R., Ultimate methane yield from beef cattle manure: Effect of temperature, ration constituents, antibiotics and manure age. *Agric. Wastes* **3**, 241–256 (1981).

15. Hashimoto, A. G., Methane production and effluent quality from fermentation of beef cattle manure and molasses. *Biotechnol. Bioeng. Symp.* **11**, 481–492 (1981).

16. Sievers, D. M., and Brune, D. E., Carbon/nitrogen ratio and anaerobic digestion of swine waste. *Trans. Am. Soc. Agric. Eng.* **21**, 537–541 (1978).

17. Maramba, F. D., "Biogas and Waste Recycling—The Philippine Experience." Maya Farms, Regal Printing Company, Manila, 1978.

18. Chen, Y. R., Varel, V. H., Hashimoto, A. G., The effect of temperature on methane fermentation kinetics of beef cattle manure. *Biotechnol. Bioeng. Symp.* **10**, 325–339 (1980).

19. Hashimoto, A. G., Effect of mixing duration and vacuum on methane production rates from beef cattle waste. *Biotechnol. Bioeng.* **24**, 9–23 (1982).

20. Hashimoto, A. G., Performance of a pilot scale, thermophilic, anaerobic fermenter treating cattle waste. *Resour. Recovery Conserv.* **8**, 3–17 (1982).

21. Converse, J. C., Zeikus, J. G., Graves, R. E., and Evans, G. W., "Dairy Manure Degradation under Mesophilic and Thermophilic Temperatures," Paper 75–4540, Annual Meeting, Am. Soc. Agric. Engrs., St. Joseph, Michigan, 1975.

22. Hashimoto, A. G., Thermophilic and mesophilic anaerobic fermentation of swine manure. *Agric. Wastes* **6**, 175–191 (1983).

23. National Research Council, "Food, Fuel and Fertilizer from Organic Wastes," Chapter 6. BOSTID, National Academy Press, Washington, D.C., 1981.

24. Schroepfer, G. J., Fullen, W. J., Johnson, A. S., Ziemke, N. R., and Anderson, J. J., The anaerobic contact process as applied to packinghouse wastes. *Sewage Ind. Wastes* **27**, 460–486 (1955).

25. Steffen, A. J., and Bedker, M., Operations of a full scale anaerobic contact treatment plant for meat packing wastes. *Proc. Purdue Ind. Waste Conf.* **16**, 423–437 (1962).

26. Switzenbaum, M. S., and Jewell, W. J. Anaerobic attached film expanded bed reactor treatment. *J. Water Pollut. Control Fed.* **52**, 1953–1965 (1980).

27. Schraa, G., and Jewell, W. J., High rate conversions of soluble organics with the thermophilic anaerobic attached film expanded bed. *J. Water Poll. Control Fed.* **56**, 226–232 (1984).
28. Jewell, W. J., Switzenbaum, M. S., and Morris, J. W., Municipal wastewater treatment with the anaerobic attached microbial film expanded bed process. *J. Water Pollut. Control Fed.* **53**, 482–490 (1981).
29. Jewell, W. J., and Morris, J. W., Influence of varying temperatures, flow rate and substrate concentration on the anaerobic attached film expanded bed process. *Proc. Purdue Ind. Waste Conf.* **36**, 621–632 (1982).
30. Kennedy, K. J., and van den Berg, L., Stability and performance of anaerobic fixed film reactors during hydraulic overloading at 10–35°C. *Water Res.* **16**, 1391–1398 (1982).
31. Tait, S. J., and Friedman, A. A., Anaerobic rotating biological contractor for carbonaceous wastewaters. *J. Water Pollut. Control Fed.* **52**, 2257–2264 (1980).
32. Lettinga, G., vanVelsen, L., DeZeeuw, W., and Habma, J. W., The application of anaerobic digestion to industrial pollution treatment. *In* "Anaerobic Digestion," (D. A. Stafford, B. I., Wheatley, and D. E. Hughes, eds.), pp. 167–185. Appl. Sci. Publ., London, 1980.
33. Hashimoto, A. G., Methane from cattle wastes: Effects of temperature, hydraulic retention time, and influent substrate concentration on the kinetic parameter (K). *Biotechnol. Bioeng.* **24**, 2039–2052 (1982).
34. Hashimoto, A. G., Methane from swine manure: Effect of temperature and influent substrate concentration on the kinetic parameter (K). *Agric. Wastes* **9**, 299–308 (1984).
35. Stevens, M. A., and Schulte, D. D. Low temperature anaerobic digestion of swine manure. *J. Environ. Eng. Div. Am. Soc. Civ. Eng.* **105**, EE1, 33–42 (1979).
36. Fischer, J. R., Sievers, D. M., and Fulhage, C. D., Anaerobic digestion of swine wastes. In "Energy, Agriculture and Waste Management" (W. J. Jewell, ed.), pp. 307–316. Ann Arbor Sci. Publ., Ann Arbor, Michigan, 1975.
37. Kennedy, K. J., and van den Berg, L., Anaerobic digestion of piggery waste using a stationary fixed film reactor. *Agric. Wastes* **4**, 151–158 (1982).
38. Martin, J. H., and Lichtenberger, P. L., Operation of a commercial farm scale plug flow manure digester plant. *In* "Energy from Biomass and Wastes—V," pp. 439–462. Institute of Gas Technology, Chicago, Illinois, 1981.
39. Schellenbach, S., Case study of a farmer-owned and -operated 1000-head feedlot anaerobic digester. "Energy from Biomass and Waste—VI," pp.545–562. Institute of Gas Technology, Chicago, Illinois, 1982.
40. Converse, J. C., Evans, G. W., Robinson, K. L., Gibbons, W., and Gibbons, M., Methane production from a large size on-farm digester for poultry manure. *In* "Livestock: A Renewable Resource," pp. 122–125. Amer. Soc. Agric. Engrs., St. Joseph, Michigan, 1980.
41. Chandler, J. A., "Development of Biogas Systems for Agriculture in California," California Energy Commission, Sacramento, California, 1982.
42. Lizdas, D. J. and Coe, W. B., Operation of a full scale, integrated manure anaerobic fermentation facility." *In* "Energy from Biomass and Wastes—V," pp. 359–378. Institute of Gas Technology, Chicago, Illinois, 1981.
43. Wellinger, A., and Kaufman, R., Psychrophilic methane generation from pig manure. *Process Biochem.* **11**, 26–30 (1982).
44. Pipyn, P., and Verstraete, W., Waste classification for digestibility in anaerobic systems. *In* "Anaerobic Digestion, (D. A. Stafford, B. I. Wheatley, and Hughes, D. E., eds.), pp. 151–166. Appl. Sci. Publ., London, 1980.
45. Switzenbaum, M. S., and Danskin, S. C., Anaerobic expanded bed treatment of whey. *Agric. Wastes* **4**, 411–426 (1982).
46. Stevens, T. G., and van den Berg, L., Anaerobic treatment of food processing wastes using a fixed film reactor. *Proc. Purdue Ind. Waste Conf.* **36**, 224–235 (1982).

47. Hudson, J. W., Pohland, F. G., and Pendergrass, R. P., Anaerobic packed column treatment of shellfish processing wastewaters. *Proc. Purdue Ind. Waste Conf.* **33,** 981–995 (1979).

48. Keenan, J. D., and Kormi, I., Anaerobic digestion of brewery by-products. *J. Water Pollut. Control Fed.* **53,** 66–67 (1981).

49. van den Berg, L., and Lentz, C. P., Anaerobic digestion of pear waste: Factors affecting performance. *Proc. Purdue Ind. Waste Conf.* **27,** 313–320 (1972).

50. Guttormsen, L., and Carlson, D. A., "Potato Processing Waste Treatment—Current Practice," Water Pollution Control Research Series, DAST-14. Federal Water Pollution Control Administration, U.S. Dept. of the Interior, Washington, D.C., 1969.

51. Colleran, E., The application of the anaerobic filter design to biogas production from solid and liquid agricultural wastes. *In* "Energy from Biomass and Wastes—VI," pp. 443–478. Institute of Gas Technology, Chicago, Illinois, 1982.

52. van den Berg, L., and Kennedy, K. J., Effect of substrate composition on methane production rates of downflow stationary fixed film reactors. *In* "Energy from Biomass and Wastes—VI," pp. 401–420. Institute of Gas Technology, Chicago, Illinois, 1982.

53. Sax, R. I., and Haltz, M. Pilot study of the biothane upflow anaerobic sludge blanket process for methane production. *In* "Energy from Biomass and Wastes—V," pp. 423–433. Institute of Gas Technology, Chicago, Illinois, 1981.

54. Chambarchac, B., Stabilization and methane production by commercial scale digestion of green vegetable cannery wastes. *In* "Energy from Biomass and Wastes—VI," pp. 483–504. Institute of Gas Technology, Chicago, Illinois, 1982.

55. Jerger, D. E., Conrad, J. R., Fannin, K. F., and Chynoweth, D. P., Biogasification of woody biomass. *In* "Energy from Biomass and Wastes—VI," pp. 341–367. Institute of Gas Technology, Chicago Illinois, 1982.

56. Yang, P. H., Methane production of Hawaiian seaweeds. *In* "Energy from Biomass and Waste—V," pp. 307–327. Institute of Gas Technology, Chicago, Illinois, 1981.

10

Utilization of Agricultural Residues

General

Agricultural residues are the excesses of production that have not been utilized. Although such residues contain material that can benefit man, their apparent value is less than the cost of collection, transportation, and processing for beneficial use. As a result, the residues are discharged as wastes, which can cause environmental problems and a loss of natural resources. If the residues can be utilized, such as to enhance food production, they are no longer nonproducts or wastes but become new resources. Residue utilization technologies rarely are a total waste management solution. Such technologies may not handle all of the available residues and can generate secondary residues, which also require careful management.

Only recently has residue utilization received broad emphasis as a component of waste management policy. Traditionally, emphasis has been on the treatment and disposal of wastes, with the subsequent loss of material and use of energy resources. This one-time use and disposal of material is a result of policies developed during earlier periods of abundant materials and energy, when there was less demand on world food production and energy resources and less concern about the quality of the environment.

The utilization of residues accomplishes important objectives: (a) better use of existing resources, (b) an increase in the base for food production, and (c) reduction of environmental problems caused by the accumulation and indiscriminate discharge of such residues.

TABLE 10.1

Basic Steps for Successful Residue Utilization

Clearly identify goal or need
Identify constraints and needed encouragement
Determine residue availability and characteristics and appropriate utilization technology
Identify capabilities of individuals who will operate the utilization process and use the resultant
    product
Determine market for resultant product
Identify relative economics

The factors pertinent to successful residue utilization approaches include a beneficial use, an adequate market, and an economical, although not necessarily profit-making, process. Many utilization processes would be satisfactory if they caused the overall cost of waste management to be less than that of other alternatives. Any additional steps in residue utilization should repay the extra storage, processing, and distribution costs that are incurred from that utilization. A return greater than the extra cost of utilization is desirable in that it reduces the total cost of waste management, but such reduction may not be sufficient to result in an overall profit for the producer.

Several basic steps (Table 10.1) must be taken when considering residue utilization possibilities. The first is to identify clearly the goal or need, such as development of energy resources, better pollution abatement, increased employment, better public health, reduction of deforestation, or better utilization of the natural resource base. The appropriate approach then can be developed to utilize the existing residues to meet that need or goal.

A second step is to delineate any impediments that exist or encouragements that may be needed to develop and implement residue utilization. These can include competing uses for the end product, land or residue ownership, availability of financing, societal structure, groups available to implement and utilize a technology, managerial needs, and availability of parts and repairs. Support services and market availability are important for successful implementation of residue utilization.

A third step is to identify and understand the existing state of knowledge concerning residue availability, residue characteristics, and technology for utilization. Although many potential technologies have been utilized throughout the world, it may be necessary to adapt the technologies to the local situation, personnel, equipment, and managerial talent and time. This may require demonstration projects to illustrate the application of a technology to meet a desired goal.

Knowledge of the characteristics and available quantities of agricultural and agro-industry residues is fundamental to successful residue utilization. The feasibility of residue utilization depends on the characteristics of the residue, the quantity that is realistically available, continuity of supply, value of the residue as a useful by-product or raw material for other products, and the cost of residue collection and transport. Information on the quantity and characteristics of agricultural residues, especially those characteristics needed for the evaluation of specific technologies, is needed. Table 10.2 identifies the characteristics that are relevant to certain utilization technologies. Generally the available information about residue characteristics relates to pollutional characteristics rather than those appropriate for utilization.

A fourth step is to recognize the capabilities of the individuals who will operate the utilization technology and use the resultant product. It must be remembered that farmers or food processors are businessmen. They are not microbiologists or chemical or sanitary engineers. Neither are those who are likely to be involved in using a specific technology. The technology to be used should be capable of being integrated with the facility producing the residue and with the intended use of any resultant by-product. Nonspecialists should be able to understand and operate the technology.

A fifth step is to understand the size and type of market that exists for the product of the utilization process and the economics that are involved. It makes no sense to take a residue that no one wants and turn it into a resource that no one wants. With residue utilization approaches, a cost-effectiveness analysis on a strict economic basis is not easy, since it is difficult to quantify the value of better public health, better pollution abatement, decreased unemployment, and other direct and indirect benefits of a residue utilization approach. Many residue utilization systems may be judged successful if they reduce the cost of current or anticipated waste treatment and disposal or if they result in an identified public goal, even if they do not make a profit in the strict economic sense.

The feasibility of residue utilization depends on the characteristics of the residue, the amount that realistically is available on a consistent basis, the cost of residue collection and transport, and the availability of a market for the resultant product. Detailed information about these items is crucial to the success of any residue utilization effort.

There is no single best approach to residue utilization. In each situation, the possible alternatives need careful evaluation, with the most appropriate technology or technologies chosen to achieve desired environmental, economic, and social objectives.

Many technologies can be considered for utilization of agricultural and agro-industry residues. These can be grouped according to the intended end use: energy, human food, animal feed, land application, and production of construction material, paper, and chemicals. Table 10.3 summarizes potential utilization technologies and the types of residue that have been used with the noted technologies.

TABLE 10.2

Relevant Residue Characteristics for Certain Utilization Technologies

Characteristic	Energy generation	Human and animal food	Fertilizers	Construction material and paper	Chemical production	Irrigation
Dry matter content	X	X	X	X	X	
Nutrient content						
N		X	X		X	X
P		X	X		X	X
K		X	X		X	
Cellulose content	X	X		X	X	
Btu of wet or dry solids	X					
Organic fraction		X			X	
Crude protein		X			X	
Digestible energy		X				
Presence of plant, animal, or human pathogens		X	X		X	X

TABLE 10.3

**Examples of Utilization Technologies That Have Been Used
with Agricultural Residues**

Technologies	Residues that have been used with the technologies
Energy generation	
Methane generation	Animal manures, food processing liquids and solids, crop residues, palm oil sludge, fish waste, meat slaughtering and processing residues
Pyrolysis	Animal manures, wood processing residue
Burning	Dried animal manures, wood, forestry residues, bagasse, nut husks, rice husk, cotton stalks
Human food	
Mushrooms	Rice straw compost, manure compost
Animal feed	
Direct refeeding, microbial or chemical processing	Corn stalk, cobs, and husk; coconut meal; cassava residues; molasses; treated straw; fish silage; edible oil refining residues; separated fruit and vegetable processing solids; ensiled manure; blood; feather meal; animal manures
Land application	
Nutrients and soil conditioners	Food processing wastes, distillery stillage, manure, rice straw ash
Irrigation	Agricultural processing wastewaters
Construction material and paper	Bagasse, rice straw, coconut stem
Chemicals	
Vinegar, glycerine, furfural, silica, organic acids, activated carbon, glue	Coconut shell, whey, rubber seed, ligno-cellulosic residues, molasses, cassava pulp, palm oil refining residues

The major agricultural residue utilization technologies are: (a) energy generation, (b) use for animal feed, and (c) recovery of nutrients, organic matter, and water by land application. The use of these and other possible approaches with agricultural residues are indicated in this chapter.

Animal Feed

Food Processing Residues

Many agro-industry residues, especially those derived from food processing, contain the same constituents as the original product. A number of such residues have been used directly as animal feeds. These include pineapple bran,

TABLE 10.4

Typical Chemical Composition of Some Crop Processing Wastes[a]

	Crude protein	Ether extract	Crude fiber	Ash
Sisal waste	9.0	2.4	23.8	12.7
Citrus pulp	7.0	3.5	13.0	7.1
Pineapple "bran"	3.5	0.5	16.2	5.2
Sugar beet pulp	9.9	0.7	20.3	3.4
Apple pomace	4.6	4.0	30.8	2.0
Tomato skin and seeds (dried)	24.8	22.0	27.6	6.6
Pea vines (silage)	14.8	5.9	25.4	17.4
Pecan dust (from kernel cleaning)	11.6	61.0	7.0	2.0

[a]Data expressed as percentage of dry matter. [Adapted from (1).]

bagasse, bananas, whey, coffee wastes, mustard wastes, citrus pulp, grape seed meal, and olive cake. Most have been fed to ruminants at levels of 5–20% of the total feed with no adverse effects. Characteristics of food processing residues are noted in Table 10.4.

Other examples include dehydrated tomato skins and seeds and corn husks, cobs, and trimmings. The solids resulting from the processing of fruits and vegetables such as peas, potatoes, beans, oranges, and peaches, as well as solids screened from food processing wastewater, have been dewatered and used as a cattle feed.

The production of potato and corn chips generates large amounts of solid material (centrifuged solids and filter cake) which can be used as an energy source for cattle. Wet potato and corn chip by-products can replace up to 30% of the dietary dry matter content in beef cattle finishing diets without decreasing animal performance (2). The economic feasibility of feeding such residues depends upon moisture content, transportation distance, and the cost and availability of alternative energy and nutrient sources.

Paunch manure is a residue that results from the slaughter of cattle and other animals. This residue is the content of the rumen and contains gastric juices, microorganisms, and partially digested food. Dried paunch manure is nearly odorless and can be used as an ingredient of animal feeds. It also is possible to use dehydrated paunch manure as a constituent of feed for the production of fish in ponds. About 10–12% paunch manure was used successfully in a feed for the production of channel catfish (3). Paunch manure and similar residues should be applicable as a feed constituent for carp and tilapia, which are important food fish in many countries.

Crop Residues

The cellulose and hemicellulose in the cell wall of plants are the most abundant carbohydrates in the world. Ruminants have digestive tracts that can ferment these carbohydrates by the action of microorganisms. However, not all such carbohydrates are digested, because they are physically and chemically bound to lignin and silica. Processing to rupture the plant cell walls can release the carbohydrates and improve the feed value of crop residues.

Chemical treatment will break the lignocellulose complex, making the cellulose more available for ruminant digestion. Many chemicals, such as potassium hydroxide, ammonium hydroxide, sodium hydroxide, and sodium formate, have been investigated. Alkali treatment has been utilized for many decades. Most of the early work was done in Germany and involved soaking straw in a dilute solution of sodium hydroxide for 4–24 hr (4). The apparent digestibility of the straw was increased from about 40 to 65%. However, the repeated washings that are necessary to remove the residual alkali require much fresh water and cause leaching of nutritionally valuable components.

A dry process using heat, alkali, and pressure can be used. With a simpler method, a 4–5% solution of sodium hydroxide is sprayed on straw and fed directly to animals after a 24-hr curing period. Ammonia also is effective in treating straw. These methods increase straw digestibility by about 10–15%. The digestibility of treated straw can be depressed if fed as part of high-concentrate diets. Thus, there does not appear to be any benefit in using alkali-treated straw in such diets (5).

Sodium hydroxide has been effective for the chemical treatment of crop residues such as barley straw, bagasse, and pea straw (6). Sodium chlorite plus ammonia was effective for sunflower hulls. Increases in digestibility in the range of 13–23% were obtained.

Animal Manures

The possibility of recycling animal manure directly as a feed ingredient for animals has received considerable attention. A common conclusion is that, if nutritional principles are followed, a portion of the animal manure can be used as an animal feed supplement. To be handled in this manner, manures and other residues are subject to feed control laws in the same manner as are other feed ingredients. In research evaluations, feeding processed animal manures has not altered the taste or quality of meat, milk, or eggs. Potential transfer of organisms can be controlled by dehydration, drying, composting, or ensiling prior to using the material as a part of animal diets.

The animal feeding operation should be located close to the source of the agricultural residues if direct refeeding is to be economically successful. It is unlikely that a residue can be transported considerable distances without adversely affecting the economic competitiveness of its use as a feedstuff.

Available information on the digestibilities and chemical characteristics of manures indicates that they may be considered best as a low-energy supplement of nitrogen and minerals, especially for manures from animals that exist primarily on forage and hay. Reported nutrient characteristics of poultry and cattle manure are summarized in Table 10.5.

The classification of animal manures as feedstuffs, based upon classic definitions and reported nutrient content of the manures, indicates that they should be considered the equivalent of silage or hay, when fed to ruminants (Fig. 10.1). Animal manures should not be classified as protein feeds, since the crude and digestible protein contents of dried poultry waste (DPW), broiler litter, and dairy cow and beef cattle manures are lower than that of typical protein feeds (soybean and cottonseed meal). The protein levels of beef cattle and dairy cow manures are comparable to protein levels in typical energy feeds (corn and sorghum grain) and silage and hays (corn silage and alfalfa, timothy and bermudagrass hay). The protein levels of DPW and broiler litter are higher than energy feeds and forages; however, DPW is lower in digestible protein than is broiler litter. Classifying animal manures as energy feeds also is inappropriate, because animal manures are lower in metabolizable energy than are energy feeds, protein feeds, and forages (7). When DPW is utilized as a feedstuff for laying hens, it is best considered as a source of minerals and amino acids.

Levels of manures in animal rations that have been reported as successful have ranged from 10 to no more than 25% of the ration dry matter content. Both DPW and broiler litter have been shown to have an economic value as a feedstuff that equaled or exceeded their value as a fertilizer. Generally, the appropriate level of incorporating these materials into ruminant rations was less than 20% (8, 9). The method of preparing and processing the broiler litter as a feed constituent (drying, composting, or ensiling) influences its value as a feedstuff.

Beef cattle and dairy cow manure, manure screenings, and anaerobically digested cattle manures appear to have little value as feedstuffs. The maximum level of incorporating these manures into animal feed rations, i.e., that level that did not adversely affect animal performance as compared to controls, was low, with many of the levels at or close to zero (10).

The conversion of animal manures to an animal feed can be done aerobically, using a waste stabilization process such as an oxidation ditch. In such aerobic units, microbial protein is synthesized as the wastes are stabilized. These microbially enhanced wastes can be fed to animals as part of their drinking water.

Significant increases in the amino acid fraction of the dry matter in aerobically stabilized manure have been found (11, 12). Both the growth of feeder

TABLE 10.5

Nutrient Composition of Manures[a]

Composition, % dry matter	Dried poultry waste		Broiler litter		Dairy cow manure		Beef cattle manure	
	Mean	Range	Mean	Range	Mean	Range	Mean	Range
Crude protein	28.0	17.0–40.4	26.8	14.4–40.0	15.3	12.0–21.9	16.5	12.2–27.0
True protein	14.6	11.3–21.8	15.8	13.6–18.0	12.5	—	—	—
Digestible protein (ruminant)	12.6	9.7–14.6	22.6	21.6–23.4	5.1	3.2–7.3	5.6	3.2–7.4
NPN × 6.25	9.7	7.8–11.6	7.6	4.8–15.1	—	—	—	—
Ether extract	2.2	1.4–3.2	2.4	0.8–3.5	3.0	2.5–3.8	2.8	1.6–7.4
Crude fiber	13.0	8.2–21.0	21.2	11.4–32.2	29.8	23.5–37.5	22.6	9.2–31.4
NFE	33.4	21.8–45.1	27.5	10.5–34.0	35.2	29.4–41.0	28.1	14.6–36.5
TDN	52.3	—	58.9	52.0–72.5	45.0	—	48.5	—
ADF	24.7	14.0–43.1	30.4	—	43.7	33.6–55.7	33.1	20.8–51.1
NDF	52.4	37.7–62.0	47.4	44.0–56.4	66.0	58.4–71.0	54.8	31.7–71.8
Lignin	1.4	—	9.7	9.4–10.4	14.4	7.2–27.1	8.6	5.0–15.0
Energy, kcal/kg								
Gross energy	3050	2200–3520	3650	3250–3860	3670	2500–4955	3940	2920–4870
Dry matter, %	84.7	78.7–89.7	80.6	72.7–89.1	15.5	10.9–20.7	21.1	15.7–25.2

[a]Adapted from (7).

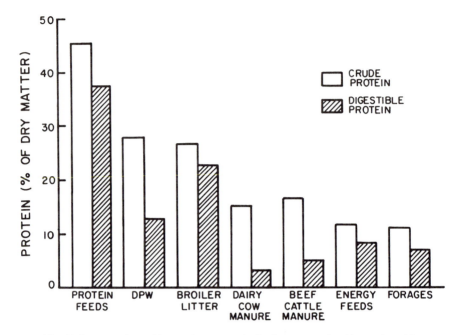

Fig. 10.1. Comparison of the protein content of animal manures and residues to that of forages, energy feeds, and protein feeds. [Adapted from (7).]

pigs and the production of eggs were enhanced when the aerobically treated wastes were fed to pigs and laying hens (12, 13).

Ensilage can be an effective, low-cost method of conserving the nutrients and organic matter in manures for use as a feedstuff. Ensiling can be conducted in simple or complex facilities and can be applied to a wide range of residues. Lactobacilli play a predominant role in ensiling since they rapidly acidify the organic matter to repress the growth of undesirable microorganisms. The lactic acid that is produced imparts a desirable odor and taste to the fermented residues. Simulation of lactic acid production can be achieved either by addition of selected organisms or by providing a suitable medium in which such organisms can develop. In terms of nutritional value of the feed produced, ensiling primarily affects the nitrogen-free substrates. The soluble carbohydrates are reduced, and the fatty acid content is increased.

Pathogenic bacteria are practically eliminated when the wastes are ensiled (4). Spore-forming bacteria do not proliferate. The ensiled manures look and smell better than the untreated manures.

An ensilage process known as "wastelage" has been used to process animal manures into animal feed. The wastelage system consists of blending wet manure, at about 70–80% moisture, with dry, standard fermentable feed ingredients, followed by ensiling the mixture for more than 10 days. Fermentation occurs in an airtight trench, bag, or silo, and the pH decreases to near 4.0. This acidic condition inhibits most biological activity. The resultant wastelage has been fed to cattle and other animals successfully (14, 15). A suitable ratio for blending cattle and swine wastes with dry feed is about 60 parts of animal waste to 40 parts of feed.

Hatchery Wastes

Within the chick hatchery industry, considerable residues, such as un-hatched eggs, dead embryos, culled chicks, and egg shells, result. Based on their nutritional characteristics, processed chick meal was identified as an animal protein concentrate equivalent to meat meal, and chick shell meal was identified as a mineral-based product equivalent to a phosphorus-deficient, steamed bone meal (16).

Fish Wastes

Fish wastage is a major problem throughout the world, with losses estimated to be about 4–5 million tons per year. Conversion of this material to animal feed by acid ensilage represents an important method of utilization.

Fish silage can be a simple and inexpensive means of preserving waste fish for animal feed. Fish silage is a liquid product made from acid and fish or fish parts. Inorganic or organic acids can be used, or the production of lactic acid can be encouraged by mixing a carbohydrate source with the fish and inoculating with lactic-acid-producing microorganisms. Optimal liquefaction can occur with 3–4% formic acid addition (17).

Compared to fish meal production, fish silage requires little capital investment and little sophisticated technology and is comparatively cheap to produce. The process can be used at both the individual and industrial level. The silage product may be used in a liquid or dried form. The process also can be applied to animal by-products such as animal and poultry offal.

Fish Production

The use of organic residues as feed for fish has occurred for centuries. Pond culture of fish has several advantages for residue utilization. Fish have a density approximately that of water and need less energy than land animals to move and support themselves. This increases the efficiency of converting residues to fish growth. Once a system is properly stocked, external requirements are a minimum of management and enough light for photosynthesis. Fish ponds can be productive with minimum labor and low capital input and can be placed on land unsuitable for agriculture.

A part of the fish food developed in a pond results from the nutrients in the residues. This food consists of small organisms, plants and animals. If fish are stocked at the correct density and species combination, the fish can harvest organisms efficiently, thus converting the nutrients to fish protein.

Many organic residues can be used as feed in aquaculture, such as oilseed cake, distillery residues, and slaughterhouse and food processing residues. Animal manures can serve as an indirect food by enhancing the production of natural aquatic food in fish ponds (18). The bacteria that decompose the manures serve as food for protozoa, zooplankton, and higher organisms, which are consumed by the fish.

Any animal manure consisting of fine particles is suitable for use in ponds. Fresh manures are dispersed and assimilated more easily than dry manure.

The rate and volume of the organic residue added to the fish ponds is important. Frequent application of small amounts of organic matter is better than periodic additions of large amounts. The latter approach can result in oxygen deficiencies and fish kills. Frequent additions permit a balanced population of bacteria, protozoa, and zooplankton. Suitable daily additions of manure to fish ponds in warm climates have been suggested as follows: cow manure, 3–4% of fish biomass; pig manure, 3–4%; chicken manure, 2–5%; and duck manure, 2% of fish biomass (4).

Carbohydrate-rich crop wastes also can be added to fish ponds. Materials such as rice bran and straw have been used. Nutrients must be added to supplement the carbohydrate material.

Intensively manured fish ponds have produced 15–30 kg of fish per hectare per day, with cow or chicken manure and nitrogen and phosphate as the only nutritional inputs to the pond. This compares to fish yields of 1–5 kg/ha/day, with no external fertilization, and 10–15 kg/ha/day, with only chemical fertilizers. A polyculture of common carp, tilapia, and silver carp was stocked in these ponds.

Polyculture is necessary to obtain high fish production levels, since each fish species feeds on different organisms. Use of polyculture also prohibits a buildup of unused food and metabolic products that can inhibit growth and require flushing or water recycling systems. Fish species used in polyculture include those that feed on: (a) aquatic vegetation and terrestrial residues, such as food processing residues, (b) zooplankton and other aquatic life, and (c) benthic material. The production achieved by polyculture is dependent upon climate; quality of the organic residues; aeration; temperature; management practices; and types, numbers, and sizes of the fishes that are stocked.

Toxicants such as insecticides or herbicides are poisonous to fish. Agricultural wastes that contain pesticide residues should not be used in fish ponds.

Algae Production

Research in several countries has shown that agricultural wastewaters can be excellent substrates for algae production. The algae are capable of photosynthesis to release oxygen, which helps support the bacteria and other microorganisms that decompose the organic matter in the wastewater (Chapter 6). This symbiotic action stabilizes the organic waste and converts the nutrients into algal proteins. Algae contain about 45–65% crude protein. In terms of protein production, algae can produce many times more kilograms of protein per hectare than can soybeans or corn.

The types of algae which can be grown depend upon local climatic conditions, wastewater characteristics, and operational characteristics. Those commonly encountered include *Chlorella, Oscillatoria,* and *Scenedesmus.* Harvesting of small, single-celled algae can be a constraint to large-scale algae production.

Postharvest processing of waste-grown algae can range from sun-drying to drum-drying. Sun-drying is a low-energy process, but the digestibility of sun-dried algae is lower than that of drum-dried algae. Algae utilization in fish, poultry, and pig feeding has been demonstrated in the Philippines, Singapore, and Israel.

Algae most suited for efficient large-scale algae production are the blue-green algae *Spirulina maxima* and *Spirulina* or *Arthospira platensis* (4, 19). These have been identified as having potential as an animal or human food since the cell protein quality is good and toxin free. These algae are large and prefer a highly alkaline medium (pH 10–11). The large size facilitates harvesting by screens or cloth. These algae have been utilized for food in Mexico and Chad for many years.

Energy Production

The processes for converting agricultural residues to energy depend upon the moisture content of the residues. In general, thermochemical processes are used with dry residues, and some form of fermentation is used with wet residues (Fig. 10.2). Many of the processes noted in Fig. 10.2 are well-established technologies in current use. Others, such as methanol production by gasification and pyrolysis and methane production from crop residues, are less well developed.

The feasibility of generating energy from agricultural residues depends upon factors such as (a) the characteristics of the residue and the quantity that is available; (b) the amount of recoverable energy in the residues and its relation to the total energy needed in the generation process; (c) compatibility of the form of

AGRICULTURAL RESIDUES	MAJOR PROCESS ALTERNATIVES	ENERGY PRODUCTS
DRY ●CROP RESIDUES ●WOOD AND WOOD PRODUCT RESIDUES	THERMOCHEMICAL ●COMBUSTION ●GASIFICATION ●PYROLYSIS	●STEAM, ELECTRICITY ●LOW VALUE GAS, METHANOL ●MEDIUM VALUE GAS, CHAR, LIQUIDS
WET ●MANURES ●FOOD PROCESSING WASTES ●HIGH MOISTURE CROP RESIDUES	FERMENTATION ●ANAEROBIC ●AEROBIC	●METHANE ●ALCOHOL

Fig. 10.2. Major energy production processes applicable to agricultural residues.

the energy to the uses for it; (d) the availability of equipment and skills needed to maintain the process; and (e) the cost of using the system, of the resultant energy, and of alternative energy sources. Typical energy values of agricultural products are noted in Table 10.6.

Thermochemical Processes

Thermochemical conversion technologies are identified by the amount of oxygen that is used and the form of energy that is produced. In direct burning processes, the oxygen that is supplied is in excess of that needed for complete oxidation of the material. Energy in the form of heat and light, as well as carbon

TABLE 10.6

Typical Energy Values of Common Fuels and Agricultural Residues (dry weight)

Energy source	Energy value (Btu/lb)[a]
Fuels	
Coal, bituminous	13,100
Gasoline	20,250
Methane	21,500
Organic materials	
Wood, pine	6,200
Charcoal	11,500
Cow manure	6,000–9,000
Leaves	7,000

[a] Btu/lb $\times$ 2.32 $\times$ 10^3 = joules per kilogram.

dioxide, water, and other combustion products, are the result of such processes.

When used directly as a fuel, the residues should be as dry as possible; otherwise, energy will be needed to remove the moisture. For example, the Btu values of wood range from 8000 to 9000 Btu per pound (18,600 to 20,900 kJ per kilogram) on a dry-weight basis. Freshly cut wood has a high moisture content, which reduces the heat value to about one-third that of dry wood. Other residues will have similar reductions in the heat value unless the moisture is decreased prior to combustion. An advantage of using agricultural residues as a fuel is the low sulfur content of the residues in comparison with most fossil fuels.

Bagasse is an example of an agricultural residue that is used as a fuel. At most sugar refineries, the sugarcane that results after the sugar has been extracted (bagasse) is used to fuel boilers at the refinery.

In gasification, the organic residues are allowed to burn in the presence of an insufficient supply of oxygen for complete combustion. The resultant product (producer gas) is a mixture of gases consisting primarily of carbon monoxide, hydrogen, methane, nitrogen, and other gases. When air is used to supply the combustion oxygen, the heating value of the gas is low. When agricultural residues are gasified, a gas with an energy content in the range of 5000–7500 kJ per cubic meter (130–200 Btu per cubic foot) can be produced (4). Natural gas has an energy content of about 37,500 kJ per cubic meter (1000 Btu per cubic foot). If pure oxygen is used, the energy content of the gas is higher since the nitrogen content is reduced substantially.

Agricultural residues should be dried to about 10–20% moisture content before being used as gasifier fuel. Many types of cellulosic agricultural residues, such as bagasse, straw, wood waste, walnut shells, and coconut residue have been used as gasifier fuel. The Philippines is using gasifier technology to power trucks and fishing boats and to reduce dependence on external fuels.

Pyrolysis involves heating the residues in the absence of air or oxygen. The resulting products include a mixture of gaseous fuels, partially oxygenated liquid fuels, and unreacted carbon (char). The relative yield depends on the rate of heating. Rapid heating promotes the production of liquids. The gaseous and liquid products are used directly as fuels or processed into other energy products. The char can be burned or gasified.

Pyrolysis has been investigated as a potential process for egg farm wastes (dead birds, litter, manure), cow manure, rice bran, and pine bark (4, 20, 21). Pyrolysis is a high-technology process requiring continuity of raw material, reasonable consistency in raw material characteristics, and adequate technical operation and maintenance.

Fermentation Processes

Fermentation processes (Fig. 10.2) can also be identified by the amount of oxygen used and the form of energy produced. Aerobic processes require the addition of oxygen; anaerobic processes do not (Chapter 5). The energy end product from the anaerobic digestion process is methane, while that from aerobic fermentation is alcohol, primarily ethyl alcohol.

A number of cellulose-, starch-, and sugar-containing agricultural residues can be converted to ethanol. These include molasses, cellulose pulp, potatoes and potato products, citrus residues, cassava residues, and food processing residues. Production of ethanol by fermentation involves (a) the availability of a suitable and continuing supply of fermentable material, (b) conversion of the raw material to a medium suitable for fermentation by yeast, (c) fermentation of the medium to alcohol, (d) recovery and purification of the ethanol, and (e) treatment, disposal, or utilization of the fermentation residue.

The availability of a suitable and continuing supply of fermentable material is necessary because there appears to be little value in operating an alcohol plant intermittently or at less than full capacity. If the raw material must be stored to assure continuity of alcohol production, it should be stored in such a manner as to avoid microbial decomposition.

Conversion of the raw material to a fermentable medium can be simple for sugar-containing residues such as molasses or cheese whey. However, starch-containing residues need to be cooked, and the starch must be hydrolyzed to sugars by acid or enzymes before fermentation. More resistant materials such as wood will require chemical treatment prior to fermentation.

Purification of the ethanol is done by distillation, which involves an energy input and cost. The alcohol production residues can be treated by anaerobic digestion to produce methane, which may be used in the production system. An energy and cost balance should be performed on any proposed ethanol production approach that uses residues to ascertain that (a) a positive energy balance results, i.e., the energy output exceeds that used in the process, and (b) the costs and energy associated with treating or utilizing the resultant fermentation by-products are considered.

The interest in nonfossil fuel energy has focused increasing attention on the production of methane from the anaerobic fermentation of organic residues. Table 10.3 indicates some of the agricultural and agro-industry residues that have been used or that have the potential for methane (biogas) generation.

Detailed information on the fundamentals of methane generation, the factors affecting the design and operation of an anaerobic digestion unit, and the use of the process with agricultural residues is presented in Chapter 9.

By using agricultural residues such as manure and food processing wastes for methane generation, the additional value of the methane can be gained while realizing other benefits. In many countries, locally collected firewood is used for heating. The collection of such wood has denuded forest and agricultural land, increased erosion, and decreased the agricultural base for crop production. The methane can reduce the need for firewood, and the digested material can be returned to the soil as a fertilizer and can improve the soil organic matter content.

The process of anaerobic digestion is a natural one which occurs whenever organic matter is microbially decomposed out of contact with air. During the anaerobic digestion of most organic wastes, gases containing 50–70% methane can be produced. Carbon dioxide (30–50%) and a small amount of other gases also are constituents of biogas. The quantity of gas produced will depend upon the biodegradability of the residues.

Key items in the successful generation of methane are (a) the biodegradability of the organic matter, (b) the ability to store the gas or use it when produced, (c) adequate raw material to meet production requirements, and (d) suitable operational control and maintenance. The biodegradable fraction of agricultural and agro-industry residues will vary, depending upon their source and how they were handled prior to digestion. For example, less than 40–50% of the organic solids in dairy manure may be biodegradable and available to produce methane. Inert material such as sand and dirt should not be included, and fresh wastes should be utilized.

Land Application

An important way to utilize liquid and solid agricultural residues is to apply them to the land and benefit from their use as fertilizers, as soil conditioners, and for irrigation. The advantages of returning organic residues to the land include (a) less dependence on increasingly costly chemical fertilizers, (b) improved soil structure, and (c) reduced need for the application of micronutrients. Additional information related to the utilization of agricultural residues by land application can be found in Chapter 11, which focuses on the use of land as a treatment process.

Fertilizers and Soil Conditioners

Before mineral fertilizers became available, animal manures, human wastes, and composts were the primary source of crop nutrients. While the availability of mineral fertilizers has resulted in a lesser interest in organic wastes as fertilizers, the principles of such use remain valid and can be employed to sustain or increase crop production.

Organic residues from agriculture contain most of the elements needed for plant growth and can supply nutrients through decomposition at a rate comparable to that needed by growing plants. In addition, application of crop residues and other organic residues to the soil helps retain moisture, reduce water and wind erosion, condition clay soils, increase the organic content of the soil, and add to soil fertility.

The feasibility of returning agricultural residues to the soil is influenced by factors such as (a) the economics of transporting the residues to the land and incorporating them in the soil, (b) the transmission of plant pathogens, and (c) the capacity of the land to assimilate the residues. The proper residue application rates must be determined by site-specific conditions such as precipitation patterns, natural soil fertility, crops to be grown, quantity of needed nutrients, and factors in the residues that may inhibit crop growth.

Crops which seem to be the most suitable for effective utilization of organic residues are those which have prolonged growth periods; crops which use large amounts of fertilizers; and fodder and grain crops, such as clover–grass mixtures, potatoes, and silage corn. Agricultural residues such as manures and food processing wastes are organic fertilizers and soil conditioners that can return nutrients and organic substances to the land, can increase soil fertility and productivity, and can improve the ecological balance.

Composting

Where residues cannot be returned immediately to the soil, composting offers another approach to recover and utilize the nutrients and soil conditioning factors in agricultural residues. The major objectives in composting are to stabilize putresible organic matter, to conserve as much of the crop nutrients and organic matter as possible, and to produce a uniform, relatively dry product suitable for use as a soil conditioner and garden supplement or for land disposal.

Composting is an aerobic, biological process in which organic residues are converted into humus by the numerous soil organisms. These include microorganisms such as bacteria, fungi, protozoa, and invertebrates such as nematodes, worms, and insects. Because composting is a microbiological process, the fundamentals of biological waste treatment noted in Chapter 5 can be applied. Important factors in the process include mixing, small particle size, oxygen for the microbial degradation of the wastes, time to accomplish the composting, and moisture. The composting can be done in open windrows or enclosed, environmentally controlled units. With the use of the controlled units, composting can be accomplished in 5–7 days, whereas in open windrows, it may take 3–8 weeks or more to produce satisfactory compost.

Composting consists of stabilization and maturation phases. During the stabilization stage, the temperature rises to a thermophilic level where the high temperature is maintained, followed by a gradual decrease in temperature to ambient levels. These conditions can be observed in batch or flow-through-type composting operations. As the temperature increases, multiplication of bacteria occurs, and the easily oxidized organic compounds are metabolized. Excess released energy results in a rapid rise in temperature. The temperature can be in the 40–60°C range. At these temperatures, the pathogenic organisms are reduced or destroyed. The ultimate rise in temperature is influenced by oxygen availability.

When the energy source is depleted, the temperature decreases gradually, and fungi and actinomycetes become active. At this stage the organic material has been stabilized but can be further matured. During maturation slow organic matter degradation continues until equilibrium conditions occur. The final product is a mixture of stable particles useful as a soil conditioner.

Small particle size is important to increase the rate of microbial decomposition. The more open-textured the mixture, the better the aeration that can be achieved. Adequate aeration provides enough air so that aerobic conditions will exist throughout most of the compost. Bulking agents such as bark can be added to the residues to promote diffusion of oxygen and movement of air through the composting material.

Water content plays an important role in composting. The proper water content of compost is about 50–60%. Compost should be damp, not wet. If insufficient water is present, the compost will dry, and decomposition will slow. With an excess of water, nutrients may be leached, and undesirable anaerobic conditions can occur.

The completion of the composting process can be noted by a marked drop in temperature and no significant increase in temperature when the mature compost is aerated. Volatile solids reductions of 40–50% can occur in a composting operation, depending upon the characteristics of the uncomposted solids.

Although ammonia nitrogen is lost during the composting process, other nutrients such as phosphorus, potassium, trace elements, and some of the organic nitrogen are conserved. Some are adsorbed on the fibrous matter in the residues and others are incorporated in the humic material formed by the organisms. When compost is applied to the soil these nutrients are released slowly and made available to plants over long periods.

The composting organisms require nitrogen for their growth. A suitable carbon-to-nitrogen (C/N) ratio for composting is about 20:1 to 30:1. If the ratio is less than 15:1, nitrogen can be lost from the process as ammonia. If the ratio is greater than 30:1, there may be insufficient nitrogen for the organisms and the composting rate will decrease.

There are many organisms that can utilize the organic matter in agricultural wastes, and these wastes have many million bacteria per gram. There is little need or value in adding enzymes or bacterial cultures to systems composting such wastes. The key to a successful composting operation is to have the environmental conditions satisfactory for the organisms that exist.

The composting of dairy, beef, swine, and poultry manure has been shown to be technically possible. Both windrow and mechanical composting of these wastes has produced suitable compost when the fundamentals of the process are followed. A number of studies have illustrated the value of adding dry, large particulate material such as bark, ground corncobs, wood shavings, straw, sawdust, and dried compost to reduce the moisture content, facilitate air movement, and reduce composting time. On-site composting of poultry manure within the poultry house has resulted in an odorless, fly-free environment. Solid meat-packing wastes also have been successfully composted (22) as have food processing waste sludges (23).

The solids from fruit processing operations such as discarded whole fruit and fruit halves and fragments have been satisfactorily composted. Some pits and a small fraction of leaves and stems were included. Separate peach and apricot wastes were mixed with municipal compost, rice hulls, and other material. Lime added to the mixture helped neutralize the fruit acids that were generated. This neutralization accelerated the process. Both bin, windrow, and forced-air composting were utilized (24).

Irrigation

Agricultural wastewaters can be a benefit when applied to land, either for crop irrigation or for groundwater recharge. Land application of wastewaters partially restores the nutrient cycle of crops, which is interrupted by the transport, treatment, and disposal of food and crop wastes to surface waters. Land application often is less expensive than treatment and disposal alternatives.

Irrigation of cropland with wastewaters results in water reclamation, utilizes the nutrients and organics in the wastewaters, and reduces the pollution load to streams. The microbial and plant action in the plant–soil filter will remove organics, most inorganics, and bacteria. Upon reaching the groundwater, the wastewater has lost its identity and has been renovated (Chapter 11). The nutrients and organics are incorporated in the crop and soil at the irrigated site. Milk processing, food processing, and milkhouse wastewaters have been used for the irrigation of pastureland and cropland used for the production of cotton, grains, forages, and trees.

Summary

The successful application of a specific residue utilization technology will depend upon the characteristics of the residue and the amount that realistically is available on a consistent basis.

Many residues can be used with several technologies. However, certain technologies can be used only if there are modifications to the residue or traditional technology. For instance, water may have to be added to dry manure to produce a suitable substrate for biogas production, or meat processing solids may have to be applied to the land at low rates to avoid both anaerobic conditions in the soil and odors.

The pressures for better resource conservation and pollution control are increasing, the need for better residue utilization exists, and many applicable technologies are available. The challenge is to change personal and public philosophies toward residue utilization and to provide the appropriate incentives to stimulate greater use of utilization technologies.

References

1. Parr, W. H., The use of crop processing waste for animal feed. *In* "Industry and Environment," Vol. 3, pp. 6–8. United Nations Environment Programme, Paris, 1980.
2. Crickenberger, R. G., and Miller, M. S., Wet potato and corn chipping by-products as feedstuffs in finishing diets for beef cattle. *Agric. Wastes* **6**, 221–234 (1983).
3. Summerfeldt, R. C., and S. C. Yin, "Paunch Manure as a Feed Supplement in Channel Catfish Farming," EPA–660/2–74–046. U.S. Environmental Protection Agency, Washington, D.C., 1974.
4. National Research Council, "Food, Fuel, Fertilizer from Organic Wastes." National Academy Press, Washington D.C., 1981.
5. Jackson, M. G., "Treating straw for animal feedings," FAO Animal Production and Health Paper 10. Food and Agriculture Organization of the United Nations, Rome, 1978.
6. Ibrahim, M. N. M., and Pearce, G. R., Effects of chemical pretreatments on the composition and in vitro digestibility of crop by-products. *Agric. Wastes* **5**, 135–156 (1983).
7. Martin, J. H., Loehr, R. C., and Pilbeam, T. E., Animal manures as feedstuffs—nutrient characteristics. *Agric. Wastes* **6**, 131–166 (1983).
8. Martin, J. H., Loehr, R. C. and Pilbeam, T. E., Animal manures as feedstuffs—broiler litter feeding trials. *Agric. Wastes* **7**, 13–38 (1983).
9. Martin, J. H., Loehr, R. C., and Pilbeam, T. E., Animal manures as feedstuffs—poultry manure feeding trials. *Agric. Wastes* **6**, 193–220 (1983).
10. Martin, J. H., Loehr, R. C., and Pilbeam, T. E., Animal manures as feedstuffs—cattle manure feeding trials. *Agric. Wastes* **7**, 81–110 (1983).
11. Martin, J. W., and Loehr, R. C., Enhancement of animal manures as feedstuffs by aerobic stabilization. *Agric. Wastes* **8**, 27–49 (1983).
12. Harmon, B. G., Recycling swine waste by aerobic fermentation. *World Anim. Rev.* **18**, 34–38 1976).
13. Martin, J. H., Performance of caged white leghorn hens fed aerobically stabilized poultry manure. *J. Poultry Sci.* **59**, 1178–1182 (1980).
14. Anthony, W. B., Cattle manure—Re-use through wastelage feeding. *In* "Animal Waste Management" pp. 105–113. Cornell University, Ithaca, New York, 1969.
15. Anthony, W. B., Cattle manure as feed for cattle. *In* "Livestock Waste Management and Pollution Abatement" pp. 293–296. Am. Soc. Agric. Engrs., St. Joseph, Michigan, 1971.
16. Tacon, A. G. S., Utilization of chick hatchery waste: The nutritional characteristics of day old chicks and egg shells. *Agric. Wastes* **4**, 335–343 (1982).
17. Disney, J. G., Parr, W. D., and Morgan, D. J., "Fish Silage: Preparation, Utilization and Prospects For Development." Tropical Products Institute, London, 1978.
18. Wohlfarth, G. W., and Schroeder, G. L., Use of manure in fish farming—A review. *Agric. Wastes*, **1**, 279–299 (1979).
19. Venkatarama, N., Devi. K. M., and Kunki, A. A. M., Utilization of rural wastes for algal biomass production with *Scenedesmus acutus* and *Spirulina platensis* in India. *Agric. Wastes*, **4**, 117–130 (1982).

20. White, R. K., and Taiganides, E. P., Pyrolysis of livestock wastes. *ASAE Publ. PROC-271 1971* 190–191 (1971).
21. Garner, W., Bricker, C. D., Ferguson, T. L., Wiegand, C. S. W., and McElroy, A. D., Pyrolysis as a method of disposal of cattle feedlot wastes. *In* "Proceedings of the Agricultural Waste Management Conference" pp. 101–124. Cornell University, Ithaca, New York, 1972.
22. Nell, J. H., and Krige, P. R., The disposal of solid abattoir waste by composting. *Water Res.* **5,** 1177–1189 (1971).
23. Hyde, M. A., and Consolazio, G. A., Composting food processing waste sludges. *Biocycle,* **23,** 58–60, 1982.
24. Rose, W. W., Chapman, J. E., Roseid, S., Katsuyama, A., Porter, V., and Mercer, W. A., Composting fruit and vegetable refuse. *Compost Sci.* **6,** 13–24 (1965).

11

Land Treatment and Stabilization of Wastes

Introduction

Of the three locations for the ultimate disposal of wastes (surface waters, atmosphere, and land) the land represents not only an appropriate treatment and stabilization medium for many wastes but also an opportunity to manage wastes with a minimum of adverse environmental effects. Application of manure, sewage sludge, municipal wastewater, and industrial wastes on land for both treatment and utilization has been practiced for centuries. The challenge is to utilize the chemical, physical, and biological properties of the soil to assimilate the wastes that are added. If the soil assimilative capacity is not exceeded, there will be minimum adverse effects to the crops that are grown, the characteristics of the soil, and the quality of the ground and surface waters. The land cannot function as a neglected waste sink. The soil and the wastes must be managed carefully as a total system.

Common waste treatment and stabilization approaches involve relocation of wastes from the point of generation to a suitable nondetrimental site. The soil can serve as a receptor of organic and inorganic residues if land treatment design is based on an understanding of the applicable scientific and engineering fundamentals and if the actual land treatment system is well designed and well managed.

The objectives of land treatment as a waste management process are (a) the treatment, stabilization, and ultimate disposal of wastes in a cost-effective, socially and environmentally acceptable manner and (b) the utilization of otherwise wasted resources while concurrently producing usable crops and minimizing disruption of the rural or agricultural community.

Land treatment is a managed treatment and ultimate stabilization process that involves the controlled application of a waste to a soil or soil–vegetative system. The process incorporates recycle and utilization of water, organics, and inorganics. The wastes are applied to the surface or mixed with the upper zone (0–1 ft) of soil. At a land treatment site, the biological degradation of organic waste constituents and the immobilization of inorganic waste constituents occurs. Controlled migration of inorganic constituents such as chloride or nitrate may be permitted if drinking water standards are met.

For many individuals, land treatment implies a spray irrigation system or a vehicle spreading sludge. Land treatment processes are considerably broader than this perception. Other terms that have been used to describe land treatment are "land farming," "sewage farm," "spray irrigation," "overland flow," "rapid infiltration," and "land application." Several of these terms also describe different approaches used to apply the wastes. Depending on the land treatment process used, different pollutant removal mechanisms may predominate.

Land treatment should not be confused with the indiscriminate dumping of waste on land or with landfills. The design goals, long-term impact, and degree of treatment of these other terrestrial systems are very different from those of land treatment.

Land treatment relies on the dynamic physical, chemical, and biological processes occurring in the soil. As a result, the applied wastes are degraded, transformed, and/or immobilized. Each soil–vegetative system will have a maximum capacity to treat, assimilate, and stabilize the applied waste. This capacity is related to the characteristics of the soil, the topography of the site, climate and other environmental conditions, and the crops to be grown. The maximum assimilative capacity of a soil therefore represents the maximum waste loading to a soil.

Many factors affecting the performance of soil systems for waste treatment and stabilization cannot be altered by the engineer or scientist. Design factors that are important include waste application or loading patterns, pretreatment of waste, drainage, vegetation, and site selection.

Land treatment can serve many useful purposes such as helping to restore the nutrient cycle that is interrupted by the disposal of wastes to streams (Chapter 2). For agricultural wastes as well as many municipal and industrial wastes, this method often is less expensive than other waste treatment and disposal options.

Land treatment can be used for untreated wastes as well as those partially treated by aeration, storage, and dewatering. There are also many opportunities to combine physical, chemical, and biological treatment methods with land treatment to obtain a least-cost method of waste management. The degree of treatment required prior to land treatment is less than that for disposal to surface waters, since the land will provide subsequent treatment and recycle. Prior treatment may range from only that required to minimize odors to that necessary to accomplish a specific reduction of BOD and nutrients.

Growth of a crop on the land treatment site can be important to the success of the process. A growing crop increases the rate of nutrient uptake and water evaporation and transpiration and decreases soil erosion. A crop will decrease the impact of water droplets and provide water storage space with the increased surface area of the crop. Plant growth results in a considerable uptake of nutrients and transpiration (loss) of water. The crop should be harvested periodically to promote crop growth and to obtain continued removal of the nutrients in the applied wastes.

Hydraulic and organic loading rates consistent with the ability of the land to accommodate the wastes are important. Intermittent application of wastes followed by a "resting" or nonapplication period is a common way to avoid overloading the soil. The upper soil must be permitted to become aerobic periodically. In addition, the soil should not become compacted by the equipment used to apply the wastes or to harvest the crop. Compaction impedes root growth, impairs crop productivity, lowers the percolation capacity of the soil, and results in more rapid soil saturation and water loss by runoff. Tillage can help correct compaction problems.

Information on the characteristics of soils is available through the Soil Survey of the United States, which describes, classifies, and maps the soils of the nation. These maps are a detailed and accurate soil inventory which can be applied to the selection of soils for land treatment. The responsibility for defining and naming different kinds of soils, for maintaining standards, and for the completion and publication of a soils inventory for the entire United States is vested by law in the Soil Conservation Service of the U.S. Department of Agriculture. The soil survey information describes texture, structure, porosity, organic matter, chemical composition, vegetation, and other important factors.

Although soil survey information has largely been used for agriculture and crop production, it can be applied for waste disposal and other environmental and sanitation developments. The soil survey enables predictions to be made of soil conditions for an area as small as 1 acre with an accuracy of 70–90%. A guide to the use of soil survey information for environmental applications such as land treatment has been developed (1).

Comparison to Other Treatment Processes

The processes and mechanisms that degrade, transform, and immobilize wastes in a land treatment site are the same as those that occur in other waste treatment systems. A major difference is that with land treatment, the processes occur in an unconfined reactor filled with soil, whereas with conventional systems the processes occur in separate tanks.

The design and operation of a land treatment site are based on sound scientific and engineering principles that are comparable to those commonly used for more conventional wastewater treatment. Both land and other treatment systems will achieve the desired performance if proper design and operation are provided. Conversely, poor performance will result for either system if suitable design and operational considerations are ignored.

For liquid wastes, a treatment system achieving secondary treatment or best practicable technologies (BPT) limitations will consist of the following processes: sedimentation, biological treatment, and sludge stabilization and disposal. If needed, tertiary treatment systems can include nutrient removal, adsorption, ion exchange, filtration, and chemical precipitation processes. For slurries and more solid wastes, anaerobic and other stabilization processes as well as dewatering and final disposal processes may be utilized.

All such processes can occur at a land treatment site (Fig. 11.1). Such a site is analogous to an unconfined biological–physical–chemical reactor that contains (a) particles (soil) that filter the applied wastewater and transform (adsorb, exchange, or precipitate) many of the applied chemicals, (b) bacteria and macroorganisms (earthworms, etc.) that degrade and stabilize the applied organics, and (c) vegetation that utilizes nutrients and inorganics during growth.

Biological treatment occurs in the upper soil layer as the applied organics are stabilized by the microorganisms in the soil. The microbial mass in the soil is large and diverse and when acclimated is capable of stabilizing organics that may

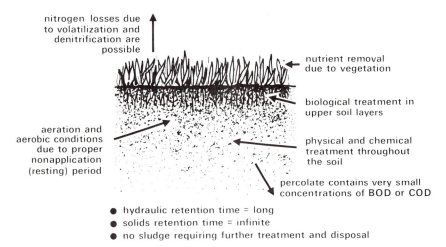

Fig. 11.1. Waste management relationships that occur with land treatment.

be in the applied wastes, especially those in agricultural wastes. One important difference is that with land treatment, there is no sludge that requires subsequent treatment and disposal. Any increase in biomass that occurs as the applied organics are stabilized remains in the soil and undergoes natural degradation until it is stabilized and becomes part of the soil humus.

Other differences of land treatment include the mechanism of oxygen transfer, particulate removal, and nutrient control. The predominant oxygen transfer in a land treatment site is gas diffusion and transfer of the oxygen across large surface areas of thin liquid films on the soil particles.

Particulates are removed at the surface or within the soil as the applied liquid or precipitation moves through the soil. Nitrogen removal can occur due to volatilization and denitrification in the soil and to vegetative uptake. Phosphorus removal results from chemical precipitation and adsorption reactions in the soil and vegetative uptake. Metals can precipitate and adsorb to soil particles and exchange sites in the soil and, to some extent, will be taken up by the vegetation. Generally, the runoff at a land treatment site is controlled, and there should be few waste constituents moving off the site with runoff.

Retention time is an important design and operational parameter for most treatment systems. In conventional wastewater treatment systems, hydraulic retention times are short—in the range of 6–24 hr—and solids retention times are about 10 days or greater, depending upon the treatment and stabilization processes used (Chapter 5). In the soil, the hydraulic retention times are much longer, many days to months, depending upon the amount of liquid that is applied.

When sludges and slurries are land treated, relatively little water is applied, and any percolation is due to precipitation. At some industrial land treatment sites there is no net leachate, as a result of (a) the small amounts of water added with the applied waste, (b) low precipitation, and (c) high evaporation.

The solids retention time at a land treatment site can be considered almost infinite since no particles should pass through the soil and few move overland with any surface flow. Because of the longer retention times, land treatment is capable of a higher degree of treatment than are most other waste treatment systems. This long retention time aids the removal of organics. Many organics not removable in conventional wastewater treatment processes are decomposed and removed in a land treatment system.

In the percolate from a properly designed and operated land treatment system the BOD is low, perhaps no more than 1–5 mg/liter, and there should be no suspended solids. In addition, because the applied phosphorus and metals will be removed in the soil, only low concentrations of these compounds should be in the percolate from a land treatment system.

Aerobic conditions must be maintained in biological treatment systems such as activated sludge or aerated lagoons, as well as land treatment systems, if the desired process performance is to be achieved and if odors are not to occur. In activated sludge or aerated lagoon systems, oxygen is added mechanically by surface or diffused aeration. In land treatment, aerobic conditions result from oxygen diffusion from the atmosphere and by maintaining an adequate nonapplication (rest) period after the wastes are applied. The rest period allows water to drain from the soil pores, oxygen to diffuse into the soil, and the organics to be stabilized.

Land treatment systems generally are less energy intensive than alternative waste treatment systems. For land treatment, energy is needed to transport the waste to the site and for application. In other systems, energy can be needed for transport, mixing, aeration, and handling of side streams (return sludge, effluent recirculation) and residuals (scum and digested sludge).

Basic Processes

Wastewater

There are three separate processes that are used for the land treatment of wastewaters: slow rate, rapid infiltration, and overland flow. The different processes accommodate different types of soils and reflect the rate of application as well as the flow rate of the wastewater. The site and design features of these processes are identified in Tables 11.1 and 11.2, respectively. The quality of the water that is expected to result from the processes (percolate for slow rate and rapid infiltration and runoff for overland flow), assuming that they are properly designed and operated, is noted in Table 11.3.

Slow Rate.   Slow-rate land treatment is the application of wastewater to a vegetated land surface, with the applied wastewater being treated as it flows through the soil. A portion of the applied wastewater percolates to the ground water, and some is used and transpired by the vegetation. Surface runoff of the applied water is minimized by proper design and operation. Surface application techniques include ridge-and-furrow and border strip flooding. Sprinkler application can be from fixed or moving irrigation equipment. Figure 11.2 illustrates a slow-rate system used to apply the effluent of an anaerobic lagoon treating meatpacking wastes to pasture. Figure 11.3 shows the fixed irrigation equipment used to apply vegetable processing wastewater from an aerated lagoon to a slow-rate site. If direct discharge effluent limitations for BOD, nitrogen, and phosphorus are stringent, a slow-rate land treatment system can be an economic and environmentally sound waste management alternative.

Many types of lands and crops have been used with slow-rate systems. These include (a) agricultural land on which corn, forages, and grasses are

TABLE 11.1

Site Features of the Processes Used for the Land Treatment of Wastewaters[a]

Feature	Slow rate	Rapid infiltration	Overland flow
Slope	Less than 20% on culti-vated land; less than 40% on noncultivated land	Not critical; excessive slopes require earthwork	2–8%
Soil permeability	Moderately slow to mod-erately rapid	Rapid (sands, loamy sands)	Low (clays, silts, and soils with impermeable barriers)
Depth to ground-water	2–3 ft (minimum)	3 ft during application cycle, 10 ft dur-ing nonapplication (lesser depths are acceptable where underdrainage is provided)	Not critical
Climatic re-strictions	Storage often needed for cold weather and when precipitation hinders application	None (possibly modifi-cation of opera-tion in cold weather)	Storage often needed for cold weather

[a] Adapted from (2).

grown; (b) forest land used for both wastewater treatment and the increased growth of forest biomass for energy and other uses; and (c) parks, golf courses, and other turfed areas.

Various factors may limit the wastewater application rate. Nitrogen often is a limiting constituent because the groundwater drinking water standards for nitrate nitrogen should not be exceeded. In arid regions, increases in chlorides and total dissolved salts in the groundwater may be the limiting factors. The hydraulic loading can be limited by the infiltration capacity of the soil. When the hydraulic capacity of the site is limited by a relatively impermeable subsurface layer or a high groundwater table, underdrains can be installed to increase the allowable loading. Grasses are usually chosen for the vegetation because of their high nitrogen uptake.

For slow-rate systems located above aquifers used as a potable water source, the nitrogen concentration in the percolate must be low enough that the groundwater quality at the project boundary can meet drinking water nitrate standards. Nitrogen removal mechanisms include crop uptake, nitrification–de-nitrification, ammonia volatilization, and storage in the soil. Nitrogen removal rates and the nitrogen application rate are important design parameters.

TABLE 11.2

Design Features for Processes Used for the Land Treatment of Wastewaters[a]

Feature	Slow rate	Rapid infiltration	Overland flow
Application techniques	Sprinkler or surface[b]	Usually surface	Sprinkler or surface
Annual application	2–20 ft	20–500 ft	10–70 ft
Typical weekly application	0.5–4 in.	4–120 in.	2.5–16 in.
Minimum preapplication treatment provided in United States	Primary sedimentation[c]	Primary sedimentation[c]	Screening and grit removal[c]
Disposition of applied wastewater	Evapotranspiration and percolation	Mainly percolation	Surface runoff and evapotranspiration with some percolation
Need for vegetation	Required	Optional	Required

[a]Adapted from (2).
[b]Includes ridge-and-furrow and border strip irrigation systems.
[c]Depends on the use of the effluent and the type of crop.

TABLE 11.3

Expected Quality of Treated Water Resulting from the Noted Land Treatment Processes[a]

Constituent (mg/liter)	Slow rate[b]		Rapid infiltration[c]		Overland flow[d]	
	Average	Upper range	Average	Upper range	Average	Upper range
BOD	<2	<5	2	<10	10	<15
Suspended solids	<1	<5	2	< 5	10	<20
Ammonia nitrogen as N	<0.5	<2	0.5	< 2	<4	< 8
Total nitrogen as N	3[e]	<8[e]	10	<20	5[f]	<10[f]
Total phosphorus as P	<0.1	<0.3	1	< 5	4	< 6

[a]Adapted from (2).
[b]Percolation of primary or secondary effluent through 1.5 m of unsaturated soil.
[c]Percolation of primary or secondary effluent through 4.5 m of unsaturated soil; phosphorus and fecal coliform removals increase with distance.
[d]Treating comminuted, screened wastewater using a slope length of 30–36 m.
[e]Concentration depends on loading rate and crop.
[f]Higher values expected when operating through a moderately cold winter or when using secondary effluent at high rates.

NOZZLE

DISTRIBUTION LINES
AND SPRAY NOZZLE

Fig. 11.2. Slow-rate distribution system used to apply lagoon effluent.

Fig. 11.3. Fixed irrigation distribution nozzle used to spray treated food processing wastewater on a slow-rate site.

Denitrification losses at slow-rate treatment sites have ranged from 3 to 70% (2) but typically are in the range of 15–25% of the applied nitrogen. The latter values can be used for design purposes in developing nitrogen application rates and are reasonably conservative. Ammonia volatilization losses of perhaps 10% of the applied nitrogen can occur if the soil pH is alkaline and the cation exchange capacity of the soil is low. Chapter 12 identifies the factors that affect ammonia losses. To be conservative, no credit should be taken for volatilization losses when developing nitrogen mass balances at a slow-rate site to identify desirable nitrogen application rates. The range of denitrification losses noted previously is broad and can be considered to include any impact of volatilization losses.

Crop uptake is the primary nitrogen removal mechanism in slow-rate systems. The amount of nitrogen removed by crop harvest depends on the nitrogen content of the crop and the crop yield. Typical annual nitrogen uptake rates for specific crops are noted in Table 11.4. Maximum nitrogen removal can be achieved by selecting crops or crop combinations with the highest nitrogen uptake potential.

TABLE 11.4

Nutrient Uptake Rates for Selected Crops[a] (kg/ha/year)[b]

	Nitrogen	Phosphorus	Potassium
Forage crops			
Alfalfa[c]	225–540	22–35	175–225
Bromegrass	130–225	40–55	245
Coastal Bermuda grass	400–675	35–45	225
Kentucky bluegrass	200–270	45	200
Quackgrass	235–280	30–45	273
Reed canary grass	335–450	40–45	315
Ryegrass	200–280	60–85	270–325
Sweet clover[c]	175	20	100
Tall fescue	150–325	30	300
Orchard grass	250–350	20–50	225–315
Field crops			
Barley	125	15	20
Corn	175–200	20–30	110
Cotton	75–110	15	40
Grain sorghum	135	15	70
Potatoes	230	20	245–325
Soybeans[c]	250	10–20	30–55
Wheat	160	15	20–45

[a] Adapted from (2).
[b] 1.0 pound/acre/year = 1.12 kg/ha/year.
[c] Legumes also will take nitrogen from the atmosphere.

Phosphorus is removed primarily by adsorption and precipitation reactions in the soil. Crop uptake can account for phosphorus removals in the range of 20–60 kg/ha/year (18–53 lb/acre/year), depending on the crop and yield (Table 11.4). The phosphorus sorption capacity of a soil depends on the amounts of clay, aluminum, iron, and calcium compounds present and the soil pH. Fine-textured mineral soils have higher phosphorus sorption capacities than coarse-textured acidic or organic soils.

Rapid Infiltration. Rapid infiltration systems are used with highly permeable soils. In these systems most of the applied wastewater percolates through the soil and drains to the groundwater or to surface waters. The purpose of a rapid infiltration system is to recharge the groundwater or to provide a treated water that can be recovered by wells or underdrains for reuse or discharge.

The wastewater is applied to permeable soils, such as sands and loamy sands, by spreading in basins or by sprinklers and is treated as it travels through the soil. Vegetation is not a normal part of the system. Emergent weeds and grasses cause few problems.

In properly designed and operated systems, removal of wastewater constituents occurs by the filtering and adsorptive reactions of the soil. Suspended solids, BOD, and microorganism removals are high, and the treated water quality is good (Table 11.3). Nitrogen removals may be no greater than 40%, unless operating procedures are used to enhance and maximize denitrification. With such procedures much greater nitrogen removals can be achieved. The factors that affect denitrification are noted in Chapter 12.

Phosphorus removals range from 50 to over 90%, depending on the residence time and the travel distance of the wastewater in the soil and on the climatic and operating conditions. As with slow-rate systems, the primary removal mechanism is adsorption with some chemical precipitation. Therefore, the long-term phosphorus removal capacity of a site is limited by the mass of soil in contact with the wastewater.

Overland Flow. The overland flow process is best suited to sites with relatively impermeable soils, although it also can be used with shallow permeable soils over impermeable subsoils. With overland flow, wastewater is applied over the upper reaches of sloped terraces and flows across the vegetated surface to runoff collection ditches. The wastewater is renovated by physical, chemical, and biological means as it flows in a thin film down the slope. Figure 11.4 shows the application of wastewater to the top of an overland flow plot.

Biological oxidation, sedimentation, and filtration are the primary removal mechanisms for organics and suspended solids in overland flow systems. Nitrogen removals result from a combination of plant uptake, denitrification, and volatilization of ammonia nitrogen. Nitrogen removal by the plants is only possible if the crop is harvested and removed from the field. Nitrogen removals can range from 70 to 90%. Less removal of nitrate and ammonium nitrogen occurs during cold weather due to reduced biological activity and plant uptake.

APPLICATION PIPE
AND NOZZLE

Fig. 11.4. Application of wastewater to the top of an overland flow site.

Phosphorus is removed by adsorption and precipitation. Treatment efficiencies generally are lower than those that occur with slow-rate systems because of the lesser contact between the wastewater and the adsorption sites within the soil. Phosphorus removals can range from 30 to 70% on a mass basis. Removals may be increased by adding alum or ferric chloride to the wastewater prior to application.

The design and performance of overland flow systems are related to the hydraulic retention time of the liquid on the slope. Kinetic relationships (3) are available to predict the removal of BOD, suspended solids (SS), ammonia, and total phosphorus in overland flow systems and to predict the hydraulic retention time of the wastewater as a function of application rate, slope, and length of the plot.

BOD removal in the first few meters of travel down the overland flow slope can be greater than the removal rate that occurs over the remaining slope. Modified first-order reactions have been shown to describe the removal of pollutants in the overland flow process (4). The removal of organic matter appears independent of slopes ranging from 2 to 6%, hydraulic loading rate, temperatures in the range of 18–28°C, and application period and appears dependent only on application rate and slope length (4).

Slurries and Dry Wastes

Animal manures and sewage sludges are examples of the types of slurries and/or dry wastes that can be applied to land. With this material the objectives of land application are stabilizing the solids; utilizing the moisture, nutrients, and microelements in wastes; and improving the soil structure, water holding capacity, and organic content. Because of the low water content of these wastes, they are either applied to the surface of the soil and subsequently plowed in or are injected directly into the soil.

Manures. Traditionally, dairy cattle manure is handled with bedding (straw or other material) and spread daily. Beef cattle wastes frequently are stockpiled or held in the feeding pens until land applied. When manure storage units are utilized the manures can be handled as a slurry. Swine and poultry manures are more liquid and are handled as a slurry. Figure 11.5 illustrates the injection of poultry manure into a cornfield. This process is used to control odors. Figure 11.6 shows the application of stockpiled cattle manure to land used for grape production in Mexico. In this case, the manure was added to increase the water-holding capacity of the soil to reduce the amount of water and energy used for irrigation.

Manure is added to agricultural fields because of its fertilizer value. Nitrogen normally is the constituent that determines the agronomic application rate. As with slow-rate wastewater treatment systems, crop uptake is the primary nitrogen removal mechanism. Agriculturally valuable crops are grown on the fields, and the land application of manure is considered a utilization rather than a disposal method.

Sludges. Several land application options for sludges are common: agricultural utilization, forest application, and land reclamation. Agricultural utilization occurs in almost every state. With this option, sludge is applied at an agronomic rate. This rate is the annual rate at which the nitrogen or phosphorus supplied by the sludge and available to the crop is equal to that needed by the crop. At such a rate, the potential of increasing the nitrate concentration of groundwater is minimized. Figure 11.7 shows the equipment that can be used to apply sludge to agricultural land.

Application of sludge to forestland is possible but is not as widespread as application to agricultural land. Sewage sludge can be a nutrient source for most tree species. Forest soils typically are low in available nitrogen and phosphorus, and many species respond dramatically to increased soil fertility from sludge

Fig. 11.5. Injection of poultry manure into a corn field.

Fig. 11.6. Application of dried cattle manure to land to increase the water-holding capacity of the soil and to supply nutrients.

Fig. 11.7. Equipment used for the injection of sewage sludge into cropland. The large tires minimize soil compaction.

Fig. 11.8. Poplar trees being grown for harvest and energy production on land to which sludge was applied to stimulate growth and provide nutrients.

additions. Heavy metal uptake in trees does not pose the same problems as those that occur with agricultural utilization. There is an emerging interest in the use of sludge as a nutrient and soil conditioning source for "energy plantations," in which fast-growing trees are grown for use as an energy source. Figure 11.8 illustrates a site used for the growth of poplar trees for potential energy production. Sludge was applied to the site to support the growth of the trees.

The use of sludge as part of a land reclamation process has occurred successfully in several states (5). The surface mining of coal, minerals, and sand has resulted in many million acres of drastically disturbed land in the United States. These lands can be reclaimed by grading the land surface to slopes that minimize erosion and by revegetation of the disturbed land. Sewage sludge has characteristics suitable for the reclamation of drastically disturbed lands. These include (a) the organic matter content, which improves the physical properties, increases the water-holding capacity, and increases the cation exchange capacity; (b) the plant nutrients; and (c) the buffer capacity of the sludge. The natural alkalinity of sludges will help neutralize the acidic conditions found in many mine spoils. This will increase the immobilization of metals and help the growth of vegetation. Metals are less soluble as the pH increases. Grasses, legumes, and trees have difficulty surviving if the pH is less than 4.5. In addition, the sludge brings an active biological community, which accelerates the recovery of soil microorganisms.

The primary purpose of the reclamation option is the reclamation of disturbed land; site management is therefore different from that of the other options. Sludge is applied only in amounts needed to reclaim the site and usually is applied only once while the site is being prepared for the seeding of grass or trees. The amount of sludge to be applied at any one time is greater for reclamation than for agricultural or forest applications because the danger of buildup of toxic materials and leaching to groundwater usually is not of concern. The sludge can be applied as a liquid or solid. In either case, the sludge should be incorporated in the soil to a depth equivalent to the plow layer. Surface applications are not as effective, since they result in poor root penetration and growth.

Vegetation

The vegetation on a land treatment site plays an important role in the use of land as an acceptor of wastes. The amount of water, organics, and minerals that can be utilized are basic considerations affecting the ability of a crop to exist when wastes are applied. Information necessary to assess the suitability of plants in a waste management system is noted in Table 11.5. Crops grown at land treatment sites should have a long growing season, a high nitrogen requirement, and, when used with wastewaters, a high water tolerance.

TABLE 11.5

**Factors Affecting the Suitability of Vegetation
at Waste Management Sites**

Water need, tolerance, and removal capability
Salinity tolerance
Nutrient need, tolerance, and uptake
Ability to withstand freezing and ice conditions
Growth and dormancy period
Resistance to diseases that may be increased by waste application
Integration with available crop management systems
Economic value

Agricultural crops most appropriate for slow-rate systems are forage and turf grasses. Forage crops that have been used successfully include reed canary grass, tall fescue, perennial ryegrass, Italian ryegrass, orchard grass, and Bermuda grass. If forage utilization and value are not a consideration, reed canary grass is often a first choice because of its high nitrogen uptake rate, winter hardiness, and persistence. Reed canary grass is slow to establish, and initially should be planted with a companion grass (ryegrass, orchard grass, or tall fescue) to provide good cover until the reed canary grass becomes established.

If high wastewater loading rates are used, orchard grass is generally more acceptable as animal feed than tall fescue or reed canary grass. Tall fescue generally is preferred as a feed over reed canary grass but is not winter-hardy.

Corn will grow satisfactorily where the water table depth is greater than 1.5–2 m. Alfalfa requires well-drained soils and water table depths of at least 3 m. The alfalfa cultivar grown should have resistance to root rot and bacterial wilt, especially if high hydraulic loading rates (>7.5 cm per week) are used.

Common agricultural crops grown for revenue at land treatment sites are corn for silage; alfalfa for silage, hay, or pasture; forage grasses for silage, hay, or pasture; grain sorghum; and grains. Many other crops also can be grown at land treatment sites if proper guidelines and management are used. Table 11.6 compares the potential of many crops for use at land treatment sites.

The nutrient removal capacity of a crop depends on the crop yield and nutrient content of the plant at harvest. Estimates of removals should be based on yields and nutrient compositions that local experience indicates can be achieved with good management. The rate of nitrogen uptake by crops changes during the growing season and is a function of the rate of dry matter accumulation and the nitrogen content of the plant. The pattern of nitrogen uptake is crop-specific.

The largest nutrient removals are achieved with perennial grasses and legumes that are cut frequently. Although legumes can fix nitrogen from the atmosphere, they are active scavengers for nitrate if it is present and therefore can be used at land treatment sites. The potential for nutrient removal with annual crops generally is less than with perennials because annuals use only part of the available growing season for growth and active uptake. Typical annual

TABLE 11.6

Comparison of Crop Characteristics Pertinent to Land Treatment Sites[a]

	Potential as revenue producer	Potential as water user	Potential as nitrogen user[b]	Moisture tolerance[c]
Field crops				
Barley	Marg	Mod	Marg	Low
Corn grain	Exc	Mod	Good	Mod
Corn silage	Exc	Mod	Exc	Mod
Grain sorghum	Good	Low	Marg	Mod
Oats	Marg	Mod	Poor	Low
Safflower	Exc	Mod	Exc	Mod
Soybeans	Good	Mod	Good–exc[d]	Mod
Wheat	Good	Mod	Good	Low
Forage crops				
Kentucky bluegrass	Good	High	Exc	Mod
Reed canary grass	Poor	High	Exc	High
Alfalfa	Exc	High	Good–exc[d]	Low
Bromegrass	Poor	High	Good	High
Clover	Exc	High	Good–exc[d]	Mod–high
Orchard grass	Good	High	Good–exc[d]	Mod
Sorghum-sudan	Good	High	Exc	Mod
Timothy	Marg	High	Good	High
Vetch	Marg	High	Exc	High
Tall fescue	Good	High	Good–exc	High
Turf crops				
Bentgrass	Exc	High	Exc	High
Bermuda grass	Good	High	Exc	High
Forest crops				
Hardwoods	Exc	High	Good–exc	High
Pine	Exc	High	Good	Mod–low
Douglas fir	Exc	High	Good	Mod

[a]Adapted from (2).
[b]Nitrogen user ratings (kg/ha): excellent (exc), >200; good, 150–200; marginal (marg), 100–150; and poor <100.
[c]Moisture tolerance ratings: high, withstands prolonged soil saturation; and low, withstands no soil saturation.
[d]Legumes also take nitrogen from the atmosphere.

uptake rates of nitrogen, phosphorus, and potassium have been presented in Table 11.4.

The amount of phosphorus in most wastes generally is higher than that needed for the growth of plants. The soils have a sorption capacity for phosphorus, and the excess rarely moves through the soil.

The potassium requirement of crops may not be met by the applied waste. In such cases, potassium fertilizers may have to be added to soils. Other macro- and micronutrients are likely to be supplied by the applied waste and the soil. With some food processing and industrial wastes, it is prudent to have the soil and waste analyzed to assure that adequate nutrients are available and that phytotoxicity due to an excess of certain micronutrients does not occur.

There is great diversity in crop tolerance, soil and climatic conditions, and characteristics of the waste to be applied. Advice of soil scientists, crop scientists, and soil and water engineers should be sought in determining the proper vegetation to use at land treatment sites.

The most common forest crops used at land treatment sites are mixed hardwoods and pines. Nutrient uptake and storage in trees depends on the species, tree density and age, length of growing season, and temperature. The understory vegetation also takes up and stores nutrients and is important in the early stages of tree establishment. During the initial stages of growth, seedlings and young trees are establishing a root system, and biomass production and nutrient uptake are relatively slow. To prevent leaching of nitrogen to groundwater during this period, nitrogen loading must be limited, or understory vegetation must be established to use and store the applied nitrogen that is in excess of the tree crop needs.

After the initial growth stage, the rates of growth and nutrient uptake increase and remain relatively constant until maturity is approached, at which time the rates decrease. When growth rates and nutrient uptake rates begin to decrease, the trees should be harvested. Maturity may be reached at 20–25 years for southern pines, 50–60 years for hardwoods, and 60–80 years for some of the western conifers. Harvesting also may be practiced in advance of maturity if short-term management is practiced.

In forestland treatment sites, the assimilation of phosphorus and metals is related more to soil properties than to plant uptake. The relatively low pH of most forest soils favors the retention of phosphorus but not metals. Addition of organic matter in forest soils improves the metal removal capacity.

Transformations in the Soil

The ability to develop economic and environmentally sound land treatment systems requires an understanding of the reactions and transformations that take place when wastes are applied to the soil. Soil is a composite medium containing inert rock, gravel and sand, reactive clay minerals, organic matter, living and dead vegetative and animal matter, plus a variety of micro- and macroorganisms. The treatment mechanisms in the soil include biological oxidation, ion exchange, chemical precipitation, adsorption, and assimilation into growing plants and animals.

The reactions and transformations that occur in a land treatment site are discussed in detail in other documents (2, 6–9). Only a limited description of the more important reactions is presented in this section.

The source of the waste (agricultural or municipal) or the form (liquid, semisolid or solid) is less important to the successful design and operation of a land treatment facility than is the nature of the waste constituents. The treatment mechanisms and rates of reaction of a constituent (e.g., metals, nitrogen, organics) are the same, regardless of source and type of waste.

Three assimilative pathways occur at a land treatment site: microbiological or chemical degradation of organic compounds; immobilization and chemical reaction of metals and cations, with little significant migration; and movement of soluble constituents, such as nitrate and chloride, with the soil water.

Degradation of organics results from the activities of indigenous microbes and other biological forms such as protozoa, mites, and earthworms. Volatile constituents and gases from biological and chemical reactions can be emitted to the atmosphere. Figure 11.9a illustrates the assimilation of organics in a land treatment system. Wastes generally are applied at discrete intervals. The slope of the line between waste applications for the concentration of an organic constituent is a measure of the biodegradability of that constituent in the soil. The decomposition rate (kilograms per hectare per year) of different organics varies substantially and ranges from very slow (half-life of months or years) to very rapid (half-life of minutes or hours). The rate of degradation depends upon the temperature and other climatic factors. The degradation normally is approximated by a first-order rate reaction.

Immobile inorganic constituents that are absorbed or precipitated but not decomposed (e.g., metals, cations) are retained and accumulated in the soil. With intermittent applications, a stepwise accumulation occurs (Fig. 11.9b). The assimilative or limiting capacity of such constituents is based upon an acceptable limit for accumulation. Examples of such limits might be cumulative amounts of soil cadmium or inhibitory concentrations of salts. When the appropriate limits are reached, application of the waste should cease.

Limits for mobile constituents are based on not exceeding groundwater quality standards. With repeated waste applications, the mobile constituents reach the groundwater or surface water quality limits in the manner noted in Fig. 11.9c. An example of a groundwater-based limit is the 10 mg/liter nitrate nitrogen standard set for drinking water.

(a) DEGRADABLE ORGANICS

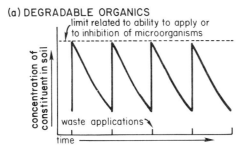

(b) IMMOBILE AND NONDEGRADABLE CONSTITUENTS

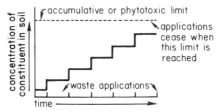

(c) MOBILE CONSTITUENTS

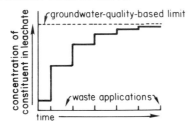

Fig. 11.9. Patterns of degradation, accumulation, and mobility that occur when wastes are added to the soil.

The diagrams in Fig. 11.9 represent separate assimilative pathways that require evaluation in the design phase of a land treatment system. Certain chemicals, such as a metal, may be evaluated with only one of these pathways. Other waste constituents may require an evaluation of all three pathways to establish the assimilative capacity. An example of such a constituent is nitrogen, which undergoes decomposition, accumulation as immobilized nitrogen, and leaching as nitrate.

A number of waste constituents are of specific interest in land treatment systems. These are oxygen-demanding material, nitrogen, phosphorus, other minerals, and microorganisms in the applied wastes. The following paragraphs describe the transformations and removals that are pertinent to the waste assimilative capacity of the soil.

Carbon

The fundamentals of carbon metabolism, as described in Chapter 5 for biological waste treatment systems, also occur in the soil. Organics in wastes are predominantly residues of plant and animal compounds and are degradable by soil microorganisms. Most of the aerobic bacterial activity occurs near the surface of the soil and decreases with depth. When wastes are added to the soil, the more easily degradable materials are metabolized rapidly, with cellulose, hemicellulose, and lignins persisting for longer periods of time.

The BOD removal efficiency of soil can be affected by the amount of vegetative cover and the infiltration capacity. Anything that adds surface area at the soil–air interface, such as litter or living plants, will increase the biological decomposition capacity of the site. The soil has a large capacity to remove organic matter from wastes.

From the environmental quality standpoint, the concern is the quantity of material that can have an adverse effect. The carbon added in a waste ultimately will exist as carbon in the protoplasm of soil microorganisms and plants, as slowly degradable soil humus, and as carbon dioxide released to the atmosphere. None of these carbon compounds has an adverse environmental effect. As in other waste treatment systems, carbon is not an element of environmental quality concern.

Oxygen

Oxygen is an important component of a soil system. If the soil is organically overloaded, oxygen can become limiting, and the biological system may become anaerobic. Anaerobic conditions will predominate if excessive flooding or waterlogged conditions prevail. Intermittent applications of wastes to soils and cultivation of the soil improve soil aeration and drainage and permit continued use of a plot of land for waste disposal.

Maintaining an unsaturated aerobic zone in the soil insures adequate filtering, biodegradation, and sorption of the waste constituents. As water moves downward in the unsaturated zone, contact occurs between the waste constituents and reactive sites on the soil's organic and inorganic particles. The aerobic unsaturated zone is more effective in pollutant removal than the saturated zone.

Nitrogen

Nitrogen is a key nutrient for protein synthesis and crop growth. Until the advent of adequate quantities of inorganic fertilizers, nitrogen management on a farm was a major factor limiting crop yields. Nitrogen undergoes transformations involving organic, inorganic, and gaseous compounds. The quantity of excess nitrogen not incorporated into plant and microbial growth or held in the soil is of concern at land treatment sites.

The nitrogen processes of interest are mineralization, immobilization, nitrification, and denitrification. The nitrogen transformations are discussed in greater detail in Chapter 12. Mineralization is the process in which nitrogen is converted to a form that is both mobile in the soil–water system and available to plants. Organic nitrogen is converted to ammonium nitrogen, which is oxidized (nitrification) to nitrite and nitrate nitrogen. Immobilization is the process in which the nitrogen is tied up as organic nitrogen, such as that in microbial cells. In the soil the processes of mineralization and immobilization occur simultaneously. If nitrogenous matter is in excess, nitrogen is mineralized. If carbon is in excess, nitrogen is converted into biological cell mass and is immobilized. Under anaerobic conditions, the oxidized nitrogen can be reduced (denitrification) to gases such as nitrous oxide and molecular nitrogen. The nitrogen cycle is summarized in Chapter 3 (Fig. 3.1).

Ammonium ions are positively charged and move slowly with the soil water because of the attractive forces between the ammonium ions and negatively charged clay and organic colloids. As long as the nitrogen stays in the ammonium form, the possibility of nitrogen loss by leaching is low. However, in normal soils, ammonium nitrogen is oxidized to nitrate, a negatively charged ion that moves freely with the soil water. Nitrate leaching can be of concern when large amounts of nitrate are present during periods before a crop is planted, when the crop does not utilize these nutrients rapidly, and when irrigation or rainfall exceeds soil water storage or crop moisture requirements. Under such conditions, excess water and soluble nitrates will move through the soil. The nitrification rate in soil is affected by the rate of nitrogen application, soil aeration, temperature, and soil moisture. Nitrification rates increase as temperature increases.

Nitrogen can be lost from the soil by leaching, denitrification, and volatilization. Once the nitrates are below the root zone, little opportunity exists for utilization by a crop or for denitrification. Denitrification can be one of the more important methods of nitrogen loss from soils. Most of the evidence for denitrification losses has originated from nitrogen balances under controlled conditions in which the amount of nitrogen added and that initially in the soil could not be accounted for by cropping, leaching, or remaining in the soil. Denitrification in the soil occurs where there is poor aeration and high oxygen demand. Volatilization of ammonia represents another loss of nitrogen from the soil. Volatilization occurs in alkaline pH ranges, with rapid rates occurring when the pH is above 9.5 and there is considerable air movement over the soil.

Soils contain many types of denitrifying bacteria that can use nitrate as hydrogen acceptor if free oxygen is not available. Nitrogen gas (N_2) is the primary end product of denitrification if the soil pH is above 6.0. Nitric or nitrous oxide may be produced in more acid soils. Denitrification occurs slowly at soil pH values less than 5.5 and at temperatures less than 10°C (50°F). For denitrification to occur in soils, nitrates or nitrites must be present (nitrification must have occurred), the nitrates must move through an area where oxygen is absent, and oxidizable organics must be in that area to serve as an energy source for the denitrifying bacteria. Vegetation enhances denitrification due to an increased carbon supply from dead root tissue and root excretions.

Although environmental quality interest is directed toward the ionic forms of nitrogen, ammonium, nitrite, and nitrate, these forms generally constitute 2% or less of the total soil nitrogen. The remainder is organic nitrogen.

The waste application rate to soil frequently is limited by the amount of nitrogen in the wastes and the relative rates of nitrogen degradation, loss, and utilization by the growing plants. The nitrogen application rates should be consistent with nitrogen needs and utilization rates so that the amount of excess nitrogen is minimized. Where heavy waste loads to the soil are contemplated because of shortage of land, reduction of the nitrogen content of the wastes prior to application may be necessary.

Migration of the nitrates is a function of the quantity of water that passes through the soil to the groundwaters. Under nonirrigated conditions, water usually moves through the soil to depths below the root zone only during the fall, winter, and early spring months. Leaching occurs during these periods because infiltration is greater than evapotranspiration and runoff. Application of wastes during these periods coincides with the increased opportunity for nitrate movement.

Planned denitrification in the soil offers opportunities to avoid excessive leaching of nitrates to groundwaters (10). Sequential nitrification and denitrification has been utilized in municipal waste treatment systems for nitrogen control but has yet to be applied to land waste disposal sites in a major way to control potential nitrate leaching to groundwaters. As the land becomes used more extensively for waste disposal, planned denitrification in the soil may provide an opportunity for relatively large nitrogen loadings per area of land, while minimizing adverse nitrogen releases to groundwater.

When wastewaters are applied to land, the nitrogen generally is readily available as degradable organic nitrogen, ammonia nitrogen, or nitrate nitrogen. However, the nitrogen in manures and sewage sludges consists primarily as slowly degradable organic nitrogen. When manure or sludge is applied to the same field every year, the availability of the nitrogen is an important factor in determining the appropriate manure or sludge application rates. The nitrogen mineralization process is rapid in the first year and decreases in subsequent years.

The nitrogen mineralization rates can be described by a series of decay constants. For example, a series of decay constants of 0.35, 0.15, 0.10, and 0.05 indicates that 35% of the nitrogen in the applied manure or sludge is available in the first year, and the amounts mineralized in years 2, 3, and 4 would be 15, 10, and 5% of the remaining nonmineralized nitrogen, respectively. After the first year the amount of the applied nitrogen in manure or sludge that is mineralized decreases by about 50% each year until the mineralization stabilizes at about 3% of the remaining organic nitrogen. This latter value (3%) is comparable to that observed for stable organic nitrogen fractions in soils.

The mineralization rate of a manure or sludge in the first year varies and is related to the degradability of the organic nitrogen and the degree of stabilization or mineralization that may have occurred before the material was applied to the soil. Table 11.7 identifies decay constants for different manures and sludges.

Fresh manures and less well stabilized sludges have more rapid initial mineralization rates. Poultry manures have a high initial decay rate because they contain uric acid and urea, which can be easily mineralized. Manures that accumulate on outdoor lots (beef cattle feedlot manure) have lower decay constants because much of the original organic nitrogen may have been mineralized and lost by leaching or volatilization. Stabilized sludges have lower decay constants than primary or undigested activated sludge, because the stabilized sludges have

TABLE 11.7

Decay Constants That Can Be Used to Estimate the Nitrogen Availability to Crops

| Item | Decay constant in year after application | | | |
	1	2	3	4
Manures[a]				
Poultry (hens)	0.90	0.10	0.05	0.05
Poultry (broilers, turkeys)	0.75	0.10	0.05	0.05
Swine	0.90	0.10	0.04	0.03
Dairy cattle (fresh)	0.50	0.15	0.05	0.05
Dairy cattle (stored)	0.30	0.10	0.07	0.05
Beef cattle (fresh)	0.75	0.15	0.10	0.05
Beef cattle (feedlots)	0.35	0.15	0.10	0.05
Beef cattle (stockpiled feedlot manure)	0.20	0.10	0.05	0.03
Sewage sludge[b]				
Primary and waste activated	0.40	0.20	0.10	0.05
Aerobically digested	0.25	0.12	0.06	0.03
Anaerobically digested	0.20	0.10	0.05	0.03
Composted	0.10	0.05	0.03	0.03

[a]Adapted from (11).
[b]Adapted from (12).

already undergone some degree of mineralization. The values noted in Table 11.7 are only indicative. Specific manures and sludges may have different decay constants.

Phosphorus

Phosphorus applied to a soil is converted to water-insoluble forms in a short time. When phosphorus in manures, wastewaters, sludge, or any material is added to a soil upon which crops are grown, a certain portion of the phosphorus is utilized by the crop, and the remainder is accumulated in the soil as organic compounds, adsorbed ions, or precipitated inorganic compounds. The phosphorus cycle is summarized in Chapter 3 (Fig. 3.2). In general less than 0.1% of the total phosphorus in soils is soluble in the water percolating through the soil. Phosphorus concentrations in groundwater and subsurface drainage systems are seldom over 0.2 mg per liter.

Phosphorus immobilization is related to the mineral constituents of the soil. Phosphorus fixation in acid soils is caused by the formation of insoluble iron and aluminum compounds. In alkaline soils the fixation is due to insoluble calcium compounds. Adsorption of the phosphorus on clay particles is caused primarily by the iron and aluminum associated with the clay. With such soils, phosphorus movement occurs primarily as the soil moves. Practices to control soil erosion help control the loss of both organic matter and the nutrients associated with the soil.

Phosphorus immobilization is less in sandy soils than in soils with a high clay content, because of the lower sorption and reactive capacities of sandy soil. The capacity of a soil to sorb phosphorus is not infinite. Each soil has a phosphorus sorptive capacity which can be exceeded by prolonged application of large amounts of phosphorus. The phosphorus sorptive capacity may not be exceeded for decades or centuries in soils having a high clay content.

Phosphorus removal by adsorption can be estimated by the use of Langmuir adsorption isotherms. Soil phosphate isotherms usually are curved rather than straight lines. This indicates that the basic assumptions of the Langmuir equation are not completely valid. Data from many soils result in isotherms that are reasonably linear over limited ranges of concentrations. Such isotherms can be used to estimate the phosphorus removals that are likely to occur.

When soils are devoid of oxygen, such as when the soil is saturated, changes in phosphate solubility may occur. Where the phosphate is immobilized as insoluble iron phosphate, anoxic conditions can result in the reduction of the ferric iron and the production of the more soluble ferrous phosphate. Where soils contain aluminum or calcium phosphates, lack of oxygen does not necessarily result in increased soluble phosphorus in the soil solution.

Metals

Wastewater and sludges can add metals as well as organic matter and nutrients to a soil. If in excess, certain metals such as cadmium, copper, nickel, selenium, and zinc have the greatest potential to contaminate the animal and human food chain. Because sludge accumulates metals during the waste treatment process, sludge has the greater potential to add excess metals to the soil. The concentrations of metals found in sludges is noted in Table 11.8. Fortunately, individuals or organizations desiring to apply sludges to land must obtain a permit from the appropriate state agency. The permit will identify (a) sludge quality suitable for land application, (b) levels of stabilization and pathogen reduction required prior to land application, (c) needed site-selection information, (d) suitable vegetation for the site, (e) desirable application processes, (f) appropriate application rates, and (g) monitoring and reporting requirements. The federal and state regulations upon which the permit is based have been developed to reduce the probability of animal or human health problems to very low levels.

The soil–plant barrier (13) can protect the plant–animal–human food chain from toxic conditions. Protection occurs when one or more of the following

TABLE 11.8

**Concentrations of Metals
in Digested Sewage Sludges[a]**

Metal	Concentration (mg/kg dry sludge)	
	Range	Median
Arsenic	1.1 to 230	10
Cadmium	1 to 3,410	10
Cobalt	11.3 to 2,490	30
Copper	84 to 17,000	800
Chromium	10 to 99,000	500
Iron	1,000 to 154,000	17,000
Lead	13 to 26,000	500
Mercury	0.6 to 56	6
Manganese	32 to 9,870	260
Molybdenum	0.1 to 214	4
Nickel	2 to 5,300	80
Selenium	1.7 to 17.2	5
Zinc	101 to 49,000	1,700

[a]Adapted from (13).

processes limit trace element transmission: (a) insolubility prevents plant uptake, (b) immobility of an element in roots prevents translocation to edible plant tissues, or (c) phytotoxicity occurs at concentrations in edible plant tissues below the concentrations that are injurious to animals. Cadmium, selenium, and molybdenum are metals which can escape the soil–plant barrier, while beryllium and cobalt are metals which may escape the barrier. The soil–plant barrier can be circumvented when animals directly ingest soil, sludge, or vegetation to which soil or sludge particles adhere.

Most microelements, with the exception of boron, are strongly adsorbed by soil particles, and their concentrations in soil solutions are very low. Adsorption is the predominant mechanism of trace metal removal by clay minerals, metal oxides, and organic matter. Soil pH is the main factor controlling soil adsorption–desorption processes and metal solubility. With the exception of molybdenum and selenium, all of the metals are more soluble at low pH values.

Sludge additions to neutral or slightly acid soil will tend to increase the soil pH. Nitrification, oxidation of organic matter, and oxidation of sulfides in anaerobic sludges increase soil acidity and decrease soil pH. Sludge additions tend to decrease the pH of calcareous soils. At agronomic rates of sludge application, the soil buffer capacity and soil pH will determine the pH of the soil–sludge mixture. The pH of the soil immediately adjacent to the roots of the plants is the pH that affects the metal uptake by plants since it is through the roots that the metals enter.

Where sludge is applied to agricultural land, the soil pH should be maintained at 6.5 or greater. For most soils, except acid soils strongly buffered by aluminum, a pH of 6.5 can be maintained by periodic addition of lime. On more acid soils, raising the pH to 6.5 may require considerable amounts of lime. Soil pH is buffered by many constituents in the soil and does not change rapidly after lime is added or sludge is applied.

Many states have limits on the total cumulative amounts of cadmium, copper, lead, nickel, and zinc that can be applied to cropland. Such limits will control the number of years that a site can be used for sludge application and the total amount of sludge that can be applied to the site.

The cation exchange capacity (CEC) of the soil is used to estimate the ability of a soil to minimize metal uptake by plants. The CEC of a soil is a measure of the net negative charge associated with both clay minerals and organic matter. The negative charge on soil organic matter and to some extent on clay minerals is pH dependent; an increase in pH increases the negative charge.

Cumulative metal recommendations were developed by joint efforts of researchers in various Agricultural Experiment Stations, the U.S. Department of Agriculture, and the U.S. Environmental Protection Agency and are noted as a function of CEC (Table 11.9). The CEC is used as a soil property that is measured easily and is related to soil components that minimize availability to

TABLE 11.9

**Maximum Amounts of Metals That Should Be Applied
to Agricultural Cropland (kg/ha)**[a]

Metal	Soil cation exchange capacity (meq/100 gm)[b]		
	<5	5–15	>15
Lead	500 (445)[c]	1000 (890)	2000 (1780)
Zinc	250 (222)	500 (445)	1000 (890)
Copper	125 (111)	250 (222)	500 (445)
Nickel	125 (111)	250 (222)	500 (445)
Cadmium	5 (4.4)	10 (8.9)	20 (17.8)

[a]Adapted from (12).
[b]Soil must be maintained at pH 6.5 or above.
[c]Values noted in parentheses are in pounds per acre.

the plant of metals added to soils. It is a general but imperfect indicator of soil components such as organic matter, clay, and metal oxides that limit the solubility of most metals. The CEC categories of <5, 5–15, and >15 meq per 100 gm (Table 11.9) correspond to sands, sandy loams and loams, and silt loams, respectively. If the soil pH is maintained at 6.5 or greater and if the application of sludge to a site ceases before the limits noted in Table 11.9 are exceeded, the soil should be usable in the future for the growth of any crop without adverse effects on yield, nor is it likely that there will be adverse food chain impacts from crops grown at the site.

Pot, greenhouse, and field studies have been used to evaluate metal uptake by crops. Significant differences in results occur with such studies (14), because the source of the added metals and the location of the experiment may affect the results.

The first difference is due to the "salt versus sludge" factor. When metals are added as soluble metal salts (sulfate, chloride, nitrate), they generally cause greater plant uptake and toxicity than when applied in sludge. Metals in sludges are in equilibrium with sludge organic matter and are not as mobile. Organic matter is added with the sludge and increases the soil CEC. The sludge also adds metal oxides, which can adsorb metals. Metal salts tend to lower the soil pH, which can increase the solubility of metals and their uptake by plants.

The second major difference is due to the "greenhouse versus field" factor. Greenhouse and growth chamber studies provide greater manageability and reproducibility than field studies. However, in greenhouse studies, the crop uptake of metals such as cadmium and zinc can be many times more than that occurring in field studies with the same soil, sludge, and crop (14). These differences result from a combination of the following factors: (a) ammonium nitrogen fertilizers which lower the pH of the soil in pots more than occurs in the field; (b) higher soluble salt levels in pots than in the field, because of a smaller soil volume; (c) confinement of plant roots to the small volume of treated soil in pots; (d) abnormal watering patterns and relative humidity in greenhouse pot studies; and (e) an inadequate supply of fertilizer nutrients to obtain maximum growth rates. The most appropriate data on which to base guidelines or actual sludge application rates result from field studies in which low metal application rates are used.

Available information indicates that when sludge is applied to land according to current federal, state, and provincial guidelines, it is unlikely that there will be deleterious uptake of metals by field crops. In one example (15), when sludge was applied according to the Province of Ontario Sludge Utilization Guidelines, there was no tendency toward high metal uptake nor toward the accumulation of PCBs by plants. It appeared that the guidelines allowed an appreciable margin of safety, which is appropriate, considering the significance of the plant–animal–human food chain.

Other Minerals

High concentrations of sodium in the applied wastes can cause substitution of sodium ions for other cations in the soil. Such wastes can include those from the lye peeling of fruits and vegetables as well as those with a high concentration of animal urine. This substitution can cause dispersion of the soil particles, can change the effective pore size, and can reduce the soil permeability. Permeability becomes low when the exchangeable sodium exceeds about 15% of the total cations in the soil exchange complex. The effect of sodium can be countered by adding a soluble calcium source to the liquid being used for irrigation. High sodium concentrations in soils can also retard seed germination and reduce crop yields.

The sodium adsorption ratio (SAR) can be used to estimate potential problems due to excess sodium in wastewater. The units of the cations in Eq. (11.1) are milliequivalents per liter.

$$SAR = \frac{Na^+}{[(Ca^{2+} + Mg^{2+})/2]^{1/2}} \qquad (11.1)$$

Wastewater with an SAR of 4 or less should result in no change to soil properties. No adverse soil impacts are expected unless the SAR exceeds about 8–10. Where high sodium concentrations are expected in agricultural wastewaters that are to be applied to land, the advice of competent soil scientists should be obtained.

High soluble salt concentrations can retard seed germination and can reduce crop yields. Salt injury to sensitive crops can occur if wastewater containing more than about 1000 per liter of dissolved salts is applied regularly during the summer months. Salt-sensitive crops include clover, orchard grass, alfalfa, and beans (2). Salt-tolerant crops include barley, cotton, Bermuda grass, and tall wheatgrass.

Microorganisms

The removal of microorganisms from wastes or wastewaters as they contact the soil is an important consideration in the use of land treatment. Bacteria and viruses are removed from wastewater as it percolates through the soil. Unless fissures exist in the soil, percolation at slow-rate sites will remove bacteria and viruses in a short distance. Most of the organisms in the applied waste are removed, generally greater than 90% removal, in the surface layer of soil. Several feet or more of soil are necessary for near-complete removal of the applied microorganisms.

Removal of bacteria, viruses, protozoa, and helminths (worms) in the applied wastes occurs by filtration, adsorption, desiccation, radiation, predation, competition for food, and exposure to other adverse conditions. Because of their relatively large size, protozoa and helminths are removed from the waste primarily by filtration at the soil surface. Bacteria are removed by both filtration and adsorption, while viruses are removed almost entirely by adsorption. Competition and predation are greatest in the surface soil layer since oxygen is more abundant, and rates of decomposition are greater. Human and animal pathogens are not adapted to the rigors of such an existence and are not expected to survive for lengthy periods. The persistence of microorganisms once they are removed by the soil depends on biological controls regulated by environmental factors such as temperature, organic matter content, and whether the soil is aerobic.

Because viruses are removed by adsorption, they can move with water moving through the soil if the soil is porous, such as at a rapid infiltration site, and where there is a high water table. The more porous the soil, the higher the wastewater application rate and the greater the possible virus migration. Appropriate preapplication techniques can reduce virus concentrations in the applied wastewater.

Sludges contain coliforms, bacterial pathogens, helminthic parasites such as *Ascaris,* and viruses. Only sludges treated by a process to significantly reduce pathogens should be applied to land growing human food chain crops. Such processes include anaerobic or aerobic digestion, air-drying, composting, and lime stabilization. When human food-chain crops are grown on land to which sludge is applied, there are additional controls (16) such as (a) site access must be controlled for 12 months after application, (b) no animals should be grazed on the site for 1 month after application if the animal product will be consumed by humans, and (c) crops consumed raw by humans should not be grown for 18 months after sludge application, unless there is no contact between the crop and sludge.

Although concerns have been expressed about the addition of human and animal pathogens when wastes are applied to soils, there have been no serious disease problems caused by the application of wastewaters or stabilized sludges on agricultural cropland when proper guidelines and permit conditions are followed. Soils that are suitable for waste disposal from the hydraulic, nutrient control, and adsorptive standpoints can provide satisfactory removal of pathogenic microorganisms. Microorganism removal is unlikely to limit the land treatment of wastes.

Land-Limiting Constituent Analysis

The development of a land treatment system includes waste characterization, site evaluation and testing, determination of the site assimilative capacity for the waste to be applied, and a determination of the controlling parameters and land requirements. The last determination involves an analysis of the land-limiting constituents (LLC) or limiting parameters.

The assimilative capacity of the soil is a quantitative measure of the potential of a given site to treat the waste constituents in an environmentally sound manner. The site assimilative capacity is related to the constituents of the waste and to the conditions at the site. The following site parameters are evaluated to determine the site assimilative capacity: (a) climatic conditions; (b) topography; (c) geology, groundwater conditions, and surface water locations; (d) soil properties such as pH, organic matter, cation exchange capacity, saturated permeability, and phosphorus adsorption capacity; and (e) vegetation use and potential.

Climatic conditions affect the water balance at the site, the growing season of the crop or vegetation, and the period of time the wastes can be applied. Some waste storage is needed for the times the waste cannot be applied because of adverse climatic conditions.

Proper determination of the site assimilative capacity for the wastes to be applied is the key component of the concept design. The assimilative capacity considers (a) the waste constituents of greatest potential environmental concern; (b) the oxidation, transformation, and immobilization reactions that will occur in the soil; and (c) the required management of the site. The assimilative capacity of the site is determined using (a) known physical, chemical, and biological reactions; (b) sound scientific and engineering principles; and (c) practical field experience.

Integration of the waste characterization and site-assimilative capacity information results in the land-limiting constituent (LLC) analysis. This concept was developed by several investigators (6, 7, 9) and was based on logic and field experience. The land-limiting constituent or limiting parameter concept is based upon the fact that soil has an assimilative capacity for inorganic and organic constituents. The constituent that results in the lowest application rate and causes the largest land area to be required is the limiting constituent. Use of this land area means that the other waste constituents are applied at a conservative rate and will not reach levels of environmental concern.

A variety of parameters can limit waste application rates. Examples include nitrate leached from the site to groundwater, cadmium in food chain crops, and salts that inhibit seed germination.

Figure 11.10 illustrates the use of the limiting constituent concept with a municipal sludge. In this example, nitrogen is the limiting constituent, thus requiring the largest land area. Nitrogen is the limiting constituent with most organic wastes such as manures and sludges. In this example, metals would be the next limiting constituent if the constraint due to nitrogen were relieved in some manner. The land-limiting constituent analysis is used to establish the required land area and the monitoring program. Monitoring during the operating life of the site will verify the reactions that were assumed and will determine if or when land application should cease.

There are two principal results from the LLC analysis. The first is a determination of how much land and which land areas are acceptable for the application of the waste under consideration. The second result is the identification of the critical constituents that limit the waste loading rate. The first result is important for land acquisition, subsequent site design, and the actual engineering of waste application rates. The second result is important because the land-

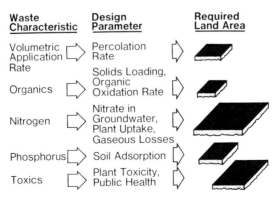

Fig. 11.10. An example of a land-limiting constituent analysis indicating that nitrogen is the limiting constituent for the waste that was analyzed.

limiting constituents determine the most appropriate parameters and locations (such as plants, soils, and groundwater) for monitoring.

Animal Manures

Animal manures are routinely and successfully applied to agricultural land to obtain the nutrient and soil-conditioning value of the material. The quantity that is applied should not cause environmental problems or nuisances. The following paragraphs provide examples that illustrate particular factors, situations, or concerns associated with the land application of manures.

The manure nutrient and salt content are factors that determine the land area required and the application rate of the manures. The characteristics of manure vary and are related to the type of feed and age and type of animal. The characteristics of the actual manures to be land-applied should be used to determine appropriate application rates. Estimates of the characteristics of manure are presented in Chapter 4.

Manure can be applied to land as a slurry or a semisolid and either injected (Fig. 11.5) or applied to the surface (Fig. 11.11). If the manure has been stored under anaerobic conditions, odorous volatile compounds will be released during spreading and as the manure lies on the land surface. These odors can be a source of complaints by adjacent land owners. Methods to control odors include injection directly in the soil, plowing shortly after the manure is surface-applied, avoiding spray distribution systems for liquids or slurries, and the use of aeration

Fig. 11.11. Land disposal of semisolid animal wastes.

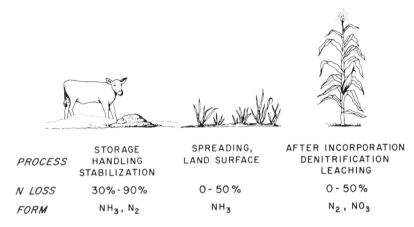

PROCESS	STORAGE HANDLING STABILIZATION	SPREADING, LAND SURFACE	AFTER INCORPORATION DENITRIFICATION LEACHING
N LOSS	30% - 90%	0 - 50 %	0 - 50 %
FORM	NH_3, N_2	NH_3	N_2, NO_3

Fig. 11.12. Estimated nitrogen losses from animal wastes.

systems to avoid the production of odors before the manure is land-applied.

Not all of the nitrogen excreted in manure is available for use by a crop. Numerous losses can take place between excretion and crop use. These include nitrogen losses (a) during collection, storage, and any treatment; (b) during transport and application on the land; (c) after application to, but before incorporation in, the soil; (d) by denitrification in the soil; and (e) by leaching (Fig. 11.12). The nitrogen losses during storage, handling, or stabilization are related to the type of manure handling system and equipment that may be used (Chapter 12, Table 12.5). It is impossible to conserve all of the nitrogen that is excreted. Ammonia losses by volatilization during storage or handling will occur. Denitrification losses can occur in aerated units and in storage after aerated units. Additional details concerning the fundamentals of nitrogen transformations and losses, as well as methods to minimize nitrogen losses from manures, are presented in Chapter 12. The decay rates of nitrogen in manures are presented in Table 11.7.

Where soil fertility has not been maintained, manure applications can result in sizable crop yield increases and should not have an adverse effect upon soil or water quality. For example, in southern Alberta, Canada, cattle manure applied annually at a rate of 70 metric tons per hectare did not cause an undesirable increase of nitrogen, phosphorus, or soluble salts in the soil (17). The plant nutrients removed from the soil through repeated cropping were replenished by the applied manure. In another study, corn forage yields were near maximum at annual beef cattle manure applications of about 100 metric tons per hectare (18).

Application of organic matter such as manures improves the tilth of the soil by increasing the soil organic matter and hydraulic conductivity and by decreasing the bulk density. Exchangeable cations (CEC) and soil nutrients also increase.

In areas with high rainfall and natural leaching, salinity in a soil is not a problem. However, in low-rainfall and irrigated areas salinity can be of concern, and the application of materials containing salts such as manures may be limited. Soil salinity is determined by measuring the electrical conductivity (EC) of a water-saturated soil paste extract. Soil saturated with 1 acre-ft of water containing 1,740 lb of salt will have an EC of 1.0 mmho per centimeter. A soil with an EC of 4 mmho per centimeter is considered saline.

When manure application rates are excessive, a decrease in crop yield may occur due to inhibitory amounts of ammonia or nitrite nitrogen or salts in the soil. Soils in the northern Great Plains of the United States frequently contain considerable quantities of salt. The addition of large quantities of manure to such soils can salinize the soil profile and adversely affect crop production. In a study in South Dakota (19), manure application rates in excess of 45 metric tons/ha/year resulted in reduced plant populations. After manure applications ceased, natural precipitation and percolation returned the soil to a nonsaline condition.

Undiluted feedlot runoff also can be a salinity hazard when land-applied or used for irrigation. The application of feedlot runoff to permeable and clay loams drastically reduced the permeability of these soils (20). This change in characteristics can be an advantage in sealing the land in a feedlot runoff lagoon to avoid groundwater infiltration, but it can be a detriment when the runoff is applied to cropland for irrigation or recovery of nutrients.

To minimize pollutant runoff from land to which manure is applied, manure should be applied to fields that (a) are well or moderately well drained; (b) are tilled shortly after the manure application; (c) are planted with crops such as corn, small grains, and grasses that will make use of the added nutrients; and (d) have a minimum slope. When manure is applied to land in the winter, snowmelt runoff from such lands may contain two to three times more nutrients and other contaminants than does runoff from rainfall in warmer times of the year.

Various forms of land treatment are effective in treating feedlot runoff. An overland flow system removed about 70% of the applied COD, 77% of the BOD, 78% of the total phosphorus, and 67% of the total nitrogen in feedlot runoff (21). Vegetative or grass filter strips also can be effective in the treatment of feedlot runoff. These filter strips are the equivalent of broad overland flow or graded terrace systems and combine infiltration, dilution, and filtration to produce an effluent suitable for discharge. With proper design, reductions of nutrients, solids, and oxygen-demanding material from feedlot runoff can be over 80% on a concentration basis and over 95% on a mass-balance basis (22). Comparable removals were obtained when farmyard runoff was treated by grass filters (23). Particulates were removed to a much greater extent than were soluble constituents.

Food Processing Wastes

Food processing wastes are well suited for land treatment because they consist primarily of easily degradable organics and because most food processing plants exist in rural areas, near the source of the crop being processed and where land is available. Hydraulic, nitrogen, and perhaps salt are the types of factors that commonly limit application rates. Land treatment systems for food processing wastes are widely used and are found throughout the United States and Canada. For instance, at any one time there may have been over 240 land treatment systems in Wisconsin (24) and 60 in the province of Ontario, Canada (25). The following paragraphs illustrate the wastes that have been land treated and the results that have been obtained. The results of these and other evaluations verify the transformations that occur in the soil and that are described in an early section of this chapter.

Fruit and vegetable canning wastewaters and meat-packing wastewaters are the food processing wastes most commonly land treated. However, poultry and milk processing wastes and whey also have been treated satisfactorily by land treatment processes.

Meat-Packing and Hatchery Wastewaters

At a meat-packing plant that processed about 10 million pounds of cattle and hogs per year liveweight, overland flow was used to treat the effluent from an anaerobic–aerobic lagoon (26). The overland flow plots consisted of clay loam over conglomerate rock, had 2–6% slopes, and were covered with native Bermuda grass. The average runoff was 50% of that applied. The percentage of removals, on a mass basis (pounds per acre per day) were significant and were 83, 70, 84 and 64% for BOD, COD, ammonia nitrogen, and total phosphorus, respectively.

Overland flow also has been used for the treatment of meat-packing wastewater that had been treated by a hydrosieve before storage and application (27). Native Bermuda grass was the cover crop, and the sites had a 6% slope. A summary of the loading rates and mass removals that occurred is presented in Table 11.10. The temperatures during the study ranged from 15 to 27°C. The loadings and the weather conditions differed widely, yet the discharges were similar, and high removals occurred. Phosphorus removals were less than the removals of other parameters, which is logical for an overland flow system. Hydraulic application rates ranged from 6500 to 16,600 gal/acre/day. The runoff ranged from 7 to 35% of the amount applied.

Both a slow-rate and an overland flow system were able to treat meat-packing wastewater in the Southwest (28). With a raw wastewater that had a COD of 9,400 mg per liter, a COD removal of 99% occurred at an application

TABLE 11.10

**Overland Flow Treatment
of Meat-Packing Wastewater**[a]

Parameter	Application rate (pounds/acre/day)[b]	Removal on a mass basis (%)
BOD_5	21.8–47.7	97–99
COD	73.2–106.2	95–99
TSS	8.7–30.5	94–99
Total N	5.0–14.6	94–98
Total P	0.5–0.9	54–92

[a]Adapted from (27).
[b]1.0 lb per acre = 1.12 kg per hectare.

DISTRIBUTION
CHANNEL

DIRECTION OF FLOW ⟶

Fig. 11.13. Distribution system used to apply wastewater at the top of a border strip land treatment site.

rate of 10 cm per week for the slow-rate system, and a 84% removal occurred with a 12.5 cm per week rate for the overland flow system. Nitrogen removals were less, averaging 58 and 44% in the slow-rate and overland flow systems, respectively. Several grasses grew very well when irrigated with meat-packing wastewater.

Border strip irrigation, which is a combination of slow-rate and overland flow and is comparable to a grass filter strip, has been used for treating meat-packing wastes. Figure 11.13 illustrates the distribution system at the top of a series of border strips, and Fig. 11.14 shows sheep grazing on border strips irrigated with the wastewater.

Hatchery wastes have been treated successfully by a combined septic tank, aerated lagoon, storage pond, and slow-rate irrigation system (29). The lagoon systems reduced the COD and nitrogen content of the initial wastes over 95%, thereby reducing significantly the amount of organics and nitrogen ultimately applied to the land. Such a combination system can be a cost-effective way of managing high-strength wastes when high organic and nutrient removals are required.

The addition of large amounts of organic matter when meat-packing wastes are applied to land will change the microbial activity. Soils to which such wastes have been added for years have higher respiratory activity, in terms of carbon dioxide production, and higher carbon mineralization rates than do control soils

DISTRIBUTION
CHANNEL AT
TOP OF SLOPE

Fig. 11.14. A border strip land treatment site used as a pasture for sheep.

to which such wastes had not been applied (30). This acclimation increases the assimilative capacity of the site.

Fruit and Vegetable Wastewaters

Land treatment has been used for fruit and vegetable canning wastewaters for many decades. The first reported application of spray irrigation (slow rate) for canning waste was in 1947 at the Hanover Cannery Company, Pennsylvania. One of the most publicized spray irrigation systems has been that used at Seabrook Farms in New Jersey. A flow of 5–10 mgd (0.22–0.44 m³ per second) of process water was sprayed over an 84-acre area of loamy sand. Each portion of the land was sprayed for 8 hr, followed by a 24-hr recovery period.

Early research on the land treatment of cannery wastes led to an improved understanding of the process, mechanisms of removal, and the engineered use of land treatment as a managed treatment process. For example, the results of the following study at Paris, Texas, provided considerable information on the fundamentals of land treatment processes and led to the development of the overland flow process as it is now used.

The results of the Paris, Texas, study on the spray irrigation of cannery wastes indicated that evaporation losses accounted for 18% of the total liquid applied, deep soil percolation accounted for 21%, and runoff accounted for 61%. Runoff from clay loam soil was greater than from sandy loam soil (31). The system removed 92–99% of the volatile solids and oxygen-demanding materials, 86–93% of the nitrogen, and 50–65% of the phosphorus. Treatment efficiency was improved by spreading the wastewater load over a larger land area and reducing the frequency of application. Phosphorus removal increased to 88% by decreasing the application schedule. A substantial fraction of the phosphorus was removed in the surface layer of soil. Surface and subsoil samples and soil water samples indicated an increase in total dissolved solids and sodium with time.

The soil was a heavy clay having an infiltration rate of about 0.05 in. per hour (32). This overland flow system was on slopes that were contoured to assure a uniform flow of water. Slopes ranged from 1 to 12%, with the results of the study indicating that slopes between 2 and 6% were preferable. The screened liquid wastes were sprayed intermittently. The liquid application rate in the test area was about 0.6 in. per day for a 5-day week. The design rate ranged from 0.25 in. per day in winter to 0.50 in. per day in summer.

The microbial action during the overland flow and the type of vegetation are important for the success of the process. In the Paris, Texas, studies (32), stands of both native and northern species of grasses initially developed. Continuous use of the overland flow system caused reed canary grass to predominate when readily available nutrients were supplied in quantity by the wastewater. The hay harvested from the disposal site ranged up to 23% crude protein and contained about double the mineral content of good quality hay. Between one and two harvests were feasible and essential to the continuous utilization of this site for nutrient removal.

The land application of tomato processing wastes on sandy loam soils with a high sodium content (33) identifies the type of site selection and preapplication treatment that can be necessary to accommodate site-specific conditions. Thirty percent of the tomatoes were peeled by a wet caustic operation. The wastewater resulted from fluming and washing tomatoes, rinsing of caustic from the tomatoes, and floor and yard drainage. Because sodium hydroxide was used in the tomato peeling operation, the wastewater SAR was about 19 and was adjusted to an SAR of 9 or less prior to irrigation by the addition of gypsum.

Border strip slow-rate irrigation was used for crop production. Prior to irrigation, the wastewater received only coarse and fine screening plus the gypsum addition. The land had not been farmed intensively and was being reclaimed by the addition of gypsum and sulfuric acid, as well as organic matter from the wastewater and solids additions. The irrigation system application rates were as follows: wastewater, 0.36–0.5 cm per day; BOD, 77–107 kg/ha/day; COD, 126–175 kg/ha/day; and total nitrogen, 0.5–0.7 kg/ha/day (33).

Few data are available to define a maximum organic application rate that can be used as part of the design basis for concentrated organic wastewaters such as vegetable processing wastewaters. A common approach is to assume that the hydraulic application rate is the limiting parameter and that the organics will be degraded more rapidly than they are applied. This assumption is reasonable for most land treatment systems.

Organic loading rates of between 200 and 600 lb of BOD_5 per acre per day (224–672 kg/ha/day) are used in many land treatment systems. A very detailed study, however, has shown that organic application rates of more than 15,000 lb per acre per day (16,800 kg/ha/day) can be completely assimilated prior to the next application after a 4-day rest period (34). This study involved field and laboratory investigations at two vegetable processing plants that had applied wastewaters to land for 10 and 25 years, respectively. As a result of the wastewater applications, about 125 tons of organics per acre had accumulated in the first foot of soil at the land treatment sites, and about 3000 lb of organic nitrogen per acre had accumulated at one of the land treatment sites. In spite of these overall increases, at the time of the study there appeared to be no net annual accumulation of nitrogen, and the BOD in the soil water approached background levels at the 1-ft depth 1 day after the site had been irrigated. Most of the nitrogen apparently was lost by the combination of nitrification and denitrification in the soil. This combination was enhanced by the resting period, which helped provide aerated conditions for the oxidation of the nitrogen, and the high oxygen demand of the applied waste when it was applied after the resting period.

A key aspect of such assimilation rates is that the microorganisms in the soil

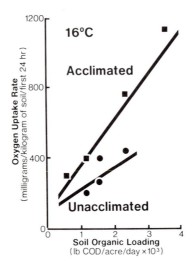

Fig. 11.15. Soils that have had wastes applied for a period of time and have the soil microorganisms acclimated to the wastes have a higher oxygen uptake rate than do soils that are not acclimated. [Adapted from (34).]

are acclimated to the applied wastes. Figure 11.15 shows the effect of acclimation on the oxygen uptake. The acclimated soil was taken from the top few inches of the land treatment site, and the unacclimated soil was taken from a comparable depth at a nearby location that had not received any waste.

At these sites, the microbial activity decreased with soil depth (Fig. 11.16). The data in this figure represent the oxygen uptake of the soils at one of the long-term land treatment sites. Although the 28°C temperature used in the laboratory test is greater than that which would occur in the field, the comparison of microbial activity remains valid. The noted decrease in microbial activity with depth is not unexpected and occurs at all land treatment sites.

Soil samples at the sites illustrate the effect of long-term wastewater application on the soil organic content (Fig. 11.17). As noted earlier, a significant increase in organic matter in the surface soils had occurred. In spite of the high application rates, groundwater data from the sites did not indicate high BOD, COD, total dissolved solids, or nitrates. This study (34) clearly indicates that land treatment is an effective process for the treatment and stabilization of vegetable processing wastewaters.

Land treatment also has been used successfully for other agricultural and food processing wastes, such as palm oil mill effluent (35) and oils used in potato processing (36). The cooking oils used in potato processing are reclaimed where possible, but certain quantities are discharged in plant wastewater. Palm and soybean oils were added to a silt loam at rates of 0.1–5.0 gm oil in 100 gm of soil

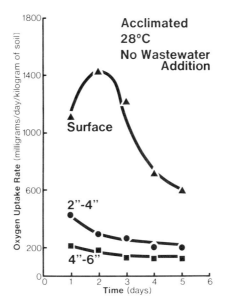

Fig. 11.16. The microbial activity at a land treatment site is greatest in the top few inches of the soil. [Adapted from (34).]

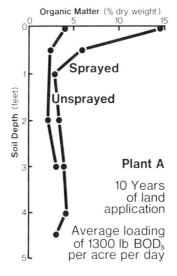

Fig. 11.17. At land treatment sites, the organic matter in the surface soil horizon can increase significantly as a result of waste additions. [Adapted from (34).]

(2.2–112 metric tons per hectare) (36). Both oils decomposed at the same rate, and decomposition proceeded very rapidly. There was no evidence of toxicity to the soil system and no evidence that there would be difficulty with land disposal of wastes containing these edible cooking oils.

Palm oil mill effluent contains high concentrations of organic matter and oil, in the range of 20,000 mg per liter of BOD and 6000 per liter of oil, respectively, plus other organic and inorganic constituents. Under managed conditions, land application of these wastes improved soil structure and soil nutrients and did not impact groundwater quality (35).

The above information is indicative of the diversity of food processing wastes that can be land treated and stabilized. Because of the degradable nature of such wastes, land treatment is an important, cost-effective, and environmentally sound process that should be considered routinely for the treatment, stabilization, and disposal of food processing wastes.

Municipal Wastewater

Early experience with land treatment for municipal wastewater in the United States dates to about the 1870s (37). However, in the first half of the twentieth century, existing land treatment systems were replaced by mechanical and biological waste treatment facilities that treat the wastewater to acceptable effluent limits before discharge to surface waters. In the last two decades, there has been a resurgence of interest in land treatment systems. This has been a result of the need to develop less costly treatment and disposal systems and because of recent federal legislation.

Much effort has been spent on developing and improving land treatment processes. The various land treatment systems have become accepted as viable wastewater management techniques that should be considered equally with any others. The regulations developed pursuant to the Federal Water Pollution Control Act Amendments of 1972 require that such consideration specifically be given for federally funded municipal wastewater projects. In the Act, the U.S. Environmental Protection Agency administrator was directed to encourage waste treatment management that results in facilities for (a) the recycling of potential pollutants through the production of agricultural, silvicultural, and aquacultural products; (b) the reclamation of wastewater; and (c) the elimination of pollutant discharges. Land treatment of municipal wastewater meets these objectives.

One of the first major land treatment studies involved an interdisciplinary team of engineers and scientists at Pennsylvania State University. The objectives of the study were to find methods of waste disposal other than stream discharge, to conserve nutrients by growing useful vegetation, and to replenish the groundwater supply by natural wastewater recharge (38). The project consisted of spraying treated sewage on croplands and forested areas. A solid set irrigation system was used.

This study indicated that the soil and the harvested crops removed from 50 to 100% of the applied phosphorus and nitrogen. The nutrients in the soil water at the 4-ft depth averaged less than 0.05 mg per liter of phosphorus and less than 9.0 mg per liter of nitrate nitrogen. The use of corn, hay, trees, and soil to purify the effluent has been described as a "living filter."

The results also indicated that a satisfactory biological and hydrologic balance could be maintained if 2 in. of wastewater were applied weekly to the land in the area of the study. The soil consisted of clay, silt, and sand mixtures that had accumulated from limestone and dolomite bedrock. About 129 acres of such soil were adequate for the treated wastes from a community of 10,000. The municipal wastes had received secondary treatment prior to the land disposal by irrigation. As discussed in subsequent paragraphs, secondary treatment is not a requirement for the application of municipal wastewaters to land.

Many full-scale land treatment systems have been designed and built for municipal wastewaters since 1970. Among the largest currently in operation are those for Muskegon, Michigan, and for Clayton County, Georgia. The experience gained from these systems has been summarized in a design manual (2) for the land treatment of wastewater.

The following paragraphs illustrate relevant results that have been obtained with these land treatment systems, especially those that have treated a combination of municipal and food processing wastes. The general site and design features associated with the land treatment processes used with wastewaters and the quality of the treated water expected from these processes have been presented in Tables 11.1, 11.2, and 11.3, respectively.

Preapplication Treatment

As noted earlier, secondary treatment is not required prior to the land treatment of wastewaters. There have been philosophical differences regarding the levels of preapplication treatment that are needed before wastewaters are applied to land. One view is that the equivalent of secondary treatment, including possibly chlorination, is necessary before the treated wastewaters are applied to the land. This view infers that land application of wastewater is a disposal mechanism that does not include wastewater treatment capabilities. Another view is that only a minimum amount of preapplication is necessary, i.e., that which minimizes nuisance conditions, increases system reliability, and reduces the risk to the public.

In identifying preapplication methods that may be needed, there are two criteria that are basic: there shall be no public health hazard, and applicable surface and ground water quality standards shall not be contravened.

There will be different preapplication needs, depending upon the degree of public access to the land treatment site, whether the crops are for human consumption, and, if so, whether they are to be eaten raw. The need for preapplication also is related to the ability of the soil to remove contaminants and to renovate the wastewater. This ability, in turn, is a function of the pollutant removal mechanisms in the soil.

The preapplication processes that should be used with land treatment systems are those that reduce risks to the public, the environment, and the equipment and permit the pollutant removal mechanisms in the soil to treat the wastewater. The use of the limiting-parameter concept in design, specific application and site management, and minimum preapplication processes should result in reliable land treatment systems that have low health and environmental risks. The relevant preapplication processes for land treatment systems are storage, odor control, removal of large solids, and possible pretreatment to reduce high concentrations of nitrogen, metals, sodium, and pathogens.

The opportunity to utilize the assimilative capacity of the soil–plant system and the need to avoid unnecessarily stringent preapplication requirements have been recognized by the Environmental Protection Agency in policy and guidance statements. The EPA guidance for assessing the appropriate degree of preapplication treatment is noted in Table 11.11.

TABLE 11.11

Guidance for Assessing the Degree of Preapplication Treatment Needed for the Land Treatment of Municipal Wastewaters[a]

Slow-rate systems
    A. Primary treatment is acceptable for isolated locations with restricted public access and when limited to crops not for direct human consumption
    B. Biological treatment by lagoons or in-plant processes plus control of the fecal coliform count to less than 1000 most probable number (MPN) per 100 ml is acceptable for controlled agriculture irrigation, except for human food crops eaten raw
    C. Biological treatment by lagoons or in-plant processes with additional BOD or SS control as needed for aesthetics plus disinfection to a log mean of 200 MPN per 100 ml[b] is acceptable in public access areas such as parks and golf courses
Rapid infiltration systems
    A. Primary treatment is acceptable for isolated locations having restricted public access
    B. Biological treatment by lagoons or in-plant processes is acceptable for urban locations having public access
Overland flow systems
    A. Screening or comminution is acceptable for isolated sites having no public access
    B. Screening or comminution plus aeration to control odors during storate or application is acceptable for urban locations having no access

[a] Adapted from (39).
[b] U.S. Environmental Protection Agency criteria for bathing waters.

Slow-Rate Systems

Slow-rate systems have been used successfully in many locations in moderate climates but, with proper design and management, can be used in more extreme climates. For example, a slow-rate system was used to treat a combination of cheese production and municipal wastewater in Thayne, Wyoming (40). Over 90% of the wastewater resulted from cheese production operations. The irrigation site is at a high elevation surrounded by mountains, with an average frost-free period of 30 days.

The wastewater flowed to an aerated lagoon with a hydraulic detention time of about 24 hr and then to a storage lagoon, from which it was pumped to a 15-ha spray field. The average flow was about 1300 m^3 per day (0.33 mgd). The wastewater was applied to different sections of the spray field each day, with each of the six sections being sprayed once per week. The treated wastewater left the system only by deep percolation and/or evapotranspiration.

The BOD and total Kjeldahl nitrogen (TKN) of the wastewater that was applied ranged from 330 to 1950 mg per liter and from 1.8 to 120 mg per liter, respectively. The wastewater was sprayed throughout the year.

In the winter, an ice pack developed and provided a means of liquid storage as an alternative to large holding lagoons. This also provided year-round use of the site, as well as additional treatment.

Even during the winter, there were significant reductions of BOD and COD in the upper 45 cm of the soil mantle. The decreases were attributed to filtration of the organic matter by the soil, as well as to the biological processes that occurred in the winter. No increase in groundwater pollutants occurred. The BOD, nitrate, ammonia, COD, phosphate, and fecal coliform concentrations in groundwater beneath the spray field were no greater than those in groundwater unaffected by the wastewater.

Creeping foxtail, reed canary grass, smooth bromegrass, and western wheatgrass were able to survive the harsh environment of the sprayfield. When the combination of nutrient uptake, forage yield, and quality of livestock feed was considered, creeping foxtail adapted best to the disposal site (41). Creeping foxtail formed a dense sod and began growth soon after the melt of the ice pack, when soil temperatures were low. This study (40, 41) clearly indicates that high-strength wastewaters can be treated by the slow-rate process on a year-round basis in very cold climates.

An evaluation of sites where municipal wastewater has been applied for many decades also indicates that the slow-rate process is a viable waste management process. For example, at Wroclaw, Poland, settled municipal wastewater has been land treated for 90 years (42). Wastewater is applied year-round.

At the Wroclaw site, the surface soils are loams with sand and sand and gravel below. Chemical and bacteriological analysis of ground waters below the irrigated fields indicated that the water quality was within the standards for drinking water. The site was grazed by cattle. The grasses are orchard grass, canary grass, and meadow grass. Some of the grasses are harvested and used for animal feed. The average hay contained 20% protein and 22% fiber on a dry-weight basis. A recent evaluation (42) at this site concluded that properly designed and operated systems treating sewage do not pose environmental hazards.

Other sites that have had long-term application of wastewaters have been evaluated. These include systems at Vineland, New Jersey; Bakersfield, California; Lubbock, Texas; Milton, Wisconsin; Mesa, Arizona; and Dickinson, North Dakota. Descriptions of these systems and their performance are included in the U.S. Environmental Protection Agency's "Process Design Manual" (2).

Concern has been expressed about the possible health risks of slow-rate land treatment. To provide a perspective on this question, the health factors associated with two wastewater treatment systems—activated sludge and slow-rate land treatment—have been compared (43). The comparison included an evaluation of removal mechanisms and the relative risk to site workers and the public by infectious organisms. The comparison indicated that under well-maintained conditions both activated sludge and land treatment provide a large measure of safety for public health. Slow-rate land treatment systems provide greater protection against parasites, viruses, metals, nitrate, and synthetic organic compounds.

Woody biomass production in forest energy plantations has increased when treated municipal wastewater has been used as a source of irrigation (44). The irrigation significantly increased total height and diameter growth, more than doubling total biomass production. The hybrid energy plantations were also very efficient in renovating the applied wastewater. The potential woody biomass production was estimated to reach 29–35 dry tons per hectare using wastewater irrigation as compared to an average of about 18 dry tons per hectare for energy plantations managed with more conventional technology.

Rapid Infiltration Systems

Because of the type of soil that is needed, rapid infiltration systems are not as widespread as slow-rate systems. Nonetheless, a number have been in operation for years, have successfully treated oxidation pond effluent and primary or secondary effluent, and have been used to renovate wastewater.

The effluent from a municipal oxidation pond was applied to a rapid infiltration site at weekly loading rates of 0.6 m per week (45). After passing through 1.3 m of soil, the groundwater generally contained less than 2 mg per liter BOD, 4 mg per liter SS, 1 mg per liter total phosphorus, 1 mg per liter ammonia nitrogen, and 10 mg per liter nitrate nitrogen. The study demonstrated that rapid infiltration can upgrade stabilization pond effluent to meet stringent effluent requirements.

Treatment of unchlorinated primary sewage effluent using rapid infiltration basins resulted in a high degree of wastewater renovation in a humid, cool northern climate (46). Inundating nine treatment basins for 7 days, followed by 14 days of rest, resulted in effluent additions totaling about 27 m. Analysis of the groundwater from the treatment site and peripheral area showed that total coliform bacteria, BOD, and COD were removed. The phosphorus concentrations in the groundwater were one-third of those in the applied effluent. Total nitrogen concentrations in the groundwater generally were greater than 10 mg per liter in wells strongly influenced by the percolate from the treatment site. Over a 6-month period, the infiltration capacity of the basins was reduced appreciably, due to accumulated organics. Oxidation of these organics rejuvenated the basin infiltration capacity.

Rapid infiltration has provided final treatment to unchlorinated Imhoff tank effluent at Fort Devens, Massachusetts, since 1942 (47). The operation involves an inundation for 2 days, followed by a 14-day recovery period. Effluent application has been about 27 m per year. Chemical analyses of soil samples obtained from the upper 3 m of the treatment beds showed that levels of organic matter, electrical conductivity, nitrogen, phosphates, calcium, magnesium, potassium, manganese, sodium, copper, iron, and zinc were higher than those of background samples. Groundwater quality in wells located 50–100 m (180–330 feet) from the infiltration basins showed that the primary effluent had been renovated after flowing through a sand and gravel formation. Total nitrogen, primarily nitrate nitrogen, in the observation wells ranged from 10–20 mg per liter. A nitrogen reduction of 60–80% had occurred probably by denitrification. Total phosphorus in the groundwater ranged from 1–2 mg per liter, slightly above the 0.7 mg per liter background level. The mean total coliform bacteria in the groundwater was less than 200 per 100 ml. The evaluation showed that the unchlorinated primary sewage effluent was being renovated by the rapid infiltration basins to a quality comparable to that achieved by conventional tertiary wastewater treatment facilities.

In Colorado, both primary and secondary effluents were treated successfully by rapid infiltration (48). The renovated water was collected 2.4–3 m below the surface of the basins. The COD of the renovated water generally was less than 10 mg per liter and was the same, whether primary or secondary effluent was applied to the basins. Virtually all of the nitrogen applied to the basins was oxidized and existed as nitrate nitrogen in the renovated water. Fecal coliform reductions ranged from 97.5 to over 99%. The coliform concentrations in the renovated water were such that disinfection probably would be required before discharge of the water to surface streams. The evaluation indicated that regulations requiring secondary treatment prior to rapid infiltration may be unnecessarily severe.

The Hollister, California, rapid infiltration site has received primary municipal effluent for 30 years. An evaluation of the groundwater and soil nitrogen and phosphorus concentrations (49) found that after the wastewater had percolated through about 7 m of unsaturated gravel and sandy loams, the total nitrogen concentrations approached those found at control locations. Effective phosphorus removal required longer travel distances, but the sorption capacity of the soil had not been exhausted, even after 30 years of continuous application.

At Lake George, New York, unchlorinated secondary effluent has been treated by rapid infiltration since 1939 (50). The basins remove BOD, COD, alkyl benzene sulfonates, coliforms, and ammonia and organic nitrogen. Soluble inorganics such as sodium, chloride, and potassium pass through the sands. Orthophosphates were removed completely in the top 3 m of the sand beds. Nitrates were removed in sand beds which were 18 m or more in depth.

Overland Flow Systems

Although overland flow has not been used as extensively as slow-rate or rapid infiltration systems, it is a sound method of waste treatment, and there is an increasing amount of information on the operation of such systems.

The performance of several full-scale overland flow systems has been evaluated (51). These systems effectively treat raw municipal wastewater to a quality better than secondary effluent. The removal kinetics of such systems appeared to be first order, and overland flow systems appear comparable to attached-growth biological systems. Phosphorus removal was lower than that for BOD, SS, and nitrogen.

TABLE 11.12

Performance of Overland Flow When Treating Raw Sewage and Lagoon Effluent[a]

	Raw sewage		Lagoon effluent	
Parameter	Site effluent concentration (mg/liter)	Average removal (%)	Site effluent concentration (mg/liter)	Average removal (%)
BOD	21 (3–50)[b]	87	14 (9–30)	55
Suspended solids	24 (2–122)	85	41 (5–152)	45
Total organic carbon	36 (20–54)	62	46 (20–61)	23
Total phosphorus	4 (0.1–8)	50	2.3 (0.6–42)	34
Total nitrogen	12 (2.7–44)	63	4.3 (3.2–14)	67

[a] Adapted from (53).

[b] The single value is the average; the numbers in the parentheses are the range of values measured.

In overland flow systems, phosphorus removal can be enhanced by the additions of coagulants before the wastewater is applied to the site. In one study (52), aluminum sulfate was added with the wastewater as it entered the distribution system. The results demonstrated that phosphorus removal can be increased to about 90% by adding 1.5–2.0 mg of aluminum for each milligram of phosphorus. With the use of coagulants, the phosphorus concentration in the overland flow runoff can be reduced to about 1.0 mg per liter.

A full-scale overland flow facility was constructed in South Carolina to treat both comminuted raw sewage and facultative lagoon effluent (53). All plots were graded to a 4% slope. The plots ranged from about 46 to 50 m in length. The average performance at this site occurred as noted in Table 11.12. The effluent from both types of wastewater equalled or exceeded that obtained by secondary treatment processes. Algal cells in the lagoon effluent were not well removed and resulted in high suspended solids concentration in the site effluent.

Summary

Based upon the preceding examples, the data and information in the "Process Design Manual" (2) and results from many other operating systems, land treatment systems can be cost-effective methods of waste treatment and stabilization. Such systems are particularly appropriate for the wastes from agricultural operations, since the wastes are generated in rural areas where land is available. Information is available to design slow-rate, rapid-infiltration, and overland flow land treatment systems, as well as systems for the application and stabilization of manures, slurries, and sludges. Such systems should be used more frequently for agricultural wastes, since the systems are environmentally and technically sound and frequently are less energy intensive than alternative waste management approaches.

References

1. Olson, G. W. "Soils and the Environment—A Guide to Soil Surveys and Their Application." Chapman and Hall, New York, 1981.
2. U.S. Environmental Protection Agency. "Process Design Manual—Land Treatment of Municipal Wastewater," EPA 625/1–81–013. Center for Environmental Research Information, Cincinnati, Ohio, 1981.
3. Martel, C. J., Jenkins, T. F., Diener, C. J., and Butler, P. L., "Development of a Rational Design Procedure for Overland Flow Systems," Report 82–2. Cold Regions Research and Engineering Laboratory, U.S. Army Corps of Engineers, Hanover, New Hampshire, 1982.
4. Smith, R. G., The Overland Flow Process. *Environ. Progress*, A.I.Ch.E., **1**, 195–205 (1982).

5. Sopper, W. E., Seaker, E. M., and Bastian, R. K., (eds), "Land Reclamation and Biomass Production with Municipal Wastewater and Sludge." Pennsylvania State University Press, University Park, Pennsylvania, 1982.

6. Overcash, M. R., and Pal, D., "Design of Land Treatment for Industrial Wastes, Theory and Practice." Ann Arbor Sci. Publ., Ann Arbor, Michigan, 1979.

7. Brown, K. W., and Associates, Inc. "Hazardous Waste Land Treatment." U.S. Environmental Protection Agency, SW–874, Municipal Environmental Research Laboratory, Cincinnati, Ohio, 1983.

8. Iskandar, I. K. (ed.), "Modeling Wastewater Renovation—Land Treatment." Wiley, New York, 1981.

9. Loehr, R. C., Jewell, W. J., Novak. J. D., Clarkson, W. W., and Friedman, G. S., "Land Application of Wastes." Van Nostrand-Reinhold, Princeton, New Jersey, 1979.

10. Lance, J. C., Fate of nitrogen in sewage effluent applied to soil. *J. Irrig. Drain. Div. Am. Soc. Civ. Eng.* **107** IR3, 131–144 (1975).

11. U.S. Department of Agriculture, "Animal Waste Utilization on Cropland and Pastureland." Science and Education Administration, Utilization Research Report No. 6. Washington, D.C., 1979.

12. U.S. Environmental Protection Agency, "Process Design Manual—Utilization of Municipal Sewage Sludge on Land." Municipal Environmental Research Laboratory, Cincinnati, Ohio, 1983.

13. Chaney, R. L., Plant Uptake of Inorganic Waste Constituents. *In* "Land Treatment of Hazardous Wastes." (J. F. Parr, P. B. March, and J. M. Kla, eds.), pp. 50–76. Noyes Data Corp., Park Ridge, New Jersey, 1983.

14. Chaney, R. L., Fate of toxic substances in sludge applied to cropland. *In* "Proceedings of the International Symposium on Land Application of Sewage Sludge." Tokyo, Japan, 1982.

15. Webber, M. D., Monteith, H. D., and Corneau, G. M., Assessment of heavy metals and PCBs at sludge application sites. *J. Water Pollut. Control Fed.* **55**, 187–195 (1983).

16. U.S. Environmental Protection Agency, Criteria for classification of solid waste disposal facilities and practices. *Fed. Regist.* **44**, 53,438–53,468 (1979).

17. Sommerfeldt, T. G., Pittman, V. J., and Milne, R. A., Effect of feedlot manure on soil and water quality. *J. Environ. Qual.* **2**, 423–427 (1973).

18. Manges, H. L., Murphy, L. S., Powers, W. L., and Schmid, L. A. "Ultimate Disposal of Beef Feedlot Wastes onto Land," EPA–600/2–78–045. Robert S. Kerr Environmental Research Laboratory, U.S. Environmental Protection Agency, Ada, Oklahoma, 1978.

19. Horton, M. L., Wiersma, J. L., Schnabel, R. R., Beyer, R. E., and Carlson, C. G., "Animal Waste Effects upon Crop Production, Soil and Runoff Waters," EPA–600/2–81–230. Robert S. Kerr Environmental Research Laboratory, U.S. Environmental Protection Agency, Ada, Oklahoma, 1981.

20. Stewart, B. A., and Mathers, A. C., Soil conditions under feedlots and on land treated with large amounts of animal wastes. *Int. Symp. Identif. Meas. Environ. Pollut. (Proc.)* 81–83 (1971).

21. Thomas, R. E., "Feasibility of Overland Flow Treatment of Feedlot Runoff," EPA–660/2–74–062. National Environmental Research Center, U.S. Environmental Protection Agency, Corvallis, Oregon, 1974.

22. Dickey, E. C., and Vanderholm, D. H., Performance and design of vegetative filters for feedlot runoff treatment. *In* "Livestock Waste: A Renewable Resource," pp. 257–261. Am. Soc. Agric. Engrs. St. Joseph, Michigan, 1981.

23. Broten, D. A., and Bubenzer, G. D., "Barnyard Sediment and Nutrient Removal By Grass Filters" ASAE Tech. Paper 78–2087, Annual Meeting, Am. Soc. Agric. Engrs., St. Joseph, Michigan, 1978.

24. O'Leary, P., Devereaux, M., Johnson, L., Lewandowski, D., and Schwerbel, D., "Guideline Document for the Design, Construction and Operation of Land Disposal Systems for Liquid Wastes." Wisconsin Department of Natural Resources, Madison, Wisconsin, (preliminary draft) November 1975.

25. Environment Canada, "Inventory of Spray Irrigation Systems in the Great Lakes Basin of Canada." Water Pollution Control Directorate, EPS 3–WP–73–5, Department of the Environment, Ottawa, Canada, 1973.

26. Witherow, J. L., Small meat packers wastes treatment systems. In "Food Processing Waste Management, Proceedings of the Cornell Agricultural Waste Management Conference" (pp. 191–221). Ithaca, New York, 1973.

27. Witherow, J. L., Rowe, M. L., and Kingery, J. L., Meatpacking wastewater treatment by spray runoff irrigation. In "Proceedings of the Sixth National Symposium of Food Processing Wastes," pp. 256–268. EPA–600/2–76–224. U.S. Environmental Protection Agency, Cincinnati, Ohio, 1976.

28. Tarquin, A. J., "Treatment of High Strength Meatpacking Plant Wastewater by Land Application," EPA–600/2–76–302. Robert S. Kerr Environmental Research Laboratory U.S. Environmental Protection Agency, Ada, Oklahoma, 1976.

29. Humenik, F. J., Overcash, M. R., Sneed, R. E., Barker, J. C., and Phillips, K., Hatchery waste management and utilization by land application. J. Water Pollut. Control Fed. **50**, 739–746 (1978).

30. Tate, K. R., Respiratory activity of soils irrigated by water and by meatworks effluent. N. Z. J. Agric. Res. **16**, 385–388 (1973).

31. Law, J. P., Thomas, R. E., and Myers, L. H., "Nutrient Removal from Cannery Wastes by Spray Irrigation of Grasslands." Water Pollut. Control Res. Ser. 16080. Robert S. Kerr Research Center, Federal Water Pollution Control Administration, Dept. of the Interior, Ada, Oklahoma, 1969.

32. Gilde, L. C., Kester, A. S., Law, J. P., Neeley, C. H., and Parmelee, D. M., A spray irrigation system for treatment of cannery wastes. J. Water Pollut. Contr. Fed. **43**, 2011–2025 (1971).

33. Crites, R. W., Lynard, W. G., and Hatch, H. L., Land treatment of cannery wastes. J. Water Pollut. Control Fed. **51**, 808–814 (1979).

34. Jewell, W. J., Kodukula, P. S., and Wujcik, W. J., "Limitations of Land Treatment of Wastes in the Vegetable Processing Industries." Department of Agricultural Engineering, Cornell University, Ithaca, New York, 1978.

35. Wood, B. J., Pillai, K. R., and Rajaratram, J. A., Palm oil mill effluent disposal on land. Agric. Wastes **1**, 103–127 (1978).

36. Smith, J. H., Decomposition in soil of waste cooking oils used in potato processing. J. Environ. Qual. **3**, 279–281 (1974).

37. Jewell, W. J., and Seabrook, B. L., "A History of Land Application as a Treatment Alternative." EPA–430/9–79–012. U.S. Environmental Protection Agency, Office of Water Program Operations, Washington D.C., 1979.

38. Kardos, L. T., Sopper, W. E., Myers, E. A., Parizek, R. R., and Nesbitt, J. B., "Renovation of Secondary Effluent for Reuse as a Water Resource," EPA–660/2–74–016. U.S. Environmental Protection Agency, Office of Research and Development, Washington, D.C., 1974.

39. Thomas, R. E., and Reed, S. C., EPA policy on land treatment and the Clean Water Act of 1977. J. Water Pollut. Control Fed. **52**, 452–460 (1980).

40. Borrelli, J., Burman, R. D., Delaney, R. H., Moyer, J. C., Hough, H. W., and Weand, B., "Land Application of Wastewater under High Altitude Conditions," EPA–600/2–78–139. Robert S. Kerr Environmental Research Laboratory, U.S. Environmental Protection Agency, Ada, Oklahoma, 1978.

41. Greene, M. C., Delaney, R. H., Moyer, J. L., and Borrelli, J., Forage production utilizing cheese plant effluent under high altitude conditions. *J. Water Pollut. Control Fed.* **52,** 2855–2864 (1980).

42. Cebula, J., and Kutera, J., Land treatment systems in Poland. *In* "Land Treatment of Waste-water" (H. L. McKim, symposium coordinator), Vol. I, pp. 257–264. U.S. Army Corps of Engineers, CRREL, Hanover, New Hampshire, 1978.

43. Uiga, A., and Crites, R. W., Relative health risks of activated sludge treatment and slow rate land treatment. *J. Water Pollut. Control Fed.* **52,** 2865–2874 (1980).

44. Sopper, W. E., and Kerr, S. N., Maximizing forest biomas energy production by municipal wastewater irrigation. *In* "Energy from Biomass and Wastes—IV" pp. 115–133. Institute of Gas Technology, Chicago, Illinois, 1980.

45. Dornbush, J. N., "Infiltration Land Treatment of Stabilization Pond Effluent," EPA–600/2–81–226. Robert S. Kerr Environmental Research Laboratory, U.S. Environmental Protection Agency, Ada, Oklahoma, 1981.

46. Satterwhite, M. B., Condike, B. J., and Stewart, G. L., "Treatment of Primary Sewage Effluent by Rapid Infiltration," Report 76–49. Cold Regions Research and Engineering Laboratory, U.S. Army Corps of Engineers, Hanover, New Hampshire, 1976.

47. Satterwhite, M. B., Stewart, G. L., Condike, B. J., and Vlach, E., "Rapid Infiltration of Primary Sweage Effluent at Fort Devens, Massachusetts," Report 76–48. Cold Regions Research and Engineering Laboratory, U.S. Army Crops of Engineers, Hanover, New Hampshire, 1976.

48. Carlson, R. R., Linstedt, K. D., Bennett, E. R., and Hartman, R. B., Rapid infiltration treatment of primary and secondary effluents. *J. Water Pollut. Control Fed.* **54,** 270–280 (1982).

49. Levine, P. E., Olson, J. V., and Crites, R. W., Nitrogen and phosphorus removal after 30 years of rapid infiltration. *In* "Land Treatment of Wastewater" (H. L. McKim, symposium coordinator), Vol. II, pp. 17–26. U.S. Army Corps of Engineers, CRREL, Hanover, New Hampshire, 1978.

50. Aulenbach, D. B., and Clesceri, N. L., The Lake George Village land application system. "Land Treatment of Wastewater" (H. L. McKim, Symposium Coordinator) Vol. II, pp. 27–35. U.S. Army Corps of Engineers, CRREL, Hanover, New Hampshire, 1978.

51. Hinrichs, D. J., Faisst, J. A., and Pivetti, D. A., "Assessment of Current Information on Overland Flow Treatment of Municipal Wastewater," EPA 430/9–80–002. U.S. Environmental Protection Agency, Office of Water Programs, Washington D.C., 1980.

52. Thomas, R. E., Bledsoe, C., and Jackson, K., "Overland Flow Treatment of Raw Wastewater with Enhanced Phosphorus Removal," EPA 600/2–76–131. Robert S. Kerr Environmental Research Laboratory, U.S. Environmental Protection Agency, Ada, Oklahoma, 1976.

53. Abernathy, R., "Overland Flow Treatment of Municipal Sewage at Easley, South Carolina," EPA 600/2–83–015. Robert S. Kerr Environmental Research Laboratory, U.S. Environmental Protection Agency, Ada, Oklahoma, 1983.

12

Nitrogen Control

Introduction

General

Major emphasis in waste treatment continues to be on the reduction of the oxygen-demanding substances released to surface waters. Early research and field experience with waste treatment plants established that ammonia oxidation exerted an oxygen demand when present in the waste or effluent. In the earlier part of this century, a large amount of nitrate in the effluent of a wastewater treatment facility was a measure of a high level of treatment and operation, and treatment plants were designed to accomplish oxidation of both carbonaceous and nitrogenous material in wastes.

The development of the standardized BOD test established that the nitrogenous oxygen demand (NOD) of untreated waste occurred a number of days, frequently greater than five, after the start of the test. The second stage of oxygen demand (NOD) of untreated wastes was considered of little consequence in receiving waters, since the oxygen demand would occur considerably downstream from the discharge of the waste. Such a lag would not occur when adequately treated wastes were discharged, since nitrifying organisms could have had an opportunity to grow in the secondary facilities. With secondary effluents, the NOD would occur in the receiving waters closer to the point of discharge of the treated wastes. If adequate nitrification were to occur in the treatment plant, the NOD could be satisfied and would not occur in the receiving waters.

Reliance on the 5-day BOD test, which does not measure NOD unless an adequate population of nitrifying organisms is in the BOD bottles, and the realization that treatment facilities could be designed and operated more economically without complete nitrification, led to an emphasis on removal of carbonaceous rather than nitrogenous oxygen demand. The lack of emphasis on meeting the NOD in treatment plant design and operation occurred from about 1930 to 1960 in the United States.

Need for Control

Environmental concerns have caused a reexamination of this position, with the result that interest is again being given to methods to control both the carbon and nitrogen content of a waste or treated effluent. The reasons nitrogen compounds are of concern include the NOD in receiving waters, ammonia toxicity to fish, increased chlorine demand due to ammonia if the water is treated as a potable source, the role of nitrogen as a controlling nutrient in eutrophication, and health problems in humans and animals. Regulatory authorities are beginning to set standards for nitrogenous materials in streams and effluents, and treatment facilities are being required to produce effluents low in both unoxidized and total nitrogen. Control and/or management of nitrogen in municipal and industrial wastes and conservation of nitrogen in agricultural wastes is receiving greater emphasis.

The oxygen demand of reduced nitrogen compounds in receiving waters frequently has been neglected or assumed to have minimum consequence. However, from the point of view of the oxygen demand on a stream, it is not insignificant. Municipal wastewaters contain about 15–20 mg per liter of nitrogen compounds, and certain agricultural wastes such as animal and fertilizer production wastewaters can have nitrogen concentrations in the 100–1000 mg per liter range. A treatment plant with 20 mg per liter of ammonia (NH_4-N) and 20 mg per liter of carbonaceous BOD in the effluent and a flow of 1 million gallons per day (mgd) would place a carbonaceous oxygen demand of about 250 lb per day and a nitrogenous oxygen demand of about 750 lb per day on the receiving stream. The concern with NOD is becoming more important as the load of oxygen-consuming wastewaters increases and as the desire for a less polluted environment continues. The NOD load must be considered in any program of stream quality improvement.

The NOD or nitrification is a biological phenomenon whose rate is a function of many factors, such as the concentration of nitrifying organisms, temperature, pH, dissolved oxygen concentration of the liquid, the type of available nitrogen compounds, and the concentration of any inhibiting compounds. The ammonia to be oxidized is that which is in excess of the nitrogen required for growth by the heterotrophic organisms in a waste treatment facility or a stream.

In untreated wastes, nitrogen will be in the form of organic nitrogen and ammonium compounds. The organic nitrogen is converted by microbial action to the ammonium ion. If adequate environmental conditions exist, nitrifying organisms can oxidize the ammonia. These organisms are autotrophs, obtaining their energy through the oxidation of ammonium ion or nitrite, as appropriate.

$$NH_4^+ + 1.5O_2 \xrightarrow{bacteria} 2H^+ + NO_2^- + H_2O \qquad (12.1)$$

This reaction requires 3.43 gm of oxygen for 1 gm of ammonia nitrogen oxidized to nitrite. The nitrite can be oxidized to nitrate.

$$NO_2^- + 0.5O_2 \xrightarrow{bacteria} NO_3^- \qquad (12.2)$$

This reaction requires 1.14 gm of oxygen for every gram of nitrite nitrogen oxidized to nitrate nitrogen. If all of the ammonium ion were oxidized, 4.57 gm of oxygen would be required per gram of ammonia nitrogen oxidized to nitrate. However, about 0.8% of the nitrogen is converted into cell material as carbon dioxide is fixed by the bacteria. Field experiments have indicated ratios of 3.22 mg of oxygen utilized per milligram of NH_4-N oxidized to NO_2-N and 1.11 mg of oxygen utilized per milligram of NO_2-N oxidized to NO_3-N (1), confirming the theoretical ratios.

High concentrations of ammonia in surface waters can be lethal to fish, and water quality criteria have been established for freshwater aquatic life (2, 3). The toxicity of ammonia is related to the quantity that enters the cell. Cell membranes are relatively impermeable to the ammonium ion (ionized ammonia, NH_4^+), whereas un-ionized ammonia (NH_3) can pass cell membranes easily. The pH of the water is important in determining the concentration of the un-ionized ammonia that exists. The proportion of the total ammonia concentration which is ionized at any time is a function of the pH of the liquid.

The two forms of ammonia exist in an equilibrium which is affected by pH:

$$NH_3 + H_2O \rightleftharpoons NH_4^+ + OH^- \qquad (12.3)$$

At high pH values, there will be a greater amount of NH_3 than at lower pH values.

The toxicity of ammonia to aquatic life in streams and wastewaters occurs because of the interaction of many factors. These include the mean ammonia concentrations to which the aquatic life are exposed, the fluctuation in the ammonia concentration, the pH and temperature of the water in which the aquatic life exists, the acclimation of the aquatic life, and other factors which may stress the aquatic life.

Prevailing methods of ammonia measurement determine only the total ammonia concentration, which is both the ionized and un-ionized forms. The proportion of the total ammonia concentration that is in the un-ionized form at any pH (F) is determined by the dissociation constants of aqueous ammonia and water and the law of mass action (3):

$$F = \frac{1}{1 + 10^{(pK_a - pH)}} \tag{12.4}$$

where pK_a is the negative logarithm of the ionization constant for Eq. (12.3).

The pK_a values of ammonia as a function of temperature are noted in Table 12.1. The relationship between F, pH, and temperature is noted in Fig. 12.1 and illustrates the impact of temperature and pH on the fraction of the total ammonia that exists as un-ionized ammonia (NH_3). An increase in NH_3 is related to an increase in pH and temperature. The presence of dissolved solids in the range of 500–1000 mg per liter will slightly lower the concentration of NH_3.

The water quality criterion for ammonia in the United States (2) is 0.02 mg per liter as un-ionized ammonia. The total ammonia ($NH_3 + NH_4^+$) concentration which contains this concentration of un-ionized ammonia at different temperatures and pH values is noted in Table 12.2.

Ammonia in waters can increase the chlorine demand of the waters if breakpoint chlorination is practiced, or it can reduce the relative effectiveness of the disinfection process (Chapter 13).

Both oxidized and reduced nitrogen compounds can serve to stimulate the growth of aquatic life in surface waters. The type of limiting nutrients can be

TABLE 12.1

Values of pK_a of Ammonia in Water as a Function of Temperature[a]

Temperature (°C)	pK_a value
5	9.90
10	9.73
15	9.56
20	9.40
25	9.25

[a]Adapted from (3). [With permission from *Water Res.* **7**, European Inland Fish Advisory Commission, Water quality criteria for European freshwater fish, Copyright 1973, Pergamon Press Ltd.]

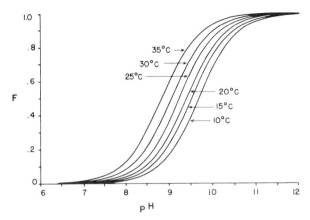

Fig. 12.1 Effect of temperature and pH on the fraction (F) of un-ionized ammonia.

different in different bodies of water. Because smaller quantities of phosphorus are necessary per unit of aquatic growth, phosphorus has received the major emphasis in eutrophication control. Where phosphorus is naturally in excess in waters, nitrogen or even carbon can be the limiting nutrient.

In the United States, a nitrate concentration of 45 mg per liter as NO_3 has been established as the maximum concentration that should be in potable waters. Maximum acceptable concentrations of nitrate reflect public health concerns and are based upon the judgment of public health officials. Acceptable concentrations vary among nations. High nitrate concentrations have caused methemoglobinemia in infants and have caused problems with animal health. In each case the problem is the same: difficulties in transport of oxygen in the blood. Nitrate-

TABLE 12.2

Total Ammonia Concentrations Which Result in an Un-Ionized Ammonia (NH₃) Concentration of 0.02 mg per Liter[a]

Temperature (°C)	pH Value				
	6.0	7.0	8.0	9.0	10.0
5	160	16	1.6	0.18	0.036
10	110	11	1.1	0.13	0.031
15	73	7.3	0.75	0.093	0.027
20	50	5.1	0.52	0.070	0.025
25	35	3.5	0.37	0.055	0.024

[a]Aqueous ammonia solutions of zero salinity. [Adapted from (6).]

reducing bacteria in the intestinal tract of humans and animals reduce nitrate to nitrite. The nitrite can oxidize hemoglobin in the blood to methemoglobin, which is unable to transport oxygen. Although nitrite is the cause of the problem, nitrites are rarely found in foods and water. The standards are placed on nitrate since it can be present in foods and water and is the precursor of nitrite. High nitrates in leafy foods such as spinach and in forage crops also can be a cause of the problem. Once high nitrate concentrations exist in surface waters and ground-waters, few practical processes exist for routine nitrate reduction or elimination. Generally the water supply is abandoned, and other potable supplies are sought.

Opportunities for Nitrogen Control

Where excess nitrogen has the potential to cause the aforementioned problems, management systems must be utilized. A summary of the possible methods is noted in Table 12.3. Few of these methods are applicable to agricultural wastes.

Traditionally the land has been the ultimate disposal medium for agricultural wastes. Where the waste applications to the land are excessive in terms of nitrogen, removal of nutrients prior to land disposal may be needed. The most feasible approaches for nitrogen control from agricultural wastewaters appear to lie in the use of nitrification and denitrification processes, either prior to, or in conjunction with land disposal. Ammonia stripping is also of interest in nitrogen control, primarily to understand the losses that can occur in treatment processes,

TABLE 12.3

Methods for Reducing the Nitrogen Content of Wastewater

Method	Nitrogen compounds removed
Physical and chemical	
Land application	NH_3, NH_4^+, organic N
Electrochemical	NH_4^+
Ammonia stripping	NH_3
Ion exchange	NO_3^-, NH_4^+
Electrodialysis	NO_3^-, NH_4^+
Reverse osmosis	NO_3^-, NH_4^+
Breakpoint chlorination	NH_4^+, organic N
Biological	
Algal utilization	All forms
Microbial denitrification	NO_3, NO_2
Land application	All forms

storage tanks, and feedlots. These losses represent a measure of nitrogen control with a specific waste but may have undesirable effects such as increased ammonia concentrations in buildings and storage units and the release of ammonia to the environment.

Nitrification

Present-day knowledge on the transformation of nitrogen and the physiology of microorganisms involved in these transformations stems primarily from the research of agronomists and soil scientists. Nitrification can be defined basically as the biological conversion of nitrogen in inorganic or organic compounds from a reduced to a more oxidized state. In the field of water pollution control, nitrification usually is referred to as a biological process in which ammonium ions are oxidized to nitrite and nitrate sequentially.

Pasteur in 1862 suggested that the oxidation of ammonia was microbiological. This suggestion was verified in classic studies with sewage and soil, which showed that oxygen was essential and that alkaline conditions favored nitrification. The ubiquity of biological nitrification in soils has been well demonstrated.

The Nitrifying Organisms

Several genera of nitrifying organisms have been reported. *Nitrosomonas, Nitrosospira, Nitrosococcus,* and *Nitrosocystis* oxidize ammonia to nitrite. *Nitrosogloea, Nitrobacter,* and *Nitrocystis* oxidize nitrite to nitrate. Of these genera, only *Nitrosomonas* and *Nitrobacter* are generally encountered in aquatic and soil ecosystems and are the nitrifying autotrophs of importance.

Although the nitrification process is largely due to autotrophic organisms, certain heterotrophic bacteria, actinomycetes, and fungi are recognized as nitrifying organisms. The genera involved include *Mycobacterium, Nocardia, Streptomyces, Agrobacterium, Bacillus,* and *Pseudomonas* (4). The first fungus to be recognized as a heterotrophic nitrifier was *Aspergillus flavus,* which forms nitrate from amino nitrogen. Heterotrophic nitrification occurs only when nitrogen is present in excess of cellular needs and when any energy source other than the oxidation of nitrogen is available. Many of these heterotrophic nitrifiers are common in lakes, streams, and soils.

The nitrifying organisms of significance in waste management are autotrophic. Organic compounds have been reported to be inhibitory to the growth of nitrifying organisms, but the claim that the organic compounds in general are inhibitors is overemphasized. The occurrence of nitrification in soils containing organic matter, in trickling filters, activated sludge tanks, and compost piles testifies that the process takes place freely in natural ecosystems containing varied degrees of organic matter.

Solids Retention Time

The concentration of nitrifying organisms is a major factor in the rate of nitrification. The quantity of nitrifiers is determined by the generation time of the organisms, which in turn is related to the amount of energy obtained in the oxidation of ammonia or nitrite. Nitrification is an exothermic reaction. The free energy change in the oxidation of ammonium to nitrite and nitrite to nitrate is reported to be in the range of -65.5 to -84 kcal per mole of ammonium and -17.5 to -20 kcal per mole of nitrite oxidized, respectively (5). The energy is utilized for the assimilation of carbon in the form of carbon dioxide or bicarbonate ion. The ammonia-oxidizing organisms obtain more energy than the nitrite-oxidizing organisms. Assuming the efficiency of cell synthesis is the same, more ammonia-oxidizing bacteria are formed than nitrite-oxidizing bacteria per unit of ammonium undergoing nitrification. In other words, the nitrite-oxidizing bacteria utilize about three times more substrate than do the ammonia-oxidizing bacteria while synthesizing the same amount of cell mass. This indicates why little significant accumulation of nitrite occurs in an ecosystem when there is no inhibition.

The energy obtained by nitrifying organisms is much lower (66 kcal per mole of substrate) than that obtained by heterotrophs in an aerobic system (686 kcal per mole of glucose as a substrate). Because of this energy difference there are low concentrations of nitrifying organisms, approximately one-tenth that of the heterotrophs, in a nitrifying biological treatment plant. The nitrifiers will be an even smaller proportion of the mixed liquor solids in a nitrifying unit.

Since the process of nitrification depends on the metabolism of a group of highly specialized aerobic organisms, it is imperative that these organisms be present in adequate numbers. The generation time for the nitrifying organisms is comparatively long, on the order of 10 or more hours, depending upon environmental conditions. The minimum solids retention time (SRT) of a nitrification treatment unit must be greater than the growth rate of the organisms under the imposed environmental conditions. The growth rate, and therefore operating SRT of the nitrification unit, is related to the temperature and the concentration of any inhibitor. The minimum SRT for nitrification increases as the temperature decreases (Fig. 12.2) because of the decreased growth rate of the microorganisms. If a biological system is to consistently obtain a required degree of nitrification, close control of the system SRT is needed to compensate for any change in environmental conditions.

Minimum SRT values are important in biological treatment units where there is the possibility of removing the nitrifying organisms faster than they can

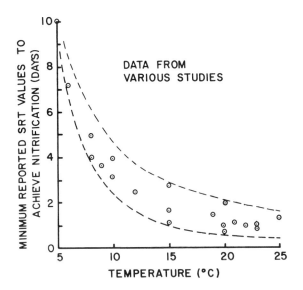

Fig. 12.2. Relationship between temperature and the minimum SRT needed to achieve nitrification. [Adapted from (6).]

grow. Minimum SRT values are of little importance when considering nitrification in a soil. Soils contain ample quantities of nitrifying bacteria, and the likelihood of removing them from the soil faster than they can reproduce is remote. Other environmental conditions have a larger effect on soil nitrification.

Dissolved Oxygen

Since the nitrifying organisms are aerobic, adequate dissolved oxygen (DO) is necessary to support nitrification, assuming other environmental conditions are satisfactory. The sensitivity of the nitrifying organisms to low dissolved oxygen concentrations is one of the reasons complete nitrification is difficult to accomplish in heavily loaded systems where the oxygen demand is significant and where adequate oxygen does not exist in the microbial floc.

A critical DO concentration exists below which nitrification does not occur. The critical DO concentration has not been precisely determined, but it appears to be around 0.3–0.5 mg per liter (7, 8). The actual limit is more dependent on the rate of oxygen diffusion to the microorganisms than on the average oxygen concentration in the system. To assure that the dissolved oxygen concentration is not limiting nitrification, the dissolved oxygen in a treatment unit generally is kept above about 1.0 mg per liter. Nitrification proceeds at a rate independent of the dissolved oxygen above the critical concentration.

Numerous studies have indicated that there was no apparent difference in the degree of nitrification when the nitrifiers were exposed to anoxic or anaerobic conditions (9–11). Although there may be an initial impact on the performance of the nitrifiers, recovery of nitrification occurs in a short time when aeration is resumed. For example, when a nitrifying sludge was kept anaerobic for 4 hr and then reaerated, its nitrifying ability increased to its original level after 20 min (10). Similar results have been obtained when nitrified poultry manure was stored without measurable DO. In one case the nitrified manure was stored for over 20 days, yet when aeration was again applied, rapid rates of nitrification occurred within 48 hr. In a submerged anaerobic filter, nitrification returned to normal conditions within 4 days of aeration after 7 days of anaerobic conditions (11). Obviously, the recovery rate of nitrification is related to the numbers of nitrifying bacteria in the system. While nitrification does not take place until adequate DO is in the biological system, when sufficient numbers of nitrifying bacteria are present, anoxic conditions of the aforementioned duration did not have a detrimental effect on the nitrifying ability of the organisms.

Temperature

Nitrification is affected by the temperature of the medium. Pure culture studies indicated that the optimum growth of nitrifiers occurred between 30 and 36°C (4).

A theta (θ) factor of 1.070 has been suggested for relating the rates of nitrification at different temperatures (12), indicating that the growth rate constant roughly doubles for each 10°C increase in the range of 6–25°C. Such temperature relationships indicate why high rates of nitrification may be difficult to obtain in the winter.

Information on nitrification at low temperatures indicates conflicting data. Different studies have indicated that nitrification did not develop below 10°C (13), that it was possible to maintain nitrification at 8°C (14), that little nitrification was achieved at temperatures below 6°C (10), and that nitrification could be accomplished at temperatures as low as 1°C (11). These studies provided little information to compare their conclusions regarding the temperature effect with the prevailing SRT values in the nitrification units. It is quite probable that when nitrification is reported as not occurring at low temperatures, the SRT of the unit was not exceeding the slower rate of the organisms at the lower temperatures. When longer SRT values are maintained at lower temperatures to compensate for the slower nitrifying organism growth rate, nitrification should occur even at low temperatures.

pH

The optimum pH for the growth of the nitrifiers is not sharply defined, but in pure cultures it has been shown to be generally on the alkaline side (Fig. 12.3).

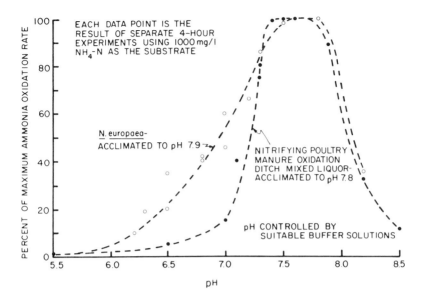

Fig. 12.3. Nitrification related to solids retention time (2).

During nitrification, hydrogen ions are produced [Eq. (12.1)], and as a result the pH drops. Theoretically, a decrease of 7.1 gm of alkalinity as $CaCO_3$ should occur for every gram of nitrate nitrogen produced. If the system has adequate buffer capacity and/or if the ammonia concentration is low, the pH will decrease only slightly. If inadequate buffer capacity exists the pH will decrease noticeably. A decrease in pH can be a practical measure of the onset of nitrification. The optimum range of pH for nitrification is between 7.5 and 8.5

Nitrification can proceed at low pH levels. In a study of the aerobic digestion of sewage sludge, the pH of the mixed liquor was in the 5.0–5.5 range during active nitrification (15). The alkalinity decreased from 700 to 100 mg per liter. Nitrification has occurred in both batch and continuous systems treating poultry manure wastewater at pH levels in the 5.0–5.5 and 6.3–6.8 ranges. These ranges occurred naturally, because of the oxidation of the high concentrations of nitrogen in this wastewater. Low pH values resulting from nitrification have also been noted in compost piles and in soils.

Nitrifying organisms can adapt to low pH levels and achieve adequate nitrification. When the pH was adjusted to a range of 5.5–6.0 in a submerged aerobic filter (8) the nitrifiers adapted to the lower pH levels, and the rate of ammonia oxidation reached that comparable to what had been achieved at a pH of 7.0.

Ammonia and Nitrite Concentrations

Although NH_4^+ is the energy source of the nitrifiers, excessive amounts can inhibit the growth of these bacteria. Ammonia is more inhibitory to *Nitrobacter* than to *Nitrosomonas*, and the inhibition of *Nitrobacter* is caused by un-ionized (free) ammonia rather than the total ammonia concentration.

High nitrite concentrations can reduce the activity of the nitrifying organisms at low pH levels. The nitrite toxicity is due to the undissociated nitrous acid (HNO_2) rather than the nitrite ion.

The concentration of free ammonia increases with an increase in the pH, and the concentration of undissociated nitrous acid increases with a decrease in pH. The reactions in the nitrification process (Fig. 12.4) illustrate the factors involved. A portion of the organic nitrogen entering an aerobic waste treatment system is converted to ammonia. The ammonia will be in solution as the ammonium ion (NH_4^+) and un-ionized ammonia (NH_3). As noted earlier, these compounds will be in an equilibrium which is affected by the pH of the liquid. When the pH increases, the concentration of the un-ionized ammonia will increase.

As nitrite oxidation occurs, there is a release of hydrogen ions, which decrease the pH to an extent related to the buffering capacity of the system. The nitrite formed will exist in equilibrium with un-ionized nitrous acid (HNO_2). As the pH decreases, the concentration of HNO_2 will increase.

Two processes work to reduce NH_3 inhibition. As the pH decreases, the ammonia equilibrium will adjust, and the concentration of NH_3 will decrease. In addition, the total ammonia concentration will decrease as it is oxidized to nitrite. These reductions tend to relieve the inhibition of the nitrobacter organisms that is caused by NH_3, promoting oxidation of nitrite to nitrate.

There are, however, situations in which nitrite is not oxidized to nitrate under conditions generally favoring compiete nitrification, i.e., in the absence of NH_3 and with adequate aeration. The continued presence of nitrite in the absence of NH_3 can be a result of HNO_2 inhibition, since as the pH decreases occur as a result of ammonia oxidation, the HNO_2 concentration increases. Certain con-

AMMONIA EQUILIBRIUM

$$\text{ORGANIC NITROGEN} \longrightarrow \boxed{NH_4^+ + OH^- \rightleftharpoons NH_3 + H_2O}$$

NITROUS ACID EQUILIBRIUM

$$NH_4 + 1.5\ O_2 \longrightarrow H_2O + H^+ + \boxed{H^+ + NO_2^- \rightleftharpoons HNO_2}$$

$$NO_2^- + 0.5\ O_2 \longrightarrow NO_3^-$$

Fig. 12.4. Nitrification reactions and ammonia and nitrous acid equilibria. [Adapted from (16).]

centrations of HNO_2 can be inhibitory to both the nitrosomonad and the nitrobacter organisms.

Free ammonia inhibits nitrite-oxidizing organisms (nitrobacters) at concentrations substantially lower than those that inhibit ammonia-oxidizing organisms (nitrosomonads). It is this differential that can cause nitrites to accumulate in a nitrifying system, without subsequent oxidation to nitrate. Nitrous acid can inhibit nitrobacters, thus also leading to an accumulation of nitrite. Inhibition by these compounds can be relieved by adjusting the conditions such that the respective concentrations are less than inhibitory concentrations. Dilution, pH adjustment, and denitrification specifically for nitrous acid can be successful in reducing inhibitory concentrations.

Many environmental factors affect the free ammonia and nitrous acid concentrations that are inhibitory. The free ammonia concentrations that begin to inhibit nitrosomonads are in the range of 10–150 mg per liter, and those that begin to inhibit nitrobacters are in the range of 0.1–1.0 mg per liter. The nitrous acid concentrations that begin to inhibit nitrobacters are in the range of 0.2–2.8 mg per liter (16).

Denitrification

General

Microbial denitrification takes place under anoxic conditions where nitrites and nitrates are used as terminal electron acceptors in place of molecular oxygen. Fundamental knowledge on this subject primarily stems from the observations reported on denitrification in soils. Denitrification rates are a function of the respiration rate of the organisms in the waste mixture or soil, the dissolved oxygen in the immediate vicinity of the organisms, and the availability of nitrite and nitrate in the liquid surrounding the organisms.

The nitrates and nitrites are reduced to gaseous nitrogen, resulting in a reduced nitrogen content of the wastewater as the gas escapes from the liquid. The composition of the gas produced is a function of the environmental conditions. Nitrogen gas (N_2) is the primary gaseous end product. Other gases produced include nitrous oxide (N_2O), nitric oxide (NO), carbon dioxide (CO_2), and hydrogen (H_2). These gases are innocuous and have no undesirable effect in the environment. Denitrification therefore offers a mechanism for removal of nitrogen from wastewaters that causes few additional problems. Uncontrolled denitrification in secondary clarifiers can hinder sedimentation.

Denitrification is brought about by facultative heterotrophic bacteria. Most of the active denitrifying organisms belong to the genera *Pseudomonas, Achromobacter, Bacillus,* and *Micrococcus.*

True dissimilatory nitrate reduction occurs when both nitrite and nitrate are used as electron acceptors, without the accumulation of potentially toxic concentrations of end products such as nitrite and ammonia. This reaction is different from the process of nitrate reduction for the purpose of cell protein synthesis, which is known as assimilatory nitrate reduction. The primary end product of this latter process is ammonium, which enters the biochemical pathways leading to protein synthesis.

The primary reactions involved in denitrification are

$$\text{Reduced organic matter} + NO_3^- \xrightarrow{\text{bacteria}}$$
$$NO_2^- + CO_2 + H_2O + \text{oxidized organic matter} \qquad (12.5)$$

which illustrates the reduction of nitrate to nitrite and

$$\text{Reduced organic matter} + NO_2^- \xrightarrow{\text{bacteria}}$$
$$N_2 + CO_2 + H_2O + \text{oxidized organic matter} \qquad (12.6)$$

which illustrates the reduction of nitrite to nitrogen gas. During dentrification, the pH will increase as hydroxide ions are produced. The degree of pH change is related to the amount of denitrification and the buffer capacity of the liquid.

To achieve high nitrogen removals by biochemical denitrification, a high degree of nitrification is a prerequisite. Denitrification as a method of nitrogen control is useful where a nitrified liquid exists (such as from irrigation drains), where a nitrified wastewater can be produced economically, and where it can be shown that nitrogen removal will have a beneficial effect. Denitrification is an important factor in soils since it reduces the amount of nitrate that can be leached to surface waters and groundwaters and because it reduces the amount of nitrogen available for plant growth.

There are many environmental factors that affect the rate of denitrification in soils and controlled biological treatment systems. The important factors governing denitrification in an ecosystem include organic matter, oxygen tension, pH, and temperature.

Organic Matter

Substrates that can act as hydrogen donors are necessary for the denitrification of oxidized nitrogen. These are oxidizable organic compounds and act as energy sources for the denitrifying population. The compounds must be available in concentrations that meet the combined requirements of organism growth and nitrite and nitrate reduction. The rate of denitrification is a function of the rate of carbon utilization and the resultant electron acceptor demand. Both endogenous and exogenous carbon sources have been explored to achieve denitrification.

Denitrification in a biological system can be accomplished by using the endogenous electron acceptor demand of the microorganisms without the addition of any exogenous electron donors. The reaction can be illustrated as

$$C_5H_7O_2N + 4NO_3^- \rightarrow 5CO_2 + NH_3 + 2N_2 + 4OH^- \qquad (12.7)$$

Based upon Eq. (12.7) approximately 1.9 mg of microbial solids are required to reduce 1.0 mg of NO_3-N. Since not all of the microbial cell is oxidized, a greater amount of microbial cells per unit of nitrate will be needed than the theoretical amount.

The endogenous electron acceptor demand of the microorganisms occurs at a low rate, and exogenous hydrogen donors are added to increase the denitrification rates. Untreated sewage and other wastes have been added for this purpose but have been undesirable, since these sources contain nitrogen in forms other than nitrate and can add unwanted amounts of organic and ammonia nitrogen to the effluent from the treatment facility. Where high nitrogen removals may not be necessary, such as where the wastes are to be applied to the land, untreated wastes can be an acceptable source of hydrogen donors.

To achieve high nitrogen removals a simple carbon source, such as sugar or alcohol, is preferred.

$$5CH_3OH + 6NO_3^- \rightarrow 5CO_2 + 3N_2 + 7H_2O + 6OH^- \qquad (12.8)$$

$$5C_6H_{12}O_6 + 24NO_3^- \rightarrow 30CO_2 + 12N_2 + 18H_2O + 24OH^- \qquad (12.9)$$

The theoretical requirement of these compounds is 1.9 mg of methanol per mg of NO_3-N reduced and 2.6 mg of glucose per mg of NO_3-N reduced. The quantity of methanol (C_m) required for denitrification is a function of the nitrate, nitrite, and DO concentrations. The requirement has been expressed as (17):

$$C_m = 2.47NO_3\text{-}N + 1.53NO_2\text{-}N + 0.87DO \qquad (12.10)$$

All the terms are in milligrams per liter.

A number of organic compounds can be used with the decision usually made on the basis of the chemical cost. In many situations methanol has been the least-cost chemical and the basis for comparing other compounds. Some 30 industrial wastes, including whey, starch processing effluent, brewery spent-grain extract, and wine sludge concentrates, were evaluated as carbon sources for denitrification (18). The majority of those wastes exhibited denitrification rates that were equal to or greater than rates observed with methanol. Sulfide and sulfur also can be sources of election donors under certain conditions (19).

Dissolved Oxygen

Denitrification does not take place until the medium is essentially oxygen free. This has been taken to be zero DO. Some studies suggest that denitrification may take place up to about 0.5 mg per liter of DO, while other reports have indicated that nitrate was not reduced at DO concentrations of about 0.2 to 0.4 mg per liter.

The role of DO in denitrification is related to the approximately 16% difference in energy yield when DO and nitrate are used as the terminal electron acceptors. The energy yield is about 570 kcal per mole of glucose metabolized when nitrate is the terminal electron acceptor and about 686 kcal per mole of glucose when oxygen is the acceptor. The facultative organisms will utilize nitrate only when adequate oxygen is not available. The difference in energy yield indicates that fewer organisms will exist in a denitrification unit than in an aerobic unit with the same organic loading.

The DO concentration surrounding the microorganisms is the important factor governing denitrification. Even in aerobic treatment systems with measurable DO in the liquid, it is possible to have some denitrification occur if the microenvironment adjacent to the organisms is devoid of oxygen. Such conditions occur in soil denitrification and wherever localized anoxic pockets occur in aerobic systems.

Aerobic denitrification can have importance in systems with high, mixed liquor suspended solids concentrations, such as aerated systems treating agricultural wastes. In oxidation ditches used for the treatment of animal wastes, nitrogen losses attributed to denitrification occurred even though bacterial nitrification took place simultaneously and high DO concentrations were in the mixed liquor.

pH

No correlation between pH and other denitrification parameters has been observed, although it is believed that active denitrification occurs under neutral or slightly alkaline conditions. The microorganisms responsible for denitrification are ubiquitous and should be able to adapt to pH levels within the broad range of microbial activity of about 5–9.5. A study of denitrification of nitrified poultry manure wastewater indicated that the pH values that resulted in the nitrification step (5.0–6.5) were not detrimental to denitrifying organisms.

Solids Retention Time

Microorganisms obtain more energy when oxygen is used as a hydrogen acceptor than when nitrate is used. Therefore, the generation time of bacteria will

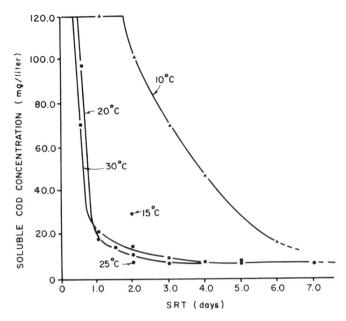

Fig. 12.5. Average steady-state effluent COD concentration during denitrification (20).

be greater and the cell mass yield will be less under denitrification conditions than under aerobic conditions. The longer generation time will require somewhat longer design SRTs in a biological denitrification reactor.

Based upon a study of suspended growth denitrification, the minimum SRT was found to be 0.5 days at 20 and 30°C and 2 days at 10°C (Fig. 12.5) (20). For practical applications a SRT of at least 3–4 days at 20 and 30°C, and 8 days at 10°C was recommended. Nitrate reduction efficiency in denitrification can be controlled by adjusting the SRT of the process to assure adequate denitrification rates as environmental conditions change.

Temperature

Microbial denitrification can occur over a wide temperature range extending from below 5 to over 60°C. Denitrification rates are negligible below about 2°C and increase as the temperature increases. Within the range of 5–27°C, an Arrhenius relationship of

$$K = K_0 e^{-E/RT} \tag{12.11}$$

was found to approximate the variation of denitrification rates with temperature (21).

Practical Application

A variety of systems have been proposed for denitrification of wastes and
wastewaters. These include anaerobic filters, denitrification using oxidation
ditches, anaerobic ponds, and modified land disposal systems.

The practice of denitrification after nitrification has been tried in laboratories, pilot plants, and large-scale plants to remove the oxidized forms of nitrogen
from the sewage effluents. Two different sets of conditions are necessary to
control the processes of nitrification and denitrification, and frequently the two
processes are separated. The majority of denitrification processes have been
applied to municipal wastewaters. Some information is available on the denitrification of wastewaters from agricultural operations.

The use of aerated units for storage and treatment of animal waste slurries
and wastewaters will minimize the odor problem normally present with anaerobic storage of the wastes and can provide some degree of treatment prior to
ultimate disposal. Nitrogen losses caused by denitrification can be obtained in
these units under specific operating conditions.

With long liquid retention times and adequate oxygen, the opportunity
exists for nitrogen removal by a nitrification–denitrification cycle. Nitrogen
losses have been determined by nitrogen balances on the units. In the majority of
cases, losses due to volatilization have been included in the balances or have
been determined to be negligible. The losses noted in Table 12.4 (22–25) represent those presumably due only to denitrification.

Uncontrolled nitrification–denitrification also has been observed in oxidation ditches for swine and aeration units for cattle feedlot wastes. Aerated units
treating animal wastes will contain ample unoxidized organic matter. Soluble
BOD levels range from 200 to 3000 mg per liter, and additional oxygen demand
occurs because of the particulate matter. No exogenous carbon compounds are
needed to achieve denitrification in these units.

TABLE 12.4

Nitrogen Losses Attributed to Denitrification—Aerated Poultry Waste Units

Liquid temperature (°C)	Dissolved oxygen (mg/liter)	Nitrogen loss (%)	Reference
12–23	2–7	36	(22)
5–21	2–6	66	(23)
12–23	0–2	50–60	(24)
11–18	0–6	70–80	(25)

Ammonia Stripping

General

Ammonia removal by desorption from a liquid waste has received close scrutiny as a method for nitrogen control. Studies with domestic sewage and industrial wastes have indicated that nitrogen removal by ammonia desorption is technically feasible. Two practical limitations have been noted: the inability to operate the process at ambient air temperatures near freezing and the buildup of calcium carbonate scale on a stripping tower.

Ammonia stripping is not an ultimate disposal process but simply transfers the nitrogen from a liquid to a gaseous phase. As such, the process is not a total solution to the excess nitrogen problem. If the stripped ammonia nitrogen falls into adjacent water bodies, it can contribute to nitrogen enrichment in these waters. Examples of such enrichment have occurred near cattle feedlots (Chapter 3).

If the stripped ammonia falls on cropland, there are minimal environmental problems, and the ammonia can be a benefit. Plants can utilize the ammonia in rain and can act as a natural sink for atmospheric ammonia. A field crop growing in air containing ammonia at normal atmospheric concentrations might satisfy as much as 10% of its nitrogen requirement by direct absorption of ammonia from the air. Annual ammonia absorption by plant canopies could be about 20 kg per hectare (26).

Ammonia is very soluble in water. However, only un-ionized ammonia (NH_3) [Eq. (12.4)] is available for desorption.

The desorption of ammonia from wastewaters involves the contacting of a liquid and gas phase in units such as spray towers, packed columns, aeration towers, or diffused-air systems. Whatever the mode of contacting, the two phases are brought together to transfer the ammonia. One phase usually flows countercurrent to the other. Ammonia being transferred from the liquid to the gas phase must pass through the interface (Fig. 12.6) and the gas and liquid film layers. During the removal of ammonia from an aqueous solution by air stripping, the greatest resistance to mass transfer occurs in the transfer from the liquid to the gas phase.

The amount of ammonia desorbed from a solution into air is directly proportional to the concentration of ammonia in the liquid, interfacial area of exposure, duration of desorption, temperature, and atmospheric pressure. If the total area of interfacial surface of a volume of liquid is A_i, then the change in the ammonia concentration in the liquid (ΔC) because of desorption from the entire surface in the duration Δt is

$$\Delta C = -k'A_i FC \, \Delta t \qquad (12.12)$$

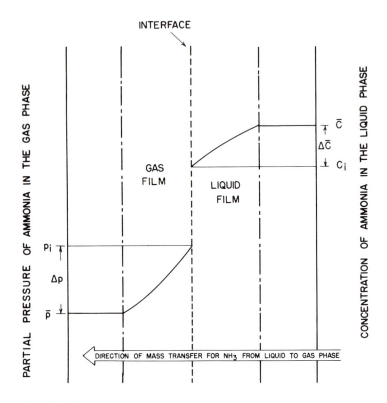

Fig. 12.6. Schematic of the transfer of ammonia from a liquid to a gas phase.

Only the desorbable ammonia concentration can be removed by stripping. The desorbable concentration is equal to the total ammonia concentration times the fraction available to be desorbed, F, [Eq. (12.4)], i.e., FC. Equation (12.12) indicates that greater quantities of ammonia can be desorbed by increasing the time of exposure, area of exposure, and the concentration of ammonia in the liquid. In this equation, k' is a desorption constant at a given temperature and pressure with units of area$^{-1} \cdot$ time^{-1}.

In ammonia desorption systems such as aeration towers or diffused aeration systems, the interfacial gas–liquid area is difficult to estimate, and the terms k' and A can be grouped into an overall desorption coefficient (K_D) for the system being utilized. K_D has the units of time^{-1}. The numerical values of K_D are system-specific and depend upon the nature of the liquid, the temperature, aeration device, rate of aeration, air-rate-to-liquid-volume or to-liquid-flow-rate ratios, and system turbulence.

Ammonia removal in batch desorption units can be expressed as

$$\log_e \frac{C_1}{C_2} = K_D F(t_2 - t_1) \tag{12.13}$$

where C_1 and C_2 are the concentrations of total ammonia at times t_1 and t_2, respectively.

Comparable equations can be written for continuous-flow desorption systems. If V is the volume of the liquid in the unit in which the desorption of ammonia is carried out, and Q is the rate of flow of the liquid into the unit, Q is the rate of outflow, assuming no losses due to evaporation. If C_1 and C_2 are the concentrations of ammonia in the influent and effluent, respectively, then a mass balance on the unit at equilibrium will show

$$Q(C_1 - C_2) = K_D F C_2 V \tag{12.14}$$

which can be written as

$$(C_1 - C_2)/C_2 = K_D F t_H \tag{12.15}$$

where t_H is the average liquid retention time in the unit and the average time of desorption in a continuous-flow desorption system.

For whatever decrease in ammonia that may be required, the time of liquid retention or desorption for which a batch or continous-flow unit should be designed may be determined if K_D and F are known. The factor F is dependent upon pH and temperature (Fig. 12.1) and can easily be determined; K_D is independent of pH but is a function of the factors noted earlier.

Equations (12.12) through (12.15) are valid in systems where the pH and temperature, and hence F, are constant. When these parameters are not constant, these equations must be modified.

Temperature

An increase in temperature will increase the desorption coefficient K_D and the removal of ammonia from a solution. Ammonia desorption from poultry and dairy manure wastewaters has provided information on the manner in which K_D changes with temperature (Fig. 12.7). The intercepts of the temperature–K_D lines have a positive intercept on the Y axis, indicating that the desorption of ammonia can occur only at temperatures greater than 3–5°C. The slope of these lines provides an estimate of the effect of temperature on the efficiency of ammonia desorption and the ability to predict the effect of changes in ambient conditions on the process. The temperature effect is greater at higher aeration

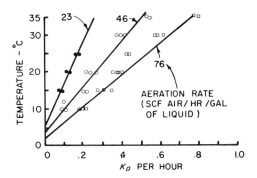

Fig. 12.7. Effect of temperature at different air flow rates on the ammonia desorption rate constant; batch units. [From (27).]

rates. The quantity of ammonia removed by a unit volume of air is directly proportional to the concentration in the liquid (Fig. 12.8) and is related to the temperature of the liquid.

Data on ammonia desorption from poultry and dairy manure wastewaters indicated that K_D values at different temperatures can be related by (27)

$$K_{D_2} = (K_{D_1})1.063^{(\theta_2 - \theta_1)} \tag{12.16}$$

where K_{D_1} and K_{D_2} are the desorption rates at the same rate of air flow and at temperatures θ_1 and θ_2 in °C.

At any given temperature, higher values of K_D can be obtained by increasing the rate of aeration.

Viscosity

Physical properties, such as viscosity, can influence the mass-transfer coefficient of gas transfer systems. Viscosity was found to affect the desorption coefficient K_D (Fig. 12.9). The data were obtained over a temperature range of 10–30°C. With an increase in the viscosity there was a proportionate decrease in K_D. This relationship was expressed as

$$K_D = a(\mu)^{-b} \tag{12.17}$$

where μ is the viscosity of the liquid, b is the slope of the lines in Fig. 12.9, and a is the value of K_D when μ is 1 centipoise. A comparison of the slopes, b, indicated that neither the type of liquid nor the air flow rate within the noted

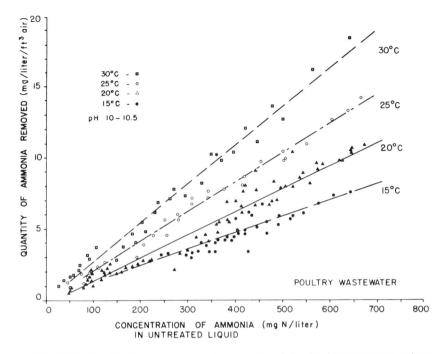

Fig. 12.8. Quantity of ammonia removed per quantity of air related to temperature and ammonia concentration. [From (27).]

range had a significant effect. The value of b ranged from 2.9 to 3.9, with an average of 3.5. The value of a represents an estimate of the maximum K_D value that can be obtained at the noted air flow rates. The fact that Eq. (12.17) was consistent over many air flows with different wastes and over a broad temperature range suggests its usefulness for general design purposes with wastes of different characteristics.

By knowing the viscosities of the suspensions of wastes and by using Eq. (12.17), it is possible to predict the value of K_D of a liquid (K_{D_2}) if the K_D of a similar liquid (K_{D_1}) and the viscosities of both liquids (μ_1 and μ_2) are known.

$$\left(\frac{K_{D_1}}{K_{D_2}} \right) = \left(\frac{\mu_1}{\mu_2} \right)^{-b} = \left(\frac{\mu_2}{\mu_1} \right)^{b} \qquad (12.18)$$

This relationship holds only when the air flows to be used for the desorption of the two liquids are the same.

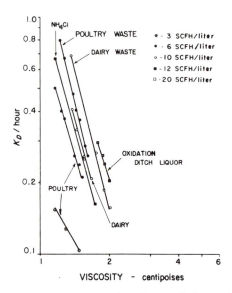

Fig. 12.9. Effect of viscosity on the desorption of ammonia from animal waste solutions; SCFH, standard cubic feet per hour. [From (27).]

Practical Application

Knowledge about the factors affecting gaseous ammonia losses can be important in the management of agricultural wastes. Under field conditions ammonia is lost from oxidation ponds, anaerobic lagoons, and manure storage systems. Waste management systems can have different nitrogen management goals, such as conservation to keep as much nitrogen as possible prior to application of wastes to cropland or loss to reduce nitrogen in runoff or leachate to the groundwater. The concepts presented in this section help in understanding the nitrogen loss phenomena that occur in natural systems as well as approaches to achieve a specific nitrogen management goal.

Considerable amounts of ammonia can be lost to the atmosphere from anaerobic lagoons. Losses ranging from 40 to 65% have been reported for anaerobic lagoons storing animal wastes (28). Increased concentrations of ammonia nitrogen have been measured in the air surrounding such lagoons.

Ammonia stripping also has occurred from oxidation ponds that have a high pH. The pH can occur naturally, such as when there is considerable sunlight and algal activity (Chapter 6), or can result from the addition of lime and other bases. In a full-scale oxidation pond, an effluent ammonia nitrogen concentration of about 1 mg per liter was achieved (29). The pond hydraulic retention time was about 11 days, the pond temperature was in the 25–30°C range, and the influent ammonia nitrogen concentration was about 25 mg per liter.

Ammonia nitrogen losses ranging from 3 to 60% can occur when manure is handled, stored, and applied to land. The estimated nitrogen, phosphorus, and potash losses that can occur with different manure management systems are noted in Table 12.5. Phosphorus and potash losses are less than those of nitrogen, since there are greater opportunities for nitrogen loss (volatilization and denitrification) than for phosphorus and potash loss. The latter two nutrients are conserved in manure storage units and losses occur only by surface runoff or by retention in settling basins and lagoons.

When manure is spread on the land surface and not incorporated for some period of time, nitrogen losses will occur especially when the soil temperatures are warm and the weather is breezy. An increase in airflow rate (wind velocity)

TABLE 12.5

Estimated Nutrient Losses When Manure Is Handled, Stored, and Applied to Land[a]

Method	Nutrient loss (%)		
	Nitrogen	Phosphorus	Potash
Daily cleaning and surface spreading of solid manure, not tilled on day of spreading	50	10	10
Stacking in covered storage with solid spreading, soil incorporated	15	10	10
Stacking in uncovered storage with solid spreading, soil incorporated	20	10	15
Stacking in uncovered storage with solid spreading, surface spread	35	10	15
Manure stored on dirt dry lot, scraped, hauled, and surface spread	65	15	15
Under-the-floor manure pits or storage tanks, liquid spreading, soil incorporated	30	5	5
Under-the-floor manure pits or storage tanks, liquid spreading, surface spread	45	10	10
Paved dry lot with runoff control to holding pond, applied by irrigation	80	40	30
Flush cleaning with liquid manure, drained to lagoon, applied by irrigation, or tank spread and injected	85	50	40
Flush cleaning with liquid manure drained to settling basin and lagoon, applied by irrigation	90	60	50

[a]Adapted from (30).

TABLE 12.6

**Estimated Nitrogen Lost from When Manure Is Spread on the
Soil Surface**[a]

Days between application and incorporation	Nitrogen loss (%)		
	Warm dry soil	Warm moist soil	Cool moist soil
1	30	10	0
4	40	20	5
7 or greater	50	30	10

[a]Adapted from (30).

above the soil surface increases ammonia losses. These conditions, plus the large surface area from which the ammonia can be released, enhance ammonia volatilization losses. Table 12.6 provides estimates of the nitrogen losses that can occur when the manure is surface applied. Application and retention of manure on the surface results in greater ammonia losses than does direct manure injection or rapid incorporation after application.

The manner in which manure is added to storage units will affect nitrogen losses. When manure is added to the surface of manure storages, the surface of the stored manure is disturbed, and the ammonia has a greater chance for volatilization. Surface-loaded storage units can have nitrogen losses that range from 25 to 40%, whereas bottom loaded units reduce nitrogen losses to the 5–10% range. All things being equal, bottom-loaded storage units conserve the greatest amount of nitrogen added to the units (31).

Chemicals also can be used to inhibit ammonia losses. Lime used at the rate of about 8 kg of CaO per ton of fresh manure will increase the pH of the manure to about 10 and inhibit urea hydrolysis and the formation of ammonia (32). Lesser amounts of lime can increase ammonia losses by increasing the pH and NH_3 concentrations, without inhibiting urea hydrolysis. The desirability of adding lime requires evaluation on a case-by-case basis.

Summary

The relationships presented in this section permit a better understanding of ammonia desorption as a nitrogen removal mechanism in both natural and treatment systems. The relationships can be used to evaluate existing facilities for their efficiency of ammonia desorption and to aid in the design of equipment suitable for ammonia conservation.

References

1. Wezernak, C. T., and Gannon, J. J., Evaluation of nitrification in streams. *J. Sanit. Eng. Div., Am. Soc. Civ. Eng.* **94,** SA5, 883–895 (1968).
2. U.S. Environmental Protection Agency, "Quality Criteria for Water," Office of Water and Hazardous Materials, Washington, D.C., 1976.
3. European Inland Fisheries Advisory Commission, Water quality criteria for European freshwater fish. *Water Res.* **7,** 1011–1022 (1973).
4. Alexander, M., Nitrification. *In* "Soil Nitrogen" (W. V. Bartholomew and F. E. Clark, eds.) Publ. No. 10, pp. 307–343. Agric. Soc. Agron., Madison, Wisconsin, 1965.
5. Gibbs, M., and Schiff, J. A., Chemosynthesis: The energy relationships of chemoautotrophic organisms. *In* "Plant Physiology" (F. C. Steward, ed.), Vol. 18, pp. 279–319. Academic Press, New York, 1960.
6. Sawyer, B., Lue-Hing, C., Zenz, D. R., and Obayashi, A. "Estimation of the Maximum Growth Rate of Ammonia Oxidizing Nitrifying Bacteria Growing in Municipal Wastewater," Report 81-5. Department of Research and Development, Metropolitan Sanitary District of Greater Chicago, Chicago, Illinois, March, 1981.
7. Wild, H. E., Sawyer, C. N., and McMahon, T. C., Factors affecting nitrification kinetics. *J. Water Pollut. Contr. Fed.* **43,** 1845–1854 (1971).
8. Stenstrom, M. K., and Poduska, R. A., The effect of dissolved oxygen concentration on nitrification. *Water Res.* **14,** 643–644 (1980).
9. Sutton, P. M., Jank, B. E., and Vachon, D., Nutrient removal in suspended growth systems without chemical addition. *J. Water Pollut. Contr. Fed.* **52,** 98–109 (1980).
10. Downing, A. L. Painter, H. A., and Knowles, G., Nitrification in the activated sludge process, *Inst. Sewage Purif., J. Proc.* **63,** 130–158 (1964).
11. Haug, R. T., and McCarty, P. L., Nitrification with the submerged filter. *J. Water Pollut. Contr. Fed.* **44,** 2086–2102 (1972).
12. Balakrishnan, S., and Eckenfelder, W. W., Discussion—recent approaches for trickling filter design. *J. Sanit. Eng. Div., Am. Soc. Civ. Eng.* **95,** SA1, 185–187 (1969).
13. Rimer, A., and Woodward, R. L., Two stage activated sludge pilot plant operations—Fitchburg, Massachusetts. *J. Water Pollut. Contr. Fed.* **44,** 101–116 (1972).
14. Mullbarger, M. C., Nitrification and denitrification in activated sludge systems. *J. Water Pollut. Contr. Fed.* **43,** 2054–2070 (1971).
15. Jaworski, N., Lawton, G. W., and Rohlich, G. A., Aerobic sludge digestion. *Int. J. Air Water Pollut.* **5,** 93–102 (1963).
16. Anthonisen, A. C., Loehr, R. C., Prakasam, T. B. S., and Srinath, E. G., Inhibition of nitrification by ammonia and nitrous acid. *J. Water Pollut. Contr. Fed.* **48,** 835–852 (1976).
17. McCarty, P. L., Beck, L., and St. Amant, P., Biological denitrification of waste water by addition of organic chemicals. *Proc. Purdue Ind. Waste Conf.* **24,** 1271–1285 (1969).
18. Monteith, H. D., "Evaluation of Industrial Waste Carbon Sources for Biological Denitrification," Environment Canada Rep. EPS 4-WP-79-9. Environmental Protection Service, Ottawa, Ontario, 1979.
19. Driscoll, C. T., and Bisogni, J. J., The use of sulfur and sulfide in packed bed reactors for autotrophic denitrification. *J. Water Pollut. Contr. Fed.* **50,** 569–577 (1978).
20. Stensel, H. D., Loehr, R. C., and Lawrence, A. W., Biological kinetics of suspended growth denitrification. *J. Water Pollut. Contr. Fed.* **45,** 244–261 (1973).
21. Dawson, R. N., and Murphy, K. L., The temperature dependency of biological denitrification. *Water Res.* **6,** 71–83 (1972).

22. Loehr, R. C., Anderson, D. F., and Anthonisen, A. C., An oxidation ditch for the handling and treatment of poultry wastes. *In* "Livestock Waste Management and Pollution Abatement," pp. 209–212. Am. Soc. Agric. Engrs., St. Joseph, Michigan, 1971.

23. Stewart, T. A., and McIlwain, R., Aerobic storage of poultry manure. *In* "Livestock Waste Management and Pollution Abatement," pp. 261–263. Am. Soc. Agric. Engrs., St. Joseph, Michigan, 1971.

24. Hashimoto, A. G., "An Analysis of a Diffused Air Aeration System under Caged Laying Hens," Paper NAR-71-428. Am. Soc. Agric. Eng., St. Joseph, Michigan, 1971.

25. Dunn, G. G., and Robinson, J. B. Nitrogen losses through denitrification and other changes in continuously aerated poultry manure. *In* "Proceedings of the Agricultural Waste Management Conference," pp. 545–554. Cornell University, Ithaca, New York, 1972.

26. Hutchinson, G. L., Millington, R. J., and Peters, D. B., Atmospheric ammonia: Adsorption by plants leaves. *Science* **175**, 771–772 (1972).

27. Srinath, E. G., and Loehr, R. C., Ammonia desorption by diffused aeration. *J. Water Pollut. Contr. Fed.* **46**, 1940–1956 (1974).

28. Koelliker, J. K., and Miner, J. R., Desorption of ammonia from anaerobic lagoons. *Trans. Am. Soc. Agric. Eng.* **16**, 148–151 (1973).

29. Idelovitch, E., and Michail, M., Nitrogen removal by free ammonia stripping from high pH ponds. *J. Water Pollut. Contr. Fed.* **53**, 1391–1401 (1981).

30. Christensen, L. A., Trierweiler, J. R., Ulrich, T. J., and Erickson, M. W., "Managing Animal Wastes—Guidelines for Decision Making," Economic Research Service, ERS-671. U.S. Dept. of Agricultre, 1981.

31. Muck, R. E., and Steenhuis, T. S., Nitrogen losses from manure storages. *Agric. Wastes* **4**, 41–54 (1982).

32. Muck, R. E., and Herndon, F. G., "Liming to Reduce Nitrogen Losses in Dairy Barns," Paper 83-4063. Presented at the Summer Meeting of the American Society of Agricultural Engineers, Montana, 1983.

13

Physical and Chemical Treatment

Introduction

Because agricultural wastes are biodegradable, some form of biological treatment generally represents an economic waste management approach. In addition, because many agricultural wastes are produced in rural areas, land treatment is an environmentally sound method of treatment, disposal, and reuse.

There are, however, a number of physical and chemical treatment methods that should receive consideration as part of an overall waste management system. These methods can (a) provide pretreatment of a wastewater prior to discharge to a municipal sewer system, (b) recover by-products, (c) reduce the size and operational costs of subsequent treatment processes, and (d) decrease the microbial concentration of effluents.

Physical and chemical treatment methods are not new. However, advances in equipment and new applications constantly occur. Examples of the use of these methods with agricultural wastes include the following. Screens and clarifiers (sedimentation) are utilized to separate solids from wastewaters. Air flotation units can separate grease from meat and poultry processing wastewaters. Wastewaters can be reclaimed for utilization of the treated effluent or the separated organic or inorganic fractions. Carbon adsorption has been used to recondition olive brines for reuse (1). A mixed-bed ion exchange unit has been used to reduce the salt content of food processing brines (2). Similar reuse opportunities exist with other brines such as those from sauerkraut and pickle processing.

369

Physical and chemical processes have had broad application in the treatment of agricultural wastes. The processes discussed in this chapter are disinfection, screening, sedimentation, flotation, and chemical precipitation. Additional discussion of these and other physical and chemical processes can be found in other texts (3–5).

Disinfection

The purpose of disinfection is to reduce the total bacterial concentration and to eliminate the pathogenic bacteria in water. The microbes of sanitary significance in water are either pathogenic (disease-producing) or nuisance-producing organisms. The production of a potable water supply requires that the bacterial concentration be zero or very low to avoid disease transmission. Disinfection has been practiced since the early 1900s for public water supplies.

There has been an increased emphasis on the disinfection of wastewaters before they are released. This emphasis is caused by increased recreational use of surface waters and by continual use of surface waters for water supply and waste disposal. Many states require disinfection of the effluent from all waste treatment facilities discharging to surface waters.

Pathogenic organisms in human and animal wastes may survive for days or more in surface waters, depending upon environmental conditions. Factors affecting microbial survival include pH, temperature, nutrient supply, competition with other organisms, ability to form spores, and resistance to inhibitors. The ability of pathogenic organisms to cause disease in man depends upon their concentration, virulence, and ingestion and resistance by their hosts. The pathogenic organisms of interest include bacteria, protozoa, and viruses.

In the disinfection of waters and wastewaters, the presence of specific pathogens is not determined. Rather, the reduction of a group of indicator organisms, coliforms, to acceptable concentrations is used as a measure of disinfection efficiency and performance. The coliform organisms are ubiquitous organisms commonly found in the intestinal tract of man and animals. There are a number of strains of coliform organisms that are measured by the traditional coliform test (6), such as *Escherichia coli,* which is a common coliform organism in the intestinal tract of man and animals. *Aerobacter aerogenes* is a coliform organism usually of nonfecal origin.

Generally, it is the entire group of coliform organisms that is used as an indicator of bacteriological quality, safety of potable water supplies, and efficiency of disinfection. The use of the coliform index is based on two premises: (a) that pathogenic organisms are equally or more susceptible to disinfection that coliform organisms and (b) the presence of coliform organisms suggests the possible presence of human or animal wastes and pathogenic organisms.

A direct count of coliform organisms in a water sample usually is not done. The most probable number (MPN) test is used to estimate the number of organisms in a sample of water. The MPN is a statistical value indicating the most probable concentration of organisms that were presented in the initial sample. A number of different size water samples are examined for the presence of coliform organisms, by use of specific fermentation tube methods. A membrane filter method, which permits more direct counts, may be preferred (6).

Disinfection of potable waters and wastewaters is used to reduce the concentration of coliform organisms to a low level. With potable waters, the level is such that the possibility of pathogenic organisms being in the water is extremely remote. Drinking water standards in the United States limit the number of coliform organisms in potable water to less than 1 per 100 ml. The combined value of coliform organisms as indicators of potential pathogens, as part of bacteriological standards for drinking water quality, and as indicators of disinfection to meet these standards has been demonstrated by the decrease in waterborne bacterial diseases since the adoption of these concepts and controls.

There are a number of chemicals and methods that can be used for disinfection, such as chlorine, iodine, ozone, quaternary ammonium compounds, and ultraviolet light. Due to low cost, efficiency of disinfection, and ease of application, chlorine is the most common chemical used for disinfection. Both gaseous chlorine and solid chlorine compounds such as calcium or sodium hypochlorite can be used.

The chlorine demand of a water is a function of many factors. Chlorine is an oxidizing agent and will react with a number of compounds, including organic waste matter, reduced inorganics, and living matter such as microorganisms. Survival of microorganisms after disinfection will depend upon the microbial species, the protection afforded by solids, and environmental conditions. The efficiency of chlorine disinfection is influenced by the amount and type of chlorine present in the solution, the concentration of other chlorine-demanding substances, time of contact, temperature, and the type and concentration of microbial life. The chlorine demand will be less in waters having low turbidity and suspended solids. The chlorine demand of filtered potable water is less than that of treated waste water effluent.

Chlorine combines with a wide variety of materials in a water, and a number of reactions occur. Many of them compete for the use of chlorine for disinfecting purposes. The chlorine demand of a water must be satisfied before chlorine residuals remain available for disinfection. The residual chlorine compounds are able to be disinfecting agents because they inactivate key enzyme systems in microorganisms after the chlorine species has penetrated the cell wall. Chlorine compounds capable of disinfection are either free available chlorine (FAC) or combined available chlorine (CAC) compounds. The FAC compounds are hypochlorous acid (HOCl) and hypochlorite ion (OCl$^-$). These can be formed by the addition of either gaseous chlorine or hypochlorite salts to water, i.e.,

$$Cl_2 + H_2O \rightarrow H^+ + Cl^- + HOCl \qquad (13.1)$$

$$NaOCl + H_2O \rightarrow Na^+ + OH^- + HOCl \qquad (13.2)$$

Use of gaseous chlorine tends to decrease the pH of a water as hydrogen ions are produced, while use of hypochlorites tends to increase the pH as hydroxyl ions are produced. The disinfecting ability of chlorine compounds increases as the temperature increases.

The distribution of hypochlorous acid and hypochlorite ion is a function of pH:

$$HOCl \rightleftharpoons H^+ + OCl^- \qquad (13.3)$$

and also is affected by the temperature of the solution. The disinfecting properties of the two FAC compounds are different, with the hypochlorous acid being the better disinfecting agent. Free available chlorine exists as HOCl at low pH levels. Below a pH of 6.5, hypochlorous acid predominates, while above a pH of about 8.5, hypochlorite ions predominate.

Chlorine also will react with other chemicals in the solution to produce additional disinfecting compounds. When ammonia is present, chloramines will be formed.

$$NH_4^+ + HOCl \rightarrow NH_2Cl \text{ (monochloramine)} + H_2O + H^+ \qquad (13.4)$$

$$NH_2Cl + HOCl \rightarrow NHCl_2 \text{ (dichloramine)} + H_2O \qquad (13.5)$$

$$NHCl_2 + HOCl \rightarrow NCl_3 \text{ (nitrogen trichloride)} + H_2O \qquad (13.6)$$

Nitrogen trichloride is not observed at neutral or greater pH values. The relative concentrations of mono-and dichloramines are a function of pH and temperature. Above a pH of about 8.5 monochloramines predominate, while below a pH of 6.5 dichloramines predominate. While the mono- and dichloramines are disinfecting agents, they require greater contact times to accomplish satisfactory disinfection. About 25 times the chloramine concentration is needed to obtain the same bacterial reduction as free available chlorine during the same contact period.

Free available chlorine compounds will not exist in a water until after the ammonia has been oxidized. The point at which this occurs is called the breakpoint. In waters containing ammonia, chlorine compounds will exist as combined available chlorine before the breakpoint and as free available chlorine after the breakpoint (Fig. 13.1). This point generally is not reached until approximately 9.5 mg per liter of Cl_2 is added per 1.0 mg per liter of NH_4-N. With treated municipal effluents having ammonia concentrations of about 20–30 mg per liter, the practice of breakpoint chlorination would require about 190–280 mg per liter of chlorine. Agricultural wastewaters can have much higher ammonia concentrations. As a result, breakpoint chlorination may be practiced in potable water production but is rarely practiced in disinfection of wastewaters. In order for wastewater effluent chlorination to be practical and effective, the wastewater must be well treated.

Chlorination of wastewaters or treated effluents also will reduce the BOD of the waters and will oxidize reduced compounds. Soluble ferrous and manganous compounds can be oxidized to insoluble ferric and manganic compounds by chlorine. Hydrogen sulfide can be oxidized by chlorine. The reaction occurs at a pH about 9. Theoretically, the oxidation of 1 mg per liter of H_2S requires about 8.5 mg per liter of chlorine. Nitrites are oxidized to nitrates by chlorine. The oxidation of 1 mg per liter of nitrite requires about 1.5 mg per liter of chlorine. Any BOD reduction or oxidation of inorganic compounds is incidental to the basic chlorination process since chlorination is not practiced solely to reduce the oxygen demand of wastewaters.

Available information on the application of chlorination to agricultural wastes is not extensive. Chlorination of effluents from duck farm wastewater facilities on Long Island is required, with the requirement being that a minimum chlorine residual of 0.5 mg per liter should persist after 15 min contact time.

Lime and chlorine have been used to suppress odors in hog wastes. There are more appropriate methods to minimize odors and to treat wastes. The most appropriate role of chlorination with agricultural wastes is to reduce the bacterial concentration of any wastes discharged to surface waters.

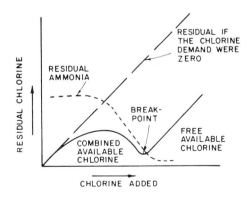

Fig. 13.1. Free and combined chlorine distribution in water.

Chlorination is an effective process for wastewater disinfection. Capital and operating costs are relatively modest when compared with some of the other processes required in wastewater treatment. Control of the process is straightforward and relatively simple. When properly applied and controlled, chlorination of wastewaters for disinfection is an effective measure for improving the bacteriological quality of the wastewater and for protecting humans and animals against transmission of enteric diseases by the water route.

Screening

Separation and concentration of the solids fraction from wastewaters is used to reduce the solids demand on subsequent handling and treatment processes and to recover solids that can be recovered for other uses such as animal feeding. Both stationary (static) and vibrating screens have been used.

Static screens intercept the wastewater, allow the liquid to pass through, and keep the solids on the surface. As the solids accumulate on the screen, they move downslope under forces caused by gravity and the fluid. A schematic of the components of a static screen is shown in Fig. 13.2. The wastewater or slurry enters a headbox, overflows a weir at the top of the screen, and flows downward over the screen, which contains three different slopes (25°, 35°, and 45°). Most of the free liquid is removed by the bottom of the first (25°) slope. More liquid is removed on the second slope as the solids roll down. The solids decelerate on the final 45° slope, which allows for final drainage. Practically all the free liquid is removed by the time the solids fall from the screen. Openings range from 0.005 to 0.10 in.

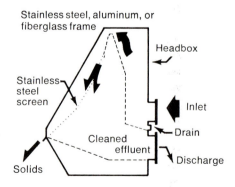

Fig. 13.2. Schematic of a static screen. [Courtesy of C-E Bauer.]

Fig. 13.3. Separation of solids from tomato processing wastewater by a static screen. [Courtesy of C-E Bauer.]

Static screens have been used successfully to separate solids from the following wastewaters: fruit and vegetable processing; seafood processing; meat and poultry processing; and brewery, winery, distillery, and feedlot slurries. Figure 13.3 illustrates the use of a static screen to separate solids from tomato processing wastewaters.

Static screens have the following advantages: (a) low initial cost, (b) compactness, (c) no motors or moving parts, (d) requirement for little if any attention, (e) performance not seriously affected by variation in flow rate or loading (f) production of a possibly useful by-product, and (g) adaptability to fit into many locations.

Vibrating screens use a motor-driven perforated screen to remove solids present in wastewater. The screen housing is supported on springs which are caused to vibrate by an eccentric drive. The screen surface may be flat or slightly funnel shaped. The vibrating motion causes the separated solids to move to the periphery of the screen where they are removed while the liquid passes through.

Vibrating screens have been used for the treatment of meat, poultry, fruit, and vegetable wastewaters. Suspended solids removals are a function of both particle size and screen size. Screen sizes can range from 0.02- to 1.0-cm openings. Suspended solids removals in the range of 20–60% have been achieved. Vibrating screens can be more sensitive to flow and solids content variations of the wastewater.

For both static and vibrating screens, maintenance to prevent screen blinding will be necessary when wastes from fish, meat, and poultry processing operations are used. These wastes can have a high grease content, which can result in screen blinding. Cleaning with detergent, caustic, and/or hot water solutions may be necessary.

Sedimentation

Sedimentation is a widely used process in waste treatment to separate settleable solids. It can be used to remove solids from a wastewater prior to discharge to a municipal treatment system (pretreatment), prior to a biological treatment process (primary clarification), and after a biological treatment process (secondary clarification). The process has been used with many agricultural wastes, particularly those from meat, poultry, fruit, and vegetable processing operations.

The principal design factor for clarifiers or sedimentation tanks is the hydraulic or surface overflow rate. The overflow rate is directly related to the settling velocity of the particulate matter that is to be removed. To remove slow-settling particles, a clarifier is designed to have a low overflow rate. To remove particles that settle rapidly, a higher overflow rate can be used. The relationship between these factors is critical to the understanding of the proper design and operation of sedimentation tanks.

Wastewaters contain a wide variety of solids having different densities and settling characteristics and ranging from discrete particles to flocculant solids. While recognizing that sedimentation of wastewater solids is not an "ideal" situation, the factors involved in sedimentation can be observed by considering the gravity settling of discrete particles in an ideal sedimentation basin. Figure 13.4a represents an ideal basin of length L, width W, and depth D. Each discrete particle will have a horizontal and vertical velocity. The horizontal velocity is determined by the vertical cross-sectional area, and the flow, i.e., $V_H = Q/DW$, where Q is the flow through the basin in terms of quantity per unit of time. The vertical velocity is a function of the settling characteristics of the particle. In an ideal basin, it is assumed that all particles are uniformly distributed at the influent

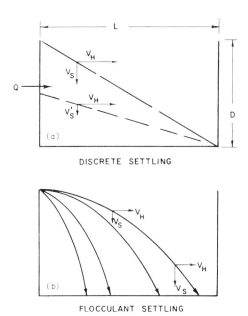

DISCRETE SETTLING

FLOCCULANT SETTLING

Fig. 13.4. Schematic illustrations of discrete and flocculant particle sedimentation.

end of the tank, that there is no effect due to density or velocity gradients throughout the tank, and that all solids remain out of the flow after they reach the bottom of the tank. Therefore in Fig. 13.4a all particles having a vertical velocity of V_S or greater would be removed from the basin. Particles having a velocity less than V_S would be removed in proportion to their settling velocity, V_S', and to the distance they were above the bottom of the tank at the influent.

The critical settling velocity of a discrete particle such that it will settle out, i.e., V_S, establishes the design of a settling tank. Engineers have chosen to use gallons per square foot-day (surface overflow rate) rather than feet per hour as the design criteria. The two items can be related by noting that when a particle has a settling velocity of V_S, the settling time of the particle, t_v, is equal to the time it takes to move in the horizontal direction through the basin t_h, i.e.,

$$t_v = t_h = \frac{D}{V_S} = \frac{L}{V_H} \tag{13.7}$$

$$V_S = \frac{V_H D}{L} = \frac{QD}{DWL} = \frac{Q}{WL} \text{ (flow/day/surface area)} \tag{13.8}$$

In an ideal sedimentation tank, depth is not an important factor controlling gravity settling, and efficiency of removal is a function of overflow rate rather than depth. With sewage and industrial wastes, idealized sedimentation does not occur, with the result that settling patterns of the type noted in Fig. 13.4b occur, especially with flocculant particles. Under these conditions, efficiency of removal is related to both overflow rate and detention time, since particle size increases with time.

Consideration of the horizontal velocity V_H is important since it can affect the removal efficiency. As the quantity of water flowing through the sedimentation tank increases, the efficiency of removal will decrease, and there will be greater solids carried over the effluent weirs.

In addition to the effective settling zone in a sedimentation basin, the tank also contains other important components such as an inlet zone, an outlet zone, and a solids removal zone. The inlet zone serves to reduce the velocity from the order of feet per second in the entering pipes to feet per hour in the tank and to distribute the influent solids as uniformly as possible. Baffled inlet zones are used to reduce the velocity and to provide both horizontal and vertical distribution of the solids.

The function of the outlet zone is to remove the clarified liquid from the settling zone. Low rates of flow are desirable to avoid carry-over of solids. Weirs are used to maintain low overflow velocities. The length of the overflow weir is a function of the solids in the sedimentation basin and the liquid flow. Overflow rates of 10,000–15,000 gal per day per linear foot of weir are common for sewage sedimentation tanks. Figure 13.5 indicates the overflow weirs on a sedimentation tank of a water treatment plant.

The design of a sedimentation tank is determined by the surface overflow rate needed to achieve a required solids removal. Rates from 600 to 1000 gal/ft²/ day have been used in conjunction with detention times of 1.5–2 hr for municipal sewage. For specific industrial or agricultural wastes, pilot plant or full-scale tests may be needed to obtain the necessary criteria for the desired solids removal. The efficiency of solids removal is affected by factors such as variable flow rates, short circuiting, turbulence, sludge density currents, and inlet and outlet conditions.

Attention must be given to the handling of the settled solids. When the settled sludge represents a significant portion of the total basin flow, such as in final clarifiers for activated sludge units, the solids loading and withdrawal methods are important. The solids loading is important in determining the area requirements of a tank if thickening is important. Settled solids may be removed by mechanical or hydraulic methods. Mechanical methods include rakes or scrapers that are driven by chains and motors. Such scrapers move the settled solids slowly along the bottom of the tank to an exit at an end of a rectangular tank (Fig. 13.6) or the center of a circular tank. Hydraulic methods involve the

Fig. 13.5. Effluent overflow weirs on a water treatment sedimentation tank.

removal of the sludge at the point of deposition (Fig. 13.7). The selection and design of the sludge removal equipment will depend upon the type of sludge to be handled. Hydraulic sludge removal methods are applied most commonly to low-density solids such as settled activated sludge. Because of the biological activity of microbial solids, the time they may be retained in a final clarifier is limited to avoid gas production and solids resuspension.

Sedimentation units can be rectangular, circular, or square. They can be classified as primary units, i.e., used before secondary units such as activated sludge or trickling filter units or as secondary or final units which are used to clarify the effluent from the secondary biological units. Many older secondary

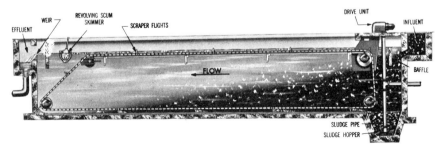

Fig. 13.6. Rectangular primary settling basin with combination chain and scraper surface skimmer and sludge collector. [Courtesy of Envirex, Inc., a Rexnord company.]

Fig. 13.7. Peripheral feed circular clarifier with hydraulic sludge removal system. [Courtesy of Envirex, Inc., a Rexnord company.]

clarifiers are not equipped with scum-scraping and removal devices. Because of the possibility of floating solids in a final clarifier, a number of states are requiring scum removal units on all final clarifiers. Without scum removal, any solids that do float in a final clarifier will go over the effluent weirs unless physically removed by the operator. Since the effluent BOD from a treatment plant is a function of the oxygen demand of any solids in the effluent, it is prudent to include scum removal units on all secondary clarifiers. Figure 13.8

Fig. 13.8. Primary sedimentation unit treating duck farm wastes.

illustrates the floating scum, the scum removal mechanism, and the effluent weirs of a primary sedimentation unit treating duck farm wastes.

As early as 1904, investigators suggested that a shallow sedimentation basin with large surface area would be superior to conventional basins. Horizontal trays have been installed in sedimentation basins to increase the net surface area. However, difficulties with sludge removal and floating material resulted in limited utilization of the tray settling concept. Low Reynolds numbers and laminar flow conditions also favor better sedimentation. Practical application of minimal sedimentation depth, increased horizontal surface area, and laminar flow conditions has led to the development of the tube settler (7), which permits shorter hydraulic retention times and better sedimentation. Tube settlers are being used at both water and wastewater treatment plants (5).

Because of the need for high quality wastewater effluents, sedimentation tanks no longer are used as a sole wastewater treatment method. They are incorporated as part of a total treatment facility as a primary and/or secondary clarifier. Surface overflow rates for secondary clarifiers usually are less than that for primary units, because of the settling characteristics of biological sludges.

Flotation

Dissolved air flotation is a process which increases the rate of removal of suspended matter from liquid wastes. The process achieves solids–liquid separation by attachment of gas bubbles to suspended particles, reducing the effective specific gravity of the particles to less than that of water. The gas is dissolved in the liquid as it flows through a pressure tank at a pressure of about 30–50 lb per square inch gauge (psig) (Fig. 13.9). The pressurized liquid is released in the influent end of a flotation tank where the dissolved gas comes out of solution and attaches to the suspended particles in the waste flow. Although air usually is used for treatment of wastes, any gas can be used. Greater quantities of carbon dioxide will go into solution than will air at a given pressure. However, unless carbon dioxide is used for other purposes or is readily available, it usually is more expensive.

Within limits, the number of air bubbles produced will be directly proportional to the product of the absolute pressure and the rate of flow of the pressurized flow. Air solubility in water increases as the temperature decreases, and the temperature of the wastewater is a factor in the effectiveness of the dissolved air flotation process. The removal rate of both floatables and settleables is a function of the size of the particle. Coagulants such as alum, ferric chloride, and clays can be used to increase the particle size. When use of chemicals is advantageous, chemical addition, coagulation, and flocculation units are utilized prior to the flotation unit. Bench-scale tests can be used to evaluate chemical type and

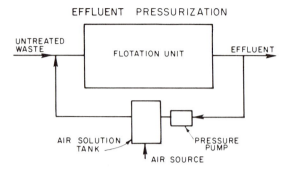

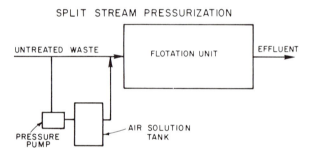

Fig. 13.9. Schematic of dissolved air flotation systems.

dosage as well as retention times, pressurized flow rate, air-to-solid or air-to-liquid ratios, and other design criteria for various wastes.

Pressurization of both influent and effluent liquid can be used (Fig. 13.9). Generally, effluent pressurization is preferred, since it avoids problems of emulsification and floc destruction that can occur when raw or flocculated wastes pass through the pressurization pump. Effluent pressurization readily allows flocculation to precede dissolved air flotation. Pressurized flow volumes are dependent on the solids concentration in the waste and generally range from 20 to 30% of the raw waste flow. Figure 13.10 illustrates a dissolved air flotation unit, which removes both settleables and floatables. Both scum and settleable solids removal equipment are needed.

Emulsified grease or oil removal by flotation is enhanced by chemical coagulation. The chemical coagulant should be in a quantity adequate to completely absorb the oil, whether free or emulsified. Good flotation properties are characterized by a tendency for the floc to float, with no tendency to settle. Excessive coagulant additions can result in a heavy floc which is only partially removed by flotation.

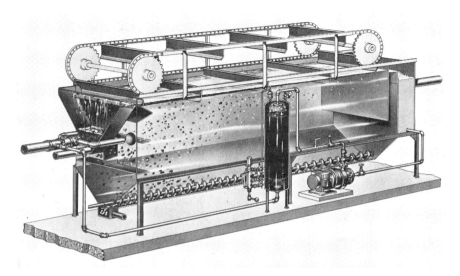

Fig. 13.10. Dissolved air flotation unit for removal of floatable and settleable solids by waste flow-effluent pressurization. [Courtesy of Envirex, Inc., a Rexnord company.]

With good operation and correct chemical addition, high treatment efficiency can be achieved; in many instances to levels comparable to biological treatment. Lower treatment efficiencies are satisfactory if flotation is used as a pretreatment step prior to physical, chemical, or biological processes.

Dissolved air flotation has been widely applied to the ore industry, to oil refinery wastes, and for thickening of activated sludge. With agricultural wastes, it has been used with food processing, meat-packing, fish processing, and edible-oil refinery waste waters. Suspended solids removal has ranged from 50 to over 90%, oil and grease from 60 to over 90%, and COD from 30 to over 80%. Different operating conditions and wastewater characteristics will result in different removal efficiencies. The design and operation of flotation units should proceed from laboratory and pilot plant studies with the actual wastes to be treated. Illustrative results with several food processing wastes are presented in Table 13.1.

Chemical Precipitation

The addition of chemicals to wastewaters offers an opportunity to precipitate particulate and colloidal material, thereby reducing the oxygen demand of the wastewater. Soluble organic compounds are poorly removed by this process. Soluble inorganic compounds, such as phosphates, can be removed if insoluble

TABLE 13.1

**Performance of Dissolved Air Flotation with Food
Processing Wastewaters**

	Removal (%)	
Wastewater	Suspended solids	BOD
Fruit and vegetable processing[a]		
Peach	84	17
Pumpkin	62	7
Peach	65–93	—
Tomato	61–84	—
Poultry processing[b]		
33% recycle		
No chemicals	54	38
Lime, 100 mg/liter	81	57
Alum, 300 mg/liter	94	33
50% recycle, no chemicals	62	38

[a]Adapted from (8).
[b]Adapted from (9).

precipitates can be formed. The amount of material precipitated from a wastewater is a function of amount of chemical added, the pH of the solution, and the type of constituents in a wastewater. Chemical precipitation has obtained suspended solids removals of up to 90+% and BOD removals of up to 50–70% from municipal wastewaters. Comparable removals can be achieved with other wastes.

Chemical precipitation of wastewater as a treatment process can be considered to be intermediate between a primary and secondary process. Chemical precipitation is useful to achieve partial treatment prior to discharge of industrial wastes to a municipal system, to achieve lower loadings on subsequent treatment units, and remove inorganic compounds such as phosphates. The quantity of chemical sludge that is generated requires careful evaluation, since it increases sludge handling and disposal costs.

The coagulants commonly used to cause precipitation in wastewaters are alum [aluminum sulfate—$Al_2(SO_4)_3$], ferric salts such as ferric sulfate [$Fe_2(SO_4)_3$] or ferric chloride (Fe Cl_3), and lime. Alum reacts with the natural alkalinity of the wastewaters, or added alkalinity if necessary, to form an insoluble aluminum hydroxide precipitate, which coagulates colloids and hastens the precipitation of other particulate matter. Lime reacts with the bicarbonate alkalinity of wastewater to form calcium carbonate, which will precipitate and enmesh other particulate matter. Insoluble calcium carbonate occurs above a pH of 9.5. Ferric salts are used to develop insoluble ferric hydroxides, which can precipitate colloids and increase the sedimentation rate of other particulates in the wastewater. Both alum and the ferric salts have the ability to precipitate negatively charged colloids in a water. Chemical precipitation of a wastewater is combined with sedimentation to remove the particulate matter.

Anionic, cationic, and nonionic organic polyelectrolytes also can be used to precipitate colloidal matter in wastewater, either separately or in combination with inorganic coagulants. A wide variety of polyelectrolytes are available from many manufacturers. Except for previous situations where specific polyelectrolytes have been proved successful, there is little to guide the choice of a polyelectrolyte to be used with a particular wastewater. Evaluation of suggested polyelectrolytes on samples of the wastewater normally is done with jar tests in the laboratory to determine proper types and dosages.

The quantity of chemical to achieve specific removals depends upon characteristics of the wastewater such as pH, alkalinity, solids content, phosphate concentration, and related factors that affect the coagulant demand. These factors vary from wastewater to wastewater, with the result that empirical relationships are used to estimate the necessary chemical dose. Laboratory jar tests with a representative sample of the specific wastewater enable an estimation of the feasible type and quantity of chemical. General predictive relationships for wastewaters remain to be developed.

The results of the jar test experiments provide a point of departure for proper chemical requirements which must be refined under actual operating conditions. The jar test procedure represents controlled conditions such as wastewater characteristics, degree of mixing, quiescent settling conditions, and time of reaction, each of which may vary in a treatment facility. Removal in practice can be poorer than estimated from jar test experiments, especially since clarification characteristics of the actual suspensions can be different from those observed in the jar tests.

While chemical precipitation can be utilized with agricultural wastes, there are few recorded cases of it being used as a major treatment process. Cow and dairy shed washings were treated with 500 mg per liter of alum and followed by sedimentation for 1 hr. The BOD and suspended solids content of these wastes were reduced by 30 and 70%, respectively (10).

Table 13.2 indicates the chemical dosages and the removals obtained when chemical precipitation was used with several vegetable processing wastewaters. As noted earlier, chemical precipitation can frequently be used with air flotation to enhance solids removal. Table 13.3 summarizes the type of results and chemical dosages that have occurred when these processes have been combined to treat fish processing wastewaters. The chemical dosages noted in Table 13.3 were able to be decreased, with a chemical cost savings of about 25%, when a polyelectrolyte also was used.

TABLE 13.2

**Performance of Chemical Precipitation with Several Vegetable
Processing Wastewaters**[a]

	Coagulant concentration used (mg/liter)		Removal (%)	
Vegetable processed	Alum	Lime	Suspended solids	BOD
Peas	44	266	81	42
Beets and corn	65	357	82	28
Lima beans	39	136	82	15
Lima beans and spinach	22	218	75	32

[a]Adapted from (9).

Acid coagulation has been used to precipitate material from meat-packing wastes. A pH between 4.0 and 4.5 was needed to obtain high removals (12). Removals that occurred in a large-scale pilot plant that used acid coagulation followed by air flotation were as follows: suspended solids, 77–94%; COD, 47–80%; fat, 98%; and total nitrogen, 38%.

Another possible use of the chemical precipitation process is for the removal of phosphorus from agricultural wastewaters. The phosphorus problem is minimized when wastewaters are discharged to land, a common method for disposal of agricultural wastes. However, for highly dilute wastes discharged to surface waters, the application of phosphorus removal methods to agricultural wastewaters becomes more critical.

TABLE 13.3

**Use of Chemical Precipitation and Air Flotation to Remove
Pollutants from Fish Processing Wastewater**[a]

Coagulant concentration (mg/liter)		Average percent removal		
Sodium hydroxide	Alum	Insoluble solids	BOD	COD
100	270	70	73	—
85	235	92	—	84

[a]Adapted from (11).

The chemicals used to precipitate phosphorus are lime, alum, and ferric salts. Lime reacts with orthophosphates in solution to precipitate hydroxylapatite. The apatite precipitate, represented by $Ca_5(OH)(PO_4)_3$, is a crystalline precipitate of variable composition. Control of pH is important, with optimum precipitation occurring at pH levels above 9.0 when lime is used. When alum is used to remove phosphates in solutions with pH values above 6.3, the phosphate removal mechanism occurs via incorporation in a complex with aluminum or via adsorption on aluminum hydroxide floc. Ferric ions and phosphate react at pH values above 7.0 to form insoluble ferric phosphate precipitates.

Chemical precipitation of phosphates in poultry, dairy, and duck farm wastewaters has indicated the type and concentration of chemicals that are feasible (13, 14). Several relationships between the characteristics of the wastewaters and phosphate removal were identified. Useful relationships occurred when the chemical dosage per remaining phosphate concentration was plotted versus the percentage of phosphate removed. A typical relationship is shown in Fig. 13.11. Both ortho and total phosphate removals and the amount remaining could be

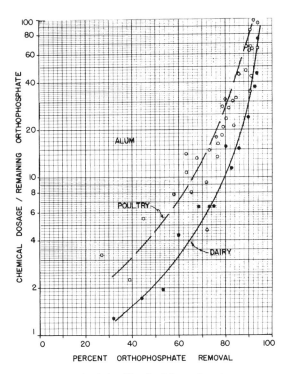

Fig. 13.11. Phosphate removal relationships for dairy and poultry manure wastewater (13).

related in this manner. If this relationship is valid for other wastewaters it offers an approach to design and operation for the removal of phosphates from wastewaters by chemical precipitation.

Decisions on the most appropriate chemical for phosphorus removal from animal wastewaters are difficult. The chemical of choice will depend upon the required dosage, the chemical cost, and the costs of ultimate solids disposal. For animal wastewaters, sludge production averaged between 0.5 and 1.0 mg per liter suspended solids increase per mg per liter of chemical used. The large chemical demand and sludge production are decided disadvantages to this method of phosphate control for concentrated animal wastewaters.

Lime has been added to the aeration tank of a system treating liquid calf manure (15). The calf manure was diluted by washwater and had a COD of about 4200 mg per liter. The phosphorus removal was related to the pH that was maintained in the aeration tank. At a pH of about 8, approximately 95% removal of the phosphorus occurred, with a resultant concentration of 10 mg phosphorus per liter in the effluent. About 10–13 kg of lime was needed to remove 1 kg of phosphorus.

References

1. Mercer, W. A., Maagdenberg, H. J., and Ralls, J. W., Reconditioning and reuse of olive processing brines. In "Proceedings of the First National Symposium on Food Processing Wastes" pp. 281–293. Pacific Northwest Water Lab., Federal Water Quality Administration, U.S. Dept. of the Interior, Washington, D.C., 1970.
2. Ralls, J. W., Mercer, W. A., and Yacoub, N. L., Reduction of salt content of food processing liquid waste effluent. In "Proceedings of the Second National Symposium on Food Processing Wastes" pp. 85–108. Pacific Northwest Water Lab., U.S. Environmental Protection Agency, Corvallis, Oregon, 1971.
3. Culp, R. L., and Culp, G. L., "Advanced Wastewater Treatment." Van Nostrand-Reinhold, Princeton, New Jersey, 1971.
4. Weber, W. J., "Physical and Chemical Treatment Processes." Wiley, New York, 1972.
5. Metcalf and Eddy, Inc., "Wastewater Engineering: Treatment, Disposal, Reuse," 2nd ed. McGraw-Hill, New York, 1979.
6. "Standard Methods for the Examination of Water and Wastewater," 15th ed. Am. Public Health Assoc., Chicago, Illinois, 1980.
7. Culp, G. L., Hsiung, K. Y., and Conley, W. R., Tube clarification process—Operating experience: J. Sanit. Eng. Div., Am. Soc. Civ. Eng. 95, SA5, 829–837 (1969).
8. Stanley Associates Engineering Ltd., "Review of Treatment Technology in the Fruit and Vegetable Processing Industry in Canada," Fisheries and Environment Canada, Rep. EPS 3-WP-77-5. Environmental Protection Service, Ottawa, Canada, 1977.
9. U.S. Environmental Protection Agency, "Pretreatment of Poultry Processing Wastes," Technology Transfer, Washington, D.C., 1973.
10. Painter, H. A., Treatment of wastewaters from farm premises. Water Waste Treat. J. 6, 352–355 (1957).

11. Claggett, F. G., "Clarification of Fish Processing Plant Effluents by Chemical Treatment and Air Flotation," Technical Report 343, Fisheries Research Board of Canada, Vancouver, British Columbia, 1972.

12. Denmead, C. F., and Cooper, R.N., "Chemical Treatment of Freezing Works Wastes," MIRINZ 448. Meat Industry Research Institute of New Zealand, Hamilton, New Zealand, 1975.

13. Loehr, R. C., Prakasam, T. B. S., Srinath, E. G., and Joo, Y. D., "Development and Demonstration of Nutrient Removal from Animal Wastes," Rep. EPA-R2-73-095. Office of Research and Monitoring, U.S. Environmental Protection Agency, Washington, D.C., 1973.

14. Loehr, R. C., and Johanson, K. J., Phosphate removal from duck farm wastes. *J. Water Pollut. Contr. Fed.* **46,** 1692–1714 (1974).

15. Kox, W. M. A., Phosphorus removal from calf manure under practical conditions. *Agric. Wastes* **3,** 7–19 (1981).

14

Nonpoint Source Control

Introduction

Pollutant sources can be classified as either point (direct) or nonpoint (diffuse) sources. Point sources discharge pollutants via pipes or ditches at identifiable locations. Examples include discharges from municipal sewage treatment plants, effluents from food processing and other industrial waste treatment facilities, and the runoff from large animal feedlots. Nonpoint source pollutants result from land-based activities, such as crop production, and generally enter surface waters as land runoff. Nonpoint discharges are intermittent and occur due to meteorological events. Point source discharges are more constant and are related to the activities in a municipality or industry. These and other differences between point and nonpoint sources are summarized in Table 14.1.

Nonpoint pollutants are conveyed by water, the source of which is uncontrolled by any individual. The water results from natural conditions such as precipitation, snowmelt, or flooding. The pollutants in nonpoint discharges are not traceable to an identifiable direct source, such as an industry or the discharge from a municipal wastewater treatment facility. The fact that land runoff may be channelled into a ditch or drain before entering surface waters does not make such runoff a point source. An important aspect of nonpoint source control options is that such control is best achieved by planning and management practices that control the source and delivery of nonpoint pollutants, thereby limiting them from reaching surface waters or groundwater.

TABLE 14.1

Characteristics of Point and Nonpoint Sources of Pollutants

Characteristic	Point source	Nonpoint source
Method of discharge to surface waters	Direct by a definable pipe or ditch	Diffuse, land-based runoff
Frequency	Relatively continuous	Intermittent, related to precipitation and climatic conditions
Ability to monitor and control	Relatively easy	Poor
General source of contaminants	Human activity in the community, industrial processing	Land-based agricultural, forestry, and recreational activities
General type of contaminants	Oxygen demanding material (BOD, COD), solids, industrial chemicals, nutrients, pathogens	Sediment, nutrients, pesticides, animal and plant pathogens, salts, organic matter
Methods of control	Effluent limitations	Best management practices
Technical approaches	Biological, physical, and chemical treatment processes for all wastes	Management, vegetative, and structural controls on selected sites
Regulatory approach for control	Discharge permits	Education, working with established agricultural and other land-oriented organizations, site demonstrations

Discharges from point sources and rural and urban nonpoint sources can affect the quality of surface waters. The relative magnitude of rural and urban nonpoint sources has been presented in Chapter 3. The present chapter identifies the factors that affect the quantity of pollutants in nonpoint sources and the approaches that can be used to control such sources. Emphasis is placed on the control of nonirrigated agricultural nonpoint sources. Information on the control of urban and irrigated nonpoint sources can be found in other documents (1, 2).

Best Management Practices

The legislative, regulatory, and technical mechanisms used for the control of nonpoint sources are different from those used for point sources. The approach used for the control of nonpoint sources is source management rather than collection and treatment of pollutants. Management practices are used as cost-effective methods to prevent or control nonpoint source pollution from agricultural activities.

The definition (3) of a best management practice (BMP) is as follows:

A practice or combination of practices that is determined by a state (or designated area-wide planning agency), after problem assessment, examination of alternative practices and appropriate public participation, to be the most effective practicable (including technological, economic, and institutional considerations) means of preventing or reducing the amount of pollution generated by nonpoint sources to a level compatible with water quality goals.

The general steps that are involved in the identification and implementation of BMPs are outlined in Fig. 14.1. The selection and application of BMPs are related to the pollutants that need to be controlled, the type of agricultural activity contributing such pollutants, and the load reduction that is required, as well as to site-specific climatic, topographical, and managerial considerations. There is no single technology or BMP that will control all nonpoint sources.

Because the intent of a BMP is to remedy a problem and to help achieve a water quality goal, the first step involved in determining appropriate BMPs is to identify the water quality problem that needs improvement. There can be different problems in different surface waters. A water quality problem can be localized, such as impairment of a local water supply by bacteria, or fish kills in a local stream, or can exist on a wider scale, such as increased nutrients causing eutrophication problems in a lake or reservoir or sediment reducing the volume of a reservoir or filling streams. The primary agricultural nonpoint source pollutants are sediment, plant nutrients, animal wastes, pesticides, and possibly salts and pathogens.

When the water quality problem is identified, attention can be given to the agricultural sources that may be contributing to the problem. The process by which potential nonpoint source pollutants are generated and transported to the surface waters also needs identification. Not all agricultural activities nor all land areas in a watershed will be contributing such pollutants, and, of those that are contributing, not all will be contributing equally.

With the problem and probable agricultural sources determined, the type of controls and the specific management practices that should be utilized will become clear. As noted in Fig. 14.1, the most appropriate BMPs will be those that are agronomically effective, environmentally effective, economically achievable, socially acceptable, and implementable. The accepted definition of these terms as used in the identification of BMPs is presented in Table 14.2.

Once the appropriate BMPs are identified, implementation can be attempted. Key aspects of implementation are (a) education of the public and agricultral community as to why the BMPs are needed and what the results will be, (b) financial or other incentives to have the BMPs put in place and utilized, (c) the institutional arrangements to assist with the implementation, and (d) field example programs to illustrate the effectiveness of the BMPs. Some form of monitoring normally is involved to see that the BMPs are in place and continue to be utilized and to determine the effectiveness of the BMPs in achieving the desired water quality improvement goals.

In implementing BMPs, emphasis should be placed on the practices, the land areas, and the agricultural activities that can achieve the greatest removal of a pollutant and hence improvement in water quality.

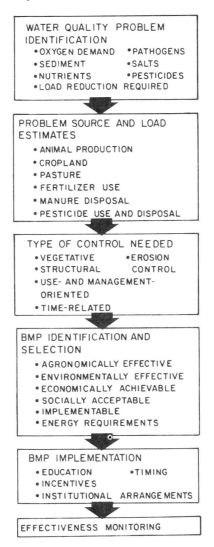

Fig. 14.1. Steps involved in the identification and implementation of best management practices for nonpoint source control.

TABLE 14.2

Definition of Terms Used to Determine BMPs[a]

Agronomic effectiveness—practices that promote productivity and commodity quality and that
  conserve, preserve, and, if possible, enhance soil productivity
Environmental effectiveness—practices that prevent, reduce, or control pollutant loads (dissolved
  or bound on particulates) entering surface water or groundwater to a level compatible with
  water quality goals; intermedia transformations and concerns should be considered
Economically feasible—practices whose economic implications are compatible with water quality
  and competing goals
Socially acceptable—practices that are reasonable and have public support
Implementable—practices that are legal; institutions are in existence or can be formulated in a
  way to provide flexibility and simplicity in administration in achieving implementation

[a]Adapted from (3).

Factors Affecting Pollutant Movement

The impact of agricultural activities on water quality results from a complex set of circumstances. Potential pollutants such as sediment, nutrients, and pesticides may leave a cropped field in runoff or percolation water, may be transported across or through the landscape to surface waters or ground water, and depending on their effect in receiving waters, may produce water quality problems (Fig. 14.2). Not all the potential pollutants that leave a field reach surface water or have an adverse water quality impact. Whether the potential pollutants reach a stream depends upon what happens on the intervening land and upon approaches to control the edge-of-field losses.

Management of agricultural nonpoint sources requires the use of practices that intercept or reduce the pollutants at one or more points in this chain. Management strategies focus on source control, delivery control, and problem mitigation.

Source control is an attempt to reduce edge-of-field loss of pollutants. Delivery control focuses on measures which prevent pollutant movement from field to stream or aquifer. Problem mitigation involves attempts to lessen the impact of pollutant discharges on potential water users. The first of these strategies (source control) is often emphasized when attempting to identify appropriate BMPs since there are a variety of well-defined management options which can reduce edge-of-field pollutant losses. These options are referred to as agricultural BMPs and include practices to reduce both pollutant availability and movement. Reduction of availability can be achieved by the management of chemical applications (pesticides, fertilizers, etc.) to limit the opportunity for loss. Management of pollutant movement results from techniques such as soil and water conservation practices (SWCPs) which prevent soil erosion and retain water.

Agricultural pollutants can be considered as two major groups: land based and management related. The land-based pollutants are associated with the soil and result from erosion and subsequent movement of soil particles to surface waters. Examples of such pollutants include sediment, animal and plant pathogens, and metals.

Management-related pollutants are those that are applied to the crop or land to enhance productivity and reduce pests. Some of the pollutants may be soluble in water and move with the runoff. Others may adsorb to the soil and move as the soil moves when eroded. Examples include pesticides and nitrogen and phosphorus in fertilizers and manures.

The management-related pollutants can be controlled by managing the input of the chemicals and fertilizers. This can be accomplished by managing the timing, amount applied, formulation and type of chemical, and application methodology. Land-based pollutants are controlled by practices that affect erosion and soil particle movement. The effectiveness of BMPs for land-based and management-related pollutants is related to how the pollutants move in and from a field.

Whether a potential pollutant moves from a field to surface waters depends on availability at the field site, detachment, and transport. The factors affecting pollutant movement are those that affect water and soil movement and the relative solubility of the pollutant. An understanding of such factors and movement helps identify the value and effectiveness of possible BMPs. The following sections attempt to provide such an understanding. Further details about these

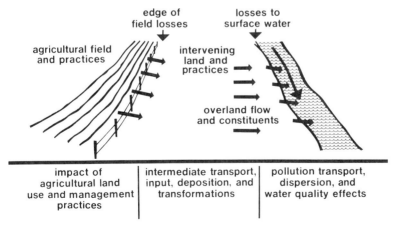

Fig. 14.2. Movement of potential pollutants from a source to a stream.

factors, the movement of sediment and agricultural pollutants, and BMPs that can control such pollutants can be found in other documents (1, 4–7).

Water Movement

The ease of water movement through the soil determines the path by which water and any constituents reach surface waters. The path of water movement (Fig. 14.3) determines the type and quantity of pollutants leaving a field.

When the rate of precipitation or snowmelt exceeds the infiltration rate of the soil and the storage capacity of the land-surface depressions, excess water becomes overland flow. Overland runoff will contain dissolved materials such as nutrients or pesticides and will carry sediment with its absorbed substances. The quality and quantity of the dissolved chemicals depend on the soil, vegetation, organic residue management, fertilizer and pesticide management, and soil conservation practices.

After water infiltrates the soil, it may percolate to the zone of saturation and move with the groundwater flow. Because rates of groundwater flow are generally slow and the subsurface paths are long, most of the water following such paths continues to flow between rainstorms. When the groundwater resurfaces in a stream channel, it is base flow.

If at some depth in the soil, the percolating water encounters a less permeable horizon, a portion of the water will move horizontally, become interflow, and reach a stream by a much shorter route. Because of the shorter route, the higher permeability of the surface soil, and generally greater hydraulic gradients, water following this path reaches the edge of the field more quickly than does base flow.

In some fields, vertical and horizontal flow may cause the soil to become saturated throughout its depth. When this happens, some of the water moving by the shallow subsurface path emerges from the soil surface and becomes overland flow.

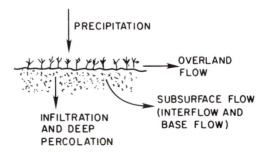

Fig. 14.3. Pathways of water movement in soil.

The physical and chemical properties of pollutants determine their path and mode of travel out of a field. Pollutants can be dissolved in the water, adsorbed to soil particles, or in a solid phase. Soluble substances will be transported in the surface and subsurface runoff, and solid and adsorbed substances will be transported in the surface runoff. The volume of runoff water generally is much greater than the quantity of sediment yielded from a field. Therefore, even if the concentration of a substance is greater in the sediment than in the water, the quantity of the substances moving off a field dissolved in the runoff might be greater than that adsorbed to the sediment. The amounts of a substance carried in the aqueous, adsorbed, and solid phases and its relative mobility in the water and sediment are important. Runoff water and the dissolved constituents will move farther in a shorter time than will sediment and the adsorbed and solid pollutants, which are subject to redeposition. The possibility of degradation and transformation is greater during the longer transport time required for sediments.

High-intensity storms increase runoff and the detachment and transport of pollutants. The kinetic energy of the rain causes larger soil particles to be detached and smaller particles to seal the soil surface. Figure 14.4 illustrates the impact that a storm can have in transporting sediment to a lake. Melting snow also can contribute significantly to the runoff and erosion process. Any increase in runoff provides greater energy to detach and transport soil and adsorbed

Fig. 14.4. Input of a stream into a lake, indicating discharge plume of silt and clay.

substances. Snowmelt also increases the volume of water transporting soluble compounds.

The infiltration rate of a soil and the ability of soil particles to adsorb pollutants depend upon the characteristics of a soil. With an increase in the infiltration rate, more water can infiltrate, and overland flow and the pollutant load associated with runoff will decrease. The amount of organic matter and clay particles influences the adsorption capacity of the soil. Soils with a high percentage of clay and organic matter will have a higher capability to adsorb pollutants than will sandy soils.

Soil Movement

Soil erosion is the detachment and transport of soil particles by precipitation and runoff. The rate of erosion ranges from a national average of about 2.8 tons/acre/year on the best land to over 50 tons/acre/year on the poorest (8).

Precipitation striking the ground detaches small particles of soil. As the rain water accumulates on the ground surface, it flows downslope, removing the loosened soil particles, thereby causing sheet erosion. As the runoff flows, it seeks the lower areas where it flows faster, which increases its capacity to carry more soil particles. The natural, small channels become larger channels or rills, resulting in rill erosion. If the size of the channels increases, gully erosion occurs.

The velocity of the water determines its ability to move and carry particles. Large, heavy particles such as sand and gravel may move only a relatively short distance. Lighter, smaller particles such as silt and clay may be carried large distances and eventually reach lakes and reservoirs (Fig. 14.4).

Factors that affect erosion and transport of pollutants are particle structure, organic matter content, and size. Regardless of the source of detachment mechanism, sediment consists of a higher percentage of smaller particles and aggregates than the original soil. This tendency is less when rills and gullies develop since bigger particles can be detached and transported. Smaller particles have a greater surface area per unit of volume and a greater adsorbing capacity than larger particles.

Sediment from surface soils will have a greater chemical enrichment than will sediment from gullies. This selective erosion process can increase overall pollutant delivery.

To quantify the increased percentage of a specific fraction in the sediment, the term enrichment ratio (ER) is used. This term indicates the ratio between the fraction of a particular constituent in the sediment and that in the original soil. The ER for clay is greater than one. The ER for clay or organic matter increases with an increase in soil detachment by raindrop splash. Enrichment ratios also increase as transport energy decreases since the smaller particles with their adsorbed substances will be transported farther.

Pollutant Movement

For the purpose of describing nonpoint source pollutant movement, potential pollutants can be classified as solid-phase pollutants, pollutants strongly adsorbed to soil particles, moderately adsorbed pollutants, and weakly adsorbed or nonadsorbed pollutants. Solid-phase pollutants, such as manure particles, and pollutants that are strongly adsorbed to soil particles, such as the pesticides DDT, toxaphene, and paraquat, move in the same manner. Transport in runoff can be high and will be related to the amount of sediment in the runoff and the amount of the substance in a soil. Transport of a strongly adsorbed substance in percolate, base flow, and interflow will be small. Because these substances are transported primarily with or on the sediment in runoff, BMPs that reduce sediment will control the transport of such pollutants.

Losses of moderately adsorbed substances, such as most pesticides, are related to both soil loss and runoff. The relative proportion of the loss will depend on the solubility of the chemical. Higher losses of atrazine and 2,4-D were found in runoff than on sediments (4). As contrasted to surface flow, transport of moderately adsorbed substances by water percolating through the soil is slow. An equilibrium will exist between the concentration of the substances dissolved in the percolate and that adsorbed to the soil.

Weakly adsorbed or soluble substances, such as nitrate, will move as the water moves. Because organic and ammonia nitrogen can be mineralized in the soil, concentrations of nitrate in base flow and interflow can be greater than those in surface runoff.

Pollutant delivery to a stream can be defined as the fraction of a pollutant leaving a defined area, such as a field, that reaches a stream. It is a function of the topographic and vegetative character of the watershed. In the case of sediment, where opportunities for redeposition are greatest, the pollutant delivery ratios are lowest.

The final form and quantity of highly soluble, conservative substances such as chlorides are not significantly affected by the particular pathway taken. For substances that are either weakly adsorbed to the soil or strongly adsorbed but degrade to more soluble constituents with time, the length and transit time of different pathways will affect actual pollutant delivery.

Pesticide Movement

Like nutrients and other potential pollutants, residual pesticides in soil or on crops can be transported from fields to surface waters. The water quality impact of a pesticide is related to its toxicity and persistence, while its ability to be transported is a function of its solubility and ability to adsorb to soil particles (Fig. 14.5). The characteristics of an individual pesticide are unique. The indi-

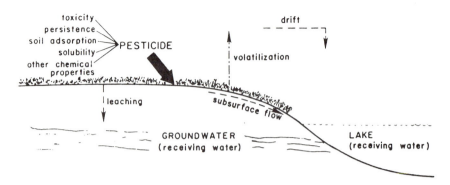

Fig. 14.5. Factors affecting the transport and water quality impact of a pesticide. [From (5).]

vidual properties of each specific pesticide to be controlled should be evaluated prior to initiating a management program to insure that the management practice that is selected is best suited to control the existing or potential problems. Detailed information about the toxicity, persistence in soil and the predominant transport mode of many herbicides, insecticides, and fungicides is available (6).

The persistence of a pesticide is a measure of the time necessary for the complete degradation of the material into harmless products. Since the primary mode of degradation is biochemical, factors which reduce biological activity decrease the rate of degradation. These factors include low soil moisture, low oxygen content, low temperature, low organic matter content, and extreme pH. The relative influence of each factor depends on the specific pesticide and site conditions.

Strongly adsorbed pesticides are relatively immobile in the soil profile. The adsorption of a pesticide to soil particles allows for degradation and reduction of its toxicity. Strong adsorption of pesticides to soil particles will generally keep pesticides close to the surface where they were applied. The soil surface is the area of greatest biological activity and where pesticide degradation rates are greater.

Organochlorines are, in general, some of the most persistent pesticides. Organophosphates, carbamates, and most herbicides are more easily degraded. The triazine herbicides are degraded primarily by chemical action but can persist for substantial periods of time. Table 14.3 lists the half-lives of some common pesticides. Half-lives are a function of the biological, chemical, and physical conditions in the soil. For example, a microorganism population may adapt to the presence of a certain pesticide, causing an increase in the degradation rate and a decrease in the pesticide half-life.

Pesticides move from agricultural fields to streams and lakes by drift or volatilization, suspended or dissolved in the surface flow, attached to eroding sediment, and dissolved in subsurface flow. Accidental spills or dumping of excess chemicals into surface waters also can contribute to pesticide concentrations in the water.

The quantity of pesticide lost by drift is dependent on the method of application. Small droplet or dust applications increase losses, whereas granular applications tend to decrease losses. Climatic conditions at the time of application are important. High winds tend to disperse pesticide molecules rapidly. Volatilization depends on the vapor pressure, water solubility, adsorption coefficient of the pesticide, and on the moisture content and temperature of the soil. Volatile pesticides frequently are regarded as more desirable, since they infiltrate soil pores more easily and more effectively reach the target pest. However, from the viewpoint of total environmental quality, volatile pesticides may be a greater environmental risk. Volatilization increases pesticide loading to the atmosphere and the possibility of eventual redeposition to streams and lakes.

Soil incorporation, cover crop, high clay and organic matter content, and deep plant penetration reduce volatilization. Surface and subsurface losses are determined by the adsorption and solubility properties of the pesticides, the characteristics of the site, and the timing, mode, placement, and rate of chemical application. The adsorption characteristics of the pesticide determine the mode of transport. Soluble and weakly adsorbed pesticides usually percolate downward through the soil profile. Although this is the most common pathway for these pesticides, intense rainfall, steep topography, or saturated soil conditions may cause such pesticides to be lost via runoff. Moderately adsorbed pesticides are primarily transported in surface flow but are also lost with eroded sediment. Strongly adsorbed pesticides such as trifluralin and toxaphene are transported primarily on eroded sediment.

TABLE 14.3

Half-Lives of Commonly Used Pesticides after Application[a]

Pesticide	Approximate half-life (weeks)
Dieldrin	100–200
Triazine herbicides	50–100
Benzoic acid herbicides	10–50
Urea herbicides	15–40
2,4-D and 2,4,5-T, herbicides	5–20
Organophosphate insecticides	1–10
Carbamate insecticides	1–5

[a]From (5).

Control Practices

The financial and technical resources available to control nonpoint source water quality problems resulting from agricultural activities are not unlimited. Thus, there is the need to focus attention on those potential sources of pollutants that have the greatest impact on water quality. In allocating program resources, the first step is to determine which agricultural activities—irrigated or nonirrigated crop production or animal production—are most critical with respect to an identified water quality problem. If, for example, cropland is identified as the most significant contributor of pollutants, it is important to locate and address these sources which are of the greatest significance and to use the most appropriate control practices.

The most appropriate BMPs are those that have the potential to (a) control or modify the use or application of a chemical to land, (b) control or modify crop or animal production management practices, (c) control the detachment mechanism of a pollutant from its site in the soil, and (d) control the transport of the pollutant. The most desirable BMPs have one or more of the following attributes:

1. Provision of optimum soil cover (vegetation, crop residues) to dissipate raindrop impact and reduce runoff velocity
2. Provision of optimum soil infiltration and flow path to minimize erosion through soil detachment and transport, and reduce runoff volume through enhanced infiltration
3. Minimization or reduction of the soil solution concentrations of soluble pesticides, plant nutrients, and other chemicals at the soil surface or within the root zone during periods of high runoff, thereby minimizing movement of such substances in runoff and percolate
4. Change in management practices so that a potential pollutant is less available for detachment and transport

Figure 14.6 indicates the major elements involved in pollutant detachment and transport from agricultural land and hence the major factors related to decisions on appropriate BMPs to control pollutants from such sources.

Control practices can be identified as management, vegetative, and structural. Management practices involve changes in timing, chemical application rates, type of chemicals used, and tillage systems. They usually do not involve other field-related activities. Vegetative practices involve changes in cropping systems and generally must be renewed annually. Vegetation plays an important role in preventing soil erosion, since it protects the soil from the impact of rain, holds the soil in place, reduces the velocity of the runoff, increases the ability of the soil to adsorb water, and can increase the rate and amount of infiltration. Structural activities require capital investment and construction or site modification and are more permanent control methods. Examples of these types of practices are noted in Table 14.4. Detailed designations of these practices and their ability to control agricultural nonpoint sources are available (5, 6).

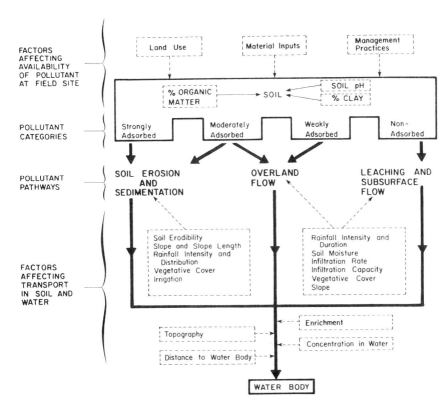

Fig. 14.6. Major factors involved in pollutant detachment and transport from agricultural land and decisions on appropriate BMPs. [Adapted from (5).]

In addition to the identified practices, common sense also plays an important role in creating and controlling nonpoint sources. Where barnyards are located very near streams (Fig. 14.7), manure runoff, increased erosion due to bare soil, and possibly direct discharge of milking parlor wastewater to the stream can occur. Beef cattle, dairy cattle, and other animal producing operations should be located away from streams and should have adequate vegetation between the operation and a stream to minimize the delivery of pollutants.

Cropland

Many of the possible BMPs noted in Table 14.4 are soil and water conservation practices (SWCPs) that can be used to reduce erosion, to control runoff, and to reduce the contribution of nonpoint pollutants from cropland. An advantage of these practices is that they have a history of nationwide use and can be implemented by existing organizations such as Soil and Water Conservation Districts.

TABLE 14.4

Examples of Managerial, Vegetative, and Structural Nonpoint Source BMPs for Nonirrigated Cropland[a]

Managerial controls
    Reducing excessive chemical application rates
    Better timing of manure, nutrient, and pesticide applications to avoid losses
    Improved methods of application
    Improved timing of field tillage operations to minimize periods of bare soil
    Use of alternative pesticides that have less of a water quality impact
    Use of insect- and disease-resistant crop varieties to reduce the need for pesticides
    Optimizing the timing of planting
    Use of mechanical weed control methods
    Reduced tillage systems
    No tillage systems
    Contour farming
Vegetative controls
    Meadowless rotation to disrupt insect cycles
    Sod-based rotations to improve soil structure and to decrease runoff
    Winter cover crops to minimize periods of bare soil
    Contour strip cropping
    Permanent vegetative cover to protect and stabilize soil
    Field borders and grass filter strips to filter sediment and associated pollutants
    Buffer strips along edge of fields and streams to reduce runoff and sediment loss
Structural controls
    Terraces to reduce erosion losses
    Diversions and interceptor drains
    Grassed waterways to reduce erosion
    Subsurface drainage to increase capacity of soil
    Detention ponds to capture sediment and adsorbed pollutants

[a]Adapted from (5).

The mechanisms by which SWCPs affect soil and water losses are identified in Table 14.5. The major impact of specific SWCPs is noted in Table 14.6. The information in Table 14.6 is suggestive only and merely indicates whether the effect is likely to be positive or negative, since the degree to which any SWCP changes a soil or field property is site-specific. In general, all of the noted SWCPs have the potential to reduce runoff velocity and transport capacity. The level of runoff reduction depends on a particular practice chosen and the location at which it is implemented. Cultural practices can control rill erosion and reduce the transport of soil detached by raindrop impact or by overland flow. Structural practices are effective in controlling gully erosion.

Each SWCP affects different stages of the sediment delivery process, i.e., raindrop impact, runoff detachment, and runoff transport. The effectiveness of each practice to reduce soil loss is site-specific. Practices that increase surface

Fig. 14.7. Barnyards near streams can be a source of pollutants.

TABLE 14.5

Effect of Soil and Water Conservation Practices in Reducing Soil and Water Loss via Runoff[a]

Mechanism	Effect on water loss	Effect on soil loss
Increase vegetative cover	Increase soil moisture storage due to increased infiltration and evapotranspiration Decrease runoff velocity and volume	Reduce raindrop impact Reduce rill erosion Reduce transport capacity
Increase surface residue	Decrease runoff velocity Increase soil moisture storage due to increased infiltration	Reduce raindrop impact Reduce rill erosion and transport capacity
Decrease slope length or steepness	Reduce runoff velocity Reduce runoff volume on permeable soils	Reduce rill and gully erosion Reduce transport capacity
Reduce tillage	Increase infiltration due to reduced surface sealing	Reduce raindrop impact Reduce soil erodibility due to improved soil structure
Increase surface storage	Increase soil moisture storage due to greater infiltration time	Reduce rill and gully erosion and transport capacity

[a]Adapted from (9).

TABLE 14.6

Major Physical Effects of Soil and Water Conservation Practices[a]

Soil and water conservation practice	Decrease runoff velocity and transport capacity	Increase surface storage	Increase soil moisture storage	Decrease raindrop impact	Improve soil structure	Reduce rill erosion	Reduce gully erosion
No tillage	+	0	+	+	+, −	+	0
Conservation tillage	+	0	+	+	+	+	0
Contour tillage	+, −	+	0	0	0	+	0
Graded rows	+	0	0	0	0	+	0
Strip cropping	+	+	+	+	+	+	0
Sod-based rotations	+[b]	0	+	+	+	+	0
Cover crops	+[c]	0	0	0	+	+	0
Contour listing	+[d]	+[d]	−	0	0	+	0
Ridge planting	+	+[d]	−	+	0	+	0
Terraces	+	+[e]	0	0	0	0	+
Diversions	+	0	−	0	0	0	+
Grassed waterways	+	0	0	+[f]	+[f]	0	+
Filter strips[g]	+	+	+	+	+	0	+

[a]Symbols are as follows: +, the effect is as indicated; 0, not considered a major effect; −, the effect is opposite of the indicated. [Adapted from (9).]
[b]During the period of sod crop only.
[c]During winter and early spring only.
[d]Only if plowing on the contour.
[e]On level raised bench terraces and tile-outlet terraces with raised outlets only.
[f]Only in the waterway.
[g]Effective only at the edge of the unit source area.

residue or vegetative cover decrease both soil detachment and transport. Other practices are effective because they reduce runoff kinetic energy.

The ability of a SWCP to control a specific sediment-related water quality problem depends on the characteristics of the sediment causing the problem. In general, clay is relatively resistant to detachment but easily transported, while the opposite is true for sand. SWCPs that control detachment of fine particles and organic matter and those that reduce soil transport control the detachment of relatively larger and heavier particles and the transport of the fine particles.

Adsorbed chemicals are associated with the clay and organic fractions of the sediment. These constituents are controlled effectively by SWCPs which provide protection from raindrop splash, such as conservation tillage. Soil and water conservation practices that rely on reducing sediment transport, such as graded terraces or contour tillage, are effective in controlling losses of organic matter and clay. Practices that control erosion by reducing raindrop impact are less influenced by storm magnitude than are practices which attempt to decrease the amount of soil detached or transported.

Detailed studies in the U.S. drainage basin to Lake Erie have shown that conservation tillage practices constituted the most cost-effective and acceptable technology for the control of soil erosion and nonpoint source phosphorus losses (10). The major objective of conversation tillage is to retain as much crop residue as possible on the land surface for the purpose of reducing the erosion impact of rainfall and the loss of detached soil particles in surface runoff.

The conservation tillage practices deemed most useful were reduced tillage and no till practices. With no till, crops are planted directly in the residue of the previous crop. Planting equipment which cuts through the residue and places the seeds at proper depth is used. With reduced tillage, the traditional moldboard plow is replaced by chisel plows and other methods which retain much of the crop residue on the soil surface.

With no tillage practices, herbicides are used to control weeds. With reduced tillage practices, weed control occurs by a combination of herbicides and tillage. Although pesticide use will increase when conservation tillage practices are used, little adverse water quality impacts are expected, since the soil runoff control that occurs with conservation tillage also prevents pesticide runoff.

It has been estimated (10) that if reduced tillage practices were used on all suitable cropland in the U.S. Lake Erie Basin, erosion potential would be reduced by about 46%. If both no till and reduced tillage were used on the cropland suitable for such practices in the basin, the overall soil erosion reduction was estimated as about 69%.

Field demonstration plots indicated that with conservation tillage practices, the yields of corn and soybeans were about the same as yields from plots receiving conventional tillage. An analysis of crop production costs indicated that reduced tillage decreased equipment and labor costs by about 8%. With no till, about 10% reduction in equipment and labor costs may occur. Thus, it was concluded (10) that although there were increased material costs for conservation tillage practices, these costs were more than offset by reduced machinery and other costs.

Because conservation practices result in no reductions in yield or loss of production and no added cost to the farmer, a zero cost can be assumed in accomplishing erosion control and resultant water quality benefits. The costs of other conservation practices were evaluated as possible nonpoint source phosphorus control measures. Except for barnyard runoff control, manure storage and controlled land application, and fertility management, all other conservation practices were expensive (greater than $5 per kilogram of total phosphorus removed) and were not cost-effective for phosphorus control (10).

Animal Production

As noted in Fig. 14.1, water quality problem identification and potential pollutant source identification are key aspects in determining appropriate BMPs. Pollutional characteristics can be used to help distinguish between water quality problems associated with animal production activities and those of other agricultural activities. Table 14.7 indicates the relative water quality parameters of animal- and crop-related activities.

TABLE 14.7

Probable Water Quality Impacts of Animal Agriculture and Cropland Runoff Nonpoint Discharges[a,b]

Parameter	Animal manures	Milkhouse wastewaters	Milking center wastewaters	Silage liquors	Cropland runoff
Dissolved oxygen	−	−	−	−	0
Biochemical oxygen demand	+	+	+	+	0
Chemical oxygen demand	+	+	+	+	0
Total nitrogen	+	+	+	+	+
Ammonia nitrogen	+	+	+	+	0
Nitrate nitrogen	*	0	0	0	+
Total phosphorus	+	+	+	+	+
Fecal coliforms and streptococci	+	0	+	0	0
Sediment	0	0	0	0	+

[a]Symbols are as follows: −, decrease in concentration when compared to background conditions; +, increase in concentration when compared to background conditions; 0, no significant impact; *, possible increase, depending upon prior oxidation of manures before runoff occurs.
[b]From (5).

TABLE 14.8

TABLE 14.8

Significance of Potential Nonpoint Sources of Water Pollutants Associated with Animal Agriculture[a,b]

Source category	Poultry	Dairy	Swine	Beef
Production facilities	−	0	0	0
Land used for manure disposal	+	+	+	+
Manure storage facilities	−	0	−	−
Pastureland and rangeland	*	+	+	+
Silos	*	+	*	+
Milkhouses and milking centers	*	+	*	*

[a]Symbols are as follows: +, a potential nonpoint source; 0, system dependent; −, not a potential nonpoint source; *, not applicable.
[b]From (5).

One water quality impact common to all nonpoint sources associated with animal production is decreased dissolved oxygen concentrations. Nonpoint discharges from animal confinement and pastures are likely to result in increases in stream BOD_5 and COD concentrations. Athough nitrogen enrichment can result from nonpoint source discharges from a variety of land-based activities, increases in ammonia nitrogen concentrations are most likely to be associated with animal production activities. Nitrate nitrogen is less useful as a parameter for source identification since nitrate nitrogen is a common constituent of cropland runoff and a result of the nitrification of the ammonia in runoff from animal production sites. The presence of fecal coliform and fecal streptococci indicator organisms in increased concentrations also suggests that animals or manures could be a source of the problem. These organisms, however, may not be present in increased concentrations where milkhouse wastewaters or silage liquors are the principal nonpoint pollutant sources. In an analysis of available water quality data to determine if animal agriculture is causing a nonpoint source problem, placing emphasis on only one water quality characteristic should be avoided.

The potential significance of nonpoint pollutant discharges associated with animal production activities is summarized in Table 14.8. Manure disposal sites are potential sources of water pollutants common to all segments of animal agriculture. Dairy farms have the greatest number of potential pollutant sources, and poultry farms have the least. The pollution potential from animal wastes is greatest when manure is spread on snow or frozen ground (Fig. 14.8). Under such conditions, there will be little opportunity for incorporation of the manure in the soil, and snowmelt and spring rains can transport the manure directly to surface waters.

The actual significance of the possible nonpoint sources of water pollutants is dependent on more than the presence of animals or manure. Factors such as

Fig. 14.8. Disposal of manure on snow-covered land.

promixity to surface waters (Fig. 14.7) and site management practices also can have a significant influence on the potential water quality impacts and the need to implement BMPs for animal production locations.

Comparative data on the unit loads from agricultural nonpoint sources were presented in Chapter 3 (Fig. 3.9). The comparisons suggest that the concentrated animal agriculture nonpoint source categories should receive priority in the allocation of available financial and technical resources for BMP implementation, with the next priority given to land used for manure disposal. As possible sources of nonpoint pollutant discharge, pastures and rangeland would have lower priority.

The objective of the comparisons in Chapter 3 and in this chapter was to establish which categories were likely to have the greatest water quality impact and what the priorities might be for allocating resources to address nonpoint water quality problems. Care should be taken in applying these generalizations to specific situations. A number of site-specific factors must be considered before one can determine the relationship of any potential pollutant source to an identified water quality problem, the need to implement BMPs, and the most appropriate BMPs.

Once it has been determined that an observed water quality problem is

TABLE 14.9

Practices to Control Nonpoint Source Problems Associated with Animal Production Activities

Open confinement facilities	Runoff collection, run-on diversion, vegetative practices—grassed outlets and buffer strips, enclosing open facilities, relocation
Land used for manure disposal	Use of application rates based on agronomic need; timing of manure application—side-dressing and top-dressing; avoidances of time periods with high runoff probability, storage, and establishment of field priorities; injection or immediate soil incorporation; and vegetative practices—field borders and buffer strips
Manure stacking facilities	Detention ponds for drainage
Pastureland and rangeland	Use of recommended stocking rates; discouraging of animal congregation in critical areas, including streams; breakup, distribution, and incorporation or removal of manure accumulations; and restricting animal access to highly erodible areas
Silage liquors	Use of recommended ensiling practices, and collection of seepage from silos
Milkhouse and milking center wastewaters	Combined storage and disposal with liquid manures; lagoons and ponds; controlled land application; and vegetative filters

related to animal agriculture, and the physical source or sources of the responsible pollutant discharges have been identified, it is then possible to select a practice or combination of practices that will effectively address the problem. Due to variables such as established water quality goals, availability of capital, management capabilities, and physical constraints, the selection of effective BMPs is a site-specific process. There is no one pollution control practice or combination of practices that can be designated as most suitable for each animal agriculture nonpoint pollutant source category.

The practices that can be used to reduce or eliminate the amount of pollutants from nonpoint animal production activities are noted in Table 14.9. Many of these also were identified as managerial, vegetative, or structural controls in Table 14.4.

Summary

The options available to control agricultural nonpoint sources are numerous. The approaches identified in this chapter are a logical progression of problem identification, identification of possible sources, and selection of BMPs to control the identified problem.

Cost-effectiveness, suitability, and likelihood of acceptance are important considerations in BMP selection. The nature of water quality management and the limit of available data suggest that an incremental program of BMP implementation can be desirable. Such a program would implement apparently appropriate practices, monitor their effectiveness, and refine these practices and/or add other practices with time.

The management, vegetative, and structural BMPs identified in Tables 14.4 and 14.9 can reduce pollutant losses in runoff. With many of the practices, reductions of solid-phase pollutants (sediment, strongly adsorbed pesticides, organic nitrogen, fixed phosphorus) are substantially greater than reductions of dissolved nutrients and pesticides. The magnitudes of pollutant reduction are site-specific, depending on local weather, soils, and crop management. For both practical and political reasons, soil and water conservation practices are likely to be the major component of programs to control nonpoint water pollution control problems caused by agriculture.

References

1. Novotny, V., and Chesters, G., "Handbook of Nonpoint Pollution—Sources and Management." Van Nostrand Reinhold, New York, 1981.
2. Evans, R. G., Walker, W. R., Skogerboe, G. V., and Smith, S. W., "Evaluation of Irrigation Methods for Salinity in Grand Valley," EPA-600/2-78-160. Robert S. Kerr Environmental Research Laboratory, U.S. Environmental Protection Agency, Ada, Oklahoma, 1978.
3. Bailey, G. W., and Waddell, T. E., Best management practices for agriculture and silviculture: An integrated overview. In "Best Management Practices for Agriculture and Silviculture" (R. C. Loehr, D. A. Haith, M. F. Walter, and C. S. Martin, eds.), pp. 35–56. Ann Arbor Sci. Pub. Inc., Ann Arbor, Michigan, 1979.
4. Haith, D. A., and Loehr, R. C. (eds.), "Effectiveness of Soil and Water Conservation Practices for Pollution Control," EPA-600/3-79-106. U.S. Environmental Protection Agency, Environmental Research Laboratory, Athens, Georgia, 1979.
5. Robillard, P. D., Walter, M. F., and Bruckner, L. M., "Planning Guide For Evaluating Agricultural Nonpoint Source Water Quality Controls," EPA-600/3-82-021. U.S. Environmental Protection Agency, Environmental Research Laboratory, Athens, Georgia, 1982.
6. Stewart, B. A., Woolhiser, D. A., Wischmeier, W. H., Caro, J. H., and Frere, M. H., "Control of Water Pollution from Cropland—Volume I—A Manual for Guideline Development," Report ARS-H-5-1. U.S. Department of Agriculture, Agricultural Research Service, Washington, D.C., 1975.
7. Stewart, B. A., Woolhiser, D. A., Wischmeier, W. H., Caro, J. H., and Frere, M. H. "Control of Water Pollution from Cropland—Volume II—An Overview," Report ARS-H-5-2. U.S. Department of Agriculture, Agricultural Research Service, Washington, D.C., 1976.
8. Conservation Foundation, "State of the Environment—1982," p. 235. Washington, D.C., 1982.
9. Walter, M. F., Steenhuis, T. S., and Haith, D. A., Nonpoint source pollution control by soil and water conservation practices. Trans. Am. Soc. Agric. Eng. 22, 834–840 (1979).
10. Yaksich, S. M., "Summary Report of the Lake Erie Wastewater Management Study," U.S. Army Corps of Engineers, Buffalo District, Buffalo, New York, 1983.

15

Management

Introduction

The theme of this book has been one of continued production of food with a minimizing of adverse environmental quality effects that may be a result of such production. A balance must result between agricultural production, profit, and environmental quality objectives. Maximum profit frequently implies no concern with environmental quality or pollution control, and maximum environmental quality control can result in severe economic constraints on agricultural production. The balance between these extremes must be the goal of agricultural producers, environmental control organizations, and the public.

Adequate information is available to avoid gross pollution from agricultural production. Good practice environmental quality guidelines have been developed by both agricultural organizations and governmental agencies. These guidelines consider waste management approaches that are integrated with agricultural production operations. There are no simple or single solutions to these problems, and it is no longer possible for a component of society to dispose of its waste by export. There is no longer any "away" into which wastes can be disposed.

The land remains the most logical point of ultimate disposal for most agricultural wastes. Other possible means of ultimate disposal include discharge of treated liquid wastes to surface waters, application of solids to the land and separate treatment of liquid wastes, solids destruction and application of residual inerts to the land, partial treatment of the wastes followed by reuse, and joint industrial–municipal cooperation on waste treatment.

413

Governmental Action

The development and application of regulations to control pollution from nonpoint sources, such as general agricultural runoff or percolation, are difficult. Few states or federal regulations have attempted this type of control. The 1972 Amendments to the Federal Water Pollution Control Act contained the first specific concern about agriculturally related nonpoint sources of pollution in federal water pollution control legislation. The Amendments indicated that each state waste treatment management plan was to identify agricultural nonpoint pollution sources, including runoff from manure disposal areas, and set forth feasible procedures to control such sources. Effluent limitations and the use of the best practicable and best available technology were applied to control the point waste discharges from agricultural operations.

Many states have water quality regulations or administrative codes or guidelines that are applicable to agricultural wastes. Agricultural wastes may not be explicitly cited as potential pollution sources in these regulations, although the regulations are sufficiently broad to be applied to agricultural pollution problems, especially those that are point sources. A number of states in which livestock production is an important factor have established additional legislation or guidelines that are specific for livestock waste management.

States generally have air pollution statutes or guidelines dealing with air pollution problems that can be associated with agricultural production facilities. Odors and particulate matter arising from feedlot dust and feed processing activities are examples of items that can be subject to control.

Agricultural producers need to be aware of the various state regulations and guidelines and any local zoning or health ordinances that may apply to their operations. The producers also should know whom to contact about specific problems. Generally, state or county extension agents and employees of the state pollution control agencies can provide information about the applicable regulations or guidelines. Advice on appropriate pollution control measures should be obtained before a problem occurs and especially before an operation is enlarged or a new operation developed.

There are many approaches utilized to control potential pollution from agricultural operations. Many states have developed registration and permit procedures for control of runoff from animal feedlots. While the procedures vary throughout the nation, they usually contain the provision that a feedlot operator shall obtain a permit from the appropriate state agency, correct any pollution hazard that exists, and assure that the operation conforms to all applicable federal, state, and local laws. Specific minimum pollution control methods such as runoff retention ponds, dikes, and distance from dwellings may be included. A map of the areas, indicating land use, streams, wells, houses, roads, topography, and other salient features, frequently is requested.

While control of potential pollution from feedlots is a recent item, pollution from food processing operations has been included in the pollution control regulations of many states for decades. Food processing wastes are point source wastes frequently discharged to surface waters, and their discharge is controlled by the prevailing pollution control regulations.

Guidelines, rather than regulations, are common approaches to minimize pollution problems from agricultural operations. A reasonable approach has been established by the Province of Ontario, Canada, which has developed a suggested code of practice for livestock buildings and disposal of animal wastes (1).

The intent of the Ontario Code of Practice is to (a) assist farmers in avoiding unnecessary and undesirable situations which could lead to disputes concerning pollution, (b) serve as guidelines for anyone concerned with the establishment of new livestock or poultry production units, (c) serve as guidelines for anyone concerned with making a major renovation or expansion of existing livestock or poultry enterprises, (d) be flexible enough in interpretation and application to cover special cases which may exist or develop from time to time, and (e) serve as a basis for a sound, considered plan of farm operation, giving due regard to waste disposal management, without being specific in design requirements. Key measures in the Code include the provisions of enough land area on which to dispose of the wastes, sufficient waste storage capacity, and sufficient distance between livestock and poultry buildings and neighboring human dwellings.

Farmers can apply for a certificate of approval, which indicates that they are following accepted guidelines for environmental quality control. Recommended land areas for disposal were based on the number of equivalent animal units contributing the wastes. The animal units were based upon the pollution potential as determined by the age, size, and feed of the animals and their period of confinement. The suggested animal unit relationship is presented in Table 15.1.

Minimum tillable land requirements for disposal of the animal wastes were suggested for both loam and sandy soils for various sizes of animal units. The recommended acreages were not necessarily the most economical, from the standpoint of efficient crop production. The minimum acreage was considered to be that required to avoid the risk of groundwater pollution by nitrogen compounds.

Manure handling and storage suggestions described in the Code were as follows: (a) sufficient storage capacity for 6 months' accumulation of wastes should be provided where a liquid manure handling system is to be used; (b) weather, cropping programs, and other local conditions may permit satisfactory operations with a shorter storage period; (c) where liquid manure is spread on land within 1000 ft of neighboring dwellings, it should be incorporated into the soil as soon as possible and preferably within 24 hr of application; (d) where solid manure is spread on land within 600 ft of neighboring dwellings, it should be incorporated into the soil as soon as possible and preferably within 24 hr of

TABLE 15.1

Equivalent Animal Units Based on Pollution Potential[a]

Animal	No. of animals equivalent to one animal unit	Basis of production
Dairy cow (plus calf)	1	Annual
Beef cow (plus calf)	1	Annual
Beef steer (400- to 1100-lb gain)	2	Annual
Bull	1	Annual
Market hog (40- to 200-lb gain)	15	As marketed
Dry sows (plus litter)	4	Annual
Laying hens	125	Annual
Chicken broilers (4–5 lb)	1000	As marketed
Pullets	300	As marketed
Turkey broilers (11–12 lb)	300	As marketed
Horse	1	Annual
Mature sheep (plus lambs)	4	Annual

[a]From (1).

application; and (e) solid and liquid manure storage units should be well managed to keep odor problems to a minimum and to prevent runoff into streams, wells, and other bodies of water.

Many states have developed guidelines to control wastes from agricultural production operations. The majority focus on animal production operations and include provisions comparable to those in the Ontario, Canada, guidelines as modified by situations and concerns within the state. Codes or guidelines are preferable to governmental regulations, since there is the danger that regulations will be applied uniformly, despite the wide variety of measures needed to protect the environment properly and the wide differences between agricultural production operations. Codes or guidelines can be changed more easily than regulations when better practices become available.

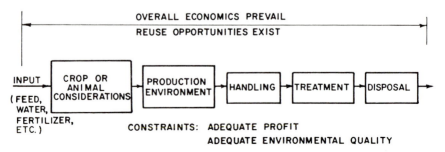

Fig. 15.1. Components of comprehensive agricultural waste management systems.

Specific state or national regulations for controlling general agricultural runoff do not seem appropriate. With the exception of food processing wastes and feedlot runoff, agricultural runoff is ubiquitous and not amenable to regulatory control. Adherence to concepts included in Chapter 14 will minimize the nonpoint source contaminants from agricultural operations. Point sources of agricultural waste discharges can be controlled by existing state and federal regulations (Chapter 1).

Decision Making

General

A satisfactory management system for environmental quality control in agriculture consists of a combination of alternatives which will provide a suitable level of environmental quality and will be economically feasible. With the broad spectrum of environmental quality effects, management interests and capabilities, legal constraints, and levels of technology, it is apparent that a separate management system must be developed for each agricultural operation or for agricultural operations that have common problems, operational methods, and environmental quality needs.

The legal, administrative, financial, and political constraints will narrow the range of alternatives for a successful agricultural waste management system. Any costs assigned to an alternative management component should be real, so that an estimate can be made of the benefits lost by using a particular alternative.

The total agricultural operation should be considered in developing suitable manure and food processing waste management systems. Waste management generally is viewed as a treatment and/or disposal problem. Superficially, this appears logical, since the problem is one of disposing of the accumulated wastes without contaminating the environment. Waste treatment and disposal are, however, only parts of an adequate waste management scheme. Consideration of waste management begins with the type and quantity of input and continues with the effect of the production environment through methods of waste handling, treatment, and disposal; the potential of waste or wastewater reuse; and a consideration of the overall economics of the agricultural production operation as it is affected by waste management (Fig. 15.1).

When methods to handle, treat, and dispose of agricultural wastes are contemplated to avoid water pollution, it is important that other types of pollution be avoided. Transfer of pollution from one sphere to another will not be a satisfactory waste management alternative.

No new facility or expansion of existing facilities for agricultural production should be considered without prior planning, which should include the probable environmental effects of the disposal of wastes from the facility. Consideration should be given to air movement patterns, rainfall and runoff relationships, ultimate waste disposal site and methods, soil characteristics, environmental quality criteria, the effects of waste treatment, and possible changes in production techniques and environmental quality restrictions. An evaluation also should be made of the probable risks of civil suits and governmental action in response to valid complaints of inadequate abatement of nuisance and pollution.

An acceptable waste management scheme will not only be technically feasible but also will be economically competitive with other approaches; will require a minimum of maintenance; may be operated on an automatic or semiautomatic schedule; will not create environmental conditions objectionable to plant workers, housed animals, and the public; and will allow for expansion and new technology.

Opportunities

Both internal and external opportunities exist in agricultural waste management systems. A number of these are outlined in Fig. 15.2. Considerable improvement can be made by "in-plant" changes. These include changes in process equipment and production processes, elimination of leaks, reduction in waste volume, separation of wastes for possible reuse or treatment and disposal according to waste strength, recovery of raw materials, change in the physical characteristics of the waste (i.e., liquid slurry or solid to accommodate waste treatment and disposal), and even a judicious choice of location for the facility.

Other opportunities exist after the wastes exit or are removed from the facility. These include the use of proper waste treatment design criteria, alternative treatment methods to obtain specified effluent requirements, treatment of wastes for acceptance on the land rather than for discharge to surface waters, runoff control, reuse of cooling water or partially treated water elsewhere in the facility such as for flushing or raw product rinsing, and a recovery of portions of the wastes for by-product utilization.

The final disposal point offers additional alternatives to be evaluated. The final residuals can be disposed of into combinations of the air, water, or land. Such disposal can be done directly or indirectly via discharge to municipal waste disposal systems. The latter approach requires that the industry pay the municipality for the discharge of its wastes, but it relieves the industry of the burden of final disposal.

A materials balance of the production facility, including (as appropriate) flow, nutrient, solids, energy, and labor balances, should be prepared as an initial step in evaluating the proper components of the combined production–waste management system. The balances will help identify if and where process changes are needed and the types of treatment and disposal that are appropriate.

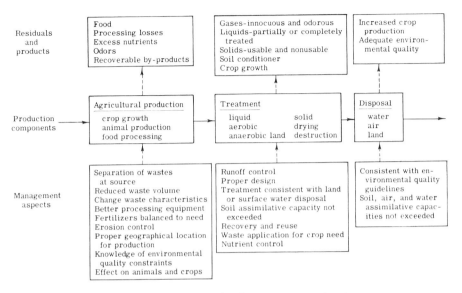

Fig. 15.2. Conceptual agricultural waste management systems.

Selection of Treatment Alternatives

For agricultural wastes, many typical management systems may be considered. Depending upon the desired complexity of the preliminary systems design and the type of waste to be treated, a number of basic components can be incorporated as in Fig. 15.3. Aspects of each of these components are outlined in Fig. 15.2.

A waste management system consists of a number of individual components. After "in-plant" processing changes have been evaluated, the unit processes for treatment can be considered. A primary consideration in evaluating unit treatment processes is the point of ultimate disposal of the treated or untreated waste. Agriculture may consider both land and water as primary points of disposal. Another equally important consideration is the degree of treatment for the disposal that is required at present and may be required in the future. For each situation, there are a number of combinations of unit processes which will satisfy the conditions. Known and possible future environmental quality constraints must be kept in mind so that unit processes selected initially can be expanded or readily modified to meet future needs. Specific unit processes applicable to animal production and food processing operations are many (Fig. 15.4), and details are provided in earlier chapters.

A major consideration for possible unit processes is their simplicity and reliability. Animal and food processers are in the business of producing food.

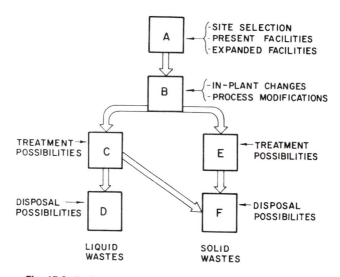

Fig. 15.3. Decisions components of a waste management system.

Waste management considerations are not among their highest priority items. Treatment and disposal processes for these operations should be as foolproof as possible, capable of being operated and maintained by individuals not trained in waste management, capable of handling varying waste loads, and consistent with equipment normally found in these operations. The simpler a process is to operate and control, the more likely it is to operate successfully and continuously. Regardless of how efficient a process may be when operating under rigidly controlled conditions, if it is sensitive to minor changes in waste charac-

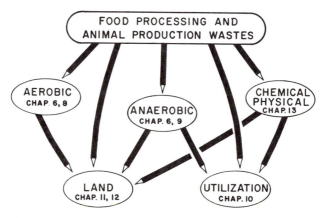

Fig. 15.4. Chapters related to treatment process and waste management decisions.

teristics or if it requires constant expert control and supervision, it is not likely to operate as intended under field conditions.

Joint Industrial–Municipal Cooperation

Discharge of untreated or partially treated wastes to a municipal treatment systems can be a possibility for both municipalities and a number of agricultural operations. The advantages include assurance that the wastes are adequately treated or disposed of, the community retains the economic base of the industry, and the industry avoids increased capital investment in waste treatment facilities and avoids the need for increased waste treatment personnel.

The details and charges regarding the use of a municipal sewerage system by an agricultural or industrial operation generally are covered by the sewer ordinances and sewer service charges of the municipality. Most municipalities have at least a general provision in their waste ordinances stipulating that harmful and objectionable materials may not be discharged to their sewerage systems. A number of items can be specifically prohibited. Examples of prohibited wastes are liquids above a certain temperature; liquids having a pH value outside a specific range; improperly shredded garbage; oils, fats, and greases above a specific concentration; any material that can cause obstructions in the sewers, such as ashes, feathers, sawdust, food processing bulk solids, paunch manure, etc.; waste toxic to the biological processes used to treat the wastes; and radioactive wastes. The logic is that an industrial operation should not place any waste into the sewers that will cause problems in the sewers or with subsequent treatment processes or that will not be removed by these processes. The burden of proof that the waste will not be detrimental to the system and can be removed by conventional treatment processes rests with the industry. In addition, many municipalities will not allow, or will allow only with special permission and surcharges, the discharge of wastes that have abnormal concentrations of BOD or suspended solids. Guidelines for the preparation of specific ordinances are available (2).

The need of complete treatment of wastes to meet more rigid effluent standards, rising costs, and the public demand for better service has led many communities to levy special charges on industrial wastes discharged to their sewerage systems. The logic behind these surcharges is that the cost of providing sewerage service should be paid directly by the users as the most equitable means of charging for the service. A number of typical service charges and examples of their use are available (3).

A properly designed service charge will distribute costs more closely related to the service provided than will any other way of raising revenue for treatment facilities. Effectively administered user charges can improve the management of industrial wastes. Charges related to the cost of treating wastes, based at least on waste volume and strength, can create an incentive for industry to pretreat, to change processes, and to manage their wastes more effectively. When the cost of the service is directly borne by the users, they have an interest in seeing that the treatment system is effectively planned and managed. User charges provide a relatively stable source of revenue to meet waste treatment costs, which allow for businesslike management of the sewerage system and provide for an orderly operation, maintenance, replacement, and expansion of the system.

The type and magnitude of the industrial waste surcharge will affect the decision of an agricultural operation to discharge its liquid wastes to a municipal system. There are several formulas in existence for industrial waste surcharges. The constant-rate formula is a common one and involves charging on the basis of a production unit such as water use, number of employees, or quantity of product produced. These formulas are simple to administer but are only a crude way of taking differences in the waste strength into account. Because testing of the effluent from an operation usually is not done with this formula, the charge is not related directly to the quality of the waste sent to the sewerage system and therefore is not likely to cause the industry to reduce its waste quantity or quality. For small municipalities where the gain in accuracy from waste testing and surveillance is small compared to the expense involved, this type of charge may be satisfactory.

A quantity–quality formula can be used where the expense of more detailed waste testing and enforcement can be justified. There are over 150 municipalities using this type of a surcharge, and the number is increasing. The formula indicates the extra charge which is to be made when the formula is applied to wastes of above-average characteristics. Surcharges based on this formula take into account both the volume and pollutional quality of the waste.

A typical surcharge of this type would be

$$C = V[Y_1 + Y_2(B - B_n) + Y_3(S - S_n) + \ldots + Y_n(N - N_n)] \quad (14.1)$$

where C = charge; V = volume; Y_1, Y_2, Y_3, Y_n = surcharge rates for specific items; B, S, N = actual concentrations of pollutants causing increased costs; and B_n, S_n, N_n = normal concentrations of pollutants. The waste characteristics usually included in surcharge calculations are volume, BOD, and suspended solids. Other characteristics such as nutrients or chlorine demand can be included as appropriate. The definition of ''normal'' concentrations varies widely among municipalities.

The surcharge rates also will vary among communities, reflecting the type of treatment and the effect a characteristic will have on a treatment process. The liquid volume affects both collection and treatment costs; BOD affects primary, secondary, and tertiary treatment costs as well as sludge disposal costs; and suspended solids affect solids handling, treatment, and disposal costs. Both capital and operating costs should be reflected in the surcharge rates.

The advantage of the quantity–quality formula is that it is an equitable charge, since it takes the characteristics of the industrial effluent and the costs associated with treating the waste directly into account. This can be important when one considers that in a small community, a continuous or seasonal agricultural operation can contribute a large percentage of the volumetric or pollutional load to the community treatment plant. Another advantage is that it provides an inducement for the agricultural or industrial operation to reduce the quantity and quality of its waste. The chief disadvantage is that the administration of the surcharge requires continued monitoring and sampling.

Pretreatment of food processing wastewater deserves consideration when constituents prohibited by, or in excess of limits permitted by, municipal regulations are present and when the processor will pay a surcharge for discharge and can effect economies by pretreatment. The type of pretreatment can include solids separation by screening, gravity, air flotation, or other means and/or biological or chemical treatment. The pretreatment processes and capacities selected depend upon the size of the processing plant, efficiency of the selected process, facilities for handling and materials removed from the wastewater, and related engineering and cost factors.

Some opportunities also exist for agricultural–municipal cooperation in the disposal of solid wastes. Examples are landfills used for dead poultry and animals and the use of animal and food processing wastes in municipal composting operations. Relationships generally are less formal than with acceptance of liquid wastes. Acceptance or denial of agricultural solid wastes by a municipality generally is based on whether the wastes can be handled with little difficulty. The agricultural producer has the responsibility of transporting the solid wastes to the disposal sites. The feasibility of such transport is a function of the proximity of the disposal site and the quality and quantity of the solid wastes.

Manure

General

Livestock producers are faced with many management decisions when they consider manure management options. The decision-making process includes the acquisition of information that will provide answers to questions such as (a) what are the environmental objectives to be achieved, (b) what other objectives such as labor reduction, energy use reduction, and minimizing costs should be considered, and (c) what legal requirements and financial commitments are involved?

The decision-making process requires technical information from diverse sources such as extension agents, universities and agricultural experiment stations, equipment dealers, the U.S. Soil Conservation Service, and other state and federal agencies. After the pertinent information is collected and evaluated, livestock producers must determine what manure management option is the most appropriate and economical for their specific operation. Factors that affect the ultimate decision are the availability and cost of capital, the type of farming operation, and the producer's planning horizon.

An important factor in the decision process is the extent to which manure can be used as a productive resource rather than as a waste for disposal. There are many conservation and utilization possibilities that can be considered (Fig. 15.5). Details about these possibilities are discussed in Chapter 10.

Many livestock producers find land application an economically viable option. Some may wish to use procedures for maximum nitrogen conservation prior to land application so as to reduce or eliminate the need to purchase inorganic fertilizers. Others may desire an anaerobic unit to produce methane, to become less dependent on fossil fuels, and to reduce energy costs.

Livestock production involves a continual series of decisions about how to organize available resources to maximize profits. The decision-making process may be structured or may be little more than intuitive judgment. Regardless of the method, economic concepts such as fixed and variable costs, opportunity costs, and the principle of diminishing returns are involved in the decision process. Manure management options are one of the components of the process.

One of the first manure management decisions is to determine whether a liquid or solid system is to be used. The decision will be related to the ability of each system to fit into an existing or new facility, labor costs, and ultimate disposal possibilities and constraints.

Any proposed manure management system must accomplish more than simply relocating a nuisance to await development for further problems and complaints. Year-round adequacy must be provided. Many manure disposal options are satisfactory during parts of the year but may be troublesome at other times. The management system must satisfactorily handle and dispose of the wastes without jeopardizing the health or well-being of the producer, the animals, or the public.

Animal production units fall into two main groups, (a) the normal unit, where disposal on land will be the prime objective, with or without the use of some waste treatment, and (b) a smaller number of operations which are single-purpose, large-scale enterprises. Many of the latter are maintained on land that is inadequate for waste disposal purposes. These units may require special treatment and disposal methods and larger capital commitments for waste management.

The four general methods of manure management are as follows: dry han-

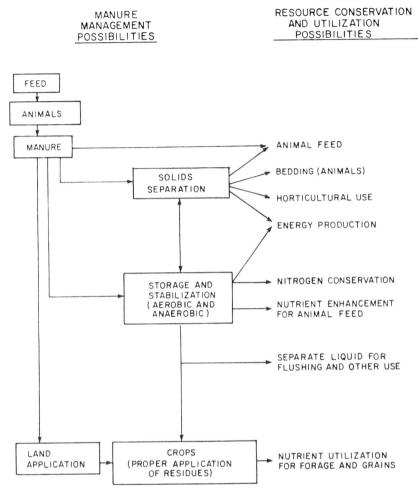

Fig. 15.5. Resource conservation and utilization possibilities that can be considered as part of manure management options.

dling—excreta handled dry or mixed with bedding, distributed on land either daily or after storage, and possibly used as part of a recovery and reuse program; semidry—a semisolid slurry disposed on the land by spreading; semiliquid— mixed with water for distribution on the land by tankers; and liquid manure— mixed with water for liquid irrigation, for treatment and disposal, or for reuse of portions of the liquid and/or solid fraction. Odors can be common with the last three methods.

Feedlots

Where animals are housed in the open, such as in feedlots, runoff from these areas has the potential for serious water pollution problems (Chapters 2 and 3). Pollution caused by runoff is reduced when animals are completely housed, as is the case in many hog, dairy, and most poultry operations.

The major considerations affecting the magnitude of the feedlot runoff problem and the possibilities for abatement include the fact that the wastes are more diffuse and not amenable to conventional waste treatment technology used for defined and controlled point source discharges; that feedlot runoff is related to rainfall and snowmelt and thus the predictability of waste flow in terms of quantity, quality, and duration is difficult; and that the characteristics of feedlot runoff are variable.

Feedlots can be divided into three basic types: (a) open feedlots, (b) open confinement, and (c) total confinement. Open feedlots are low-density lots (50–400 ft^2 per animal) with little or no cover or protection for the animals. Open confinement feedlots are higher-density lots (30–75 ft^2 per animal) which offer some protection for the animals, ranging from partially roofed to totally roofed pen areas enclosed on three sides. Total confinement feedlots are high-density structures (12–30 ft^2 per animal) with controlled environment. The waste management alternatives are different for each type. The need for runoff control is greatest with the open and open confinement feedlots.

Key runoff and waste management factors at open feedlots and open confinement feedlots include diversion of extraneous runoff; lot size and surface; lot topography and drainage; collection, retention, and disposal of feedlot runoff; and solid manure removal and disposal. In total confinement feedlots, the waste management approaches focus on removal and proper disposal of the accumulated manure. Regardless of the type of feedlot, the items in Table 15.2 should be considered in the design and construction of feedlots.

Site selection and feedlot location can reduce the potential of water pollution from feedlots. The number of animals and the surface of the lot will determine the area of the lot that is required. A site should be chosen that is away from a stream or waterway. Animals in confined feedlots should not have free access to a stream, since this would permit a natural stream to run through the lot and be the receptor of feedlot runoff. Drainage of feedlots through adjacent property, road ditches, or directly into natural streams or lakes should not be permitted. Land area should be available so that drainage from the lots can be retained on property owned by the feedlot.

Drainage from the feedlots will be affected by the slope of the lots. Slopes of 4–6% will keep the lots drier. Feedlot runoff contains greater pollutants when the lot has been permitted to remain wet. Where inadequate slopes are available, internal drainage and manure mounds should be provided. New lots should be

TABLE 15.2

Pollution Control Factors to be Considered in the Design and Construction of Feedlots

1. The minimum design of all runoff collection retention facilities is a 25-year, 24-hr storm for all new feedlots[a]
2. Runoff collection and storage structures, treatment lagoons and ponds, and other structures for the storage of liquid, solid, semisolid wastes, and ensilage should be constructed to eliminate seepage to surface or groundwaters
3. All feedlot runoff and manure should be applied to available land in a manner and at a rate that will not cause subsequent water or soil contamination or damage to crops
4. All tailwater from irrigation of feedlot runoff should be retained and recycled
5. Techniques for handling, storage, and disposal of runoff and manure should be such that odors, noise, and other nuisances are prevented

[a]From (4).

properly drained during construction. Snow buildup should be avoided, since it can increase runoff and pollution potential. Animal shelters should be placed where snow buildup can be minimized.

If a feedlot is continuously stocked so as to maintain an intact manure layer, there is little likelihood of nitrate or other pollutants moving through the soil profile to the water table. An anaerobic layer develops at the interface of the manure and soil, which helps prevent the movement of nitrate and water. Caution should be exercised in cleaning a feedlot so as not to destroy this layer.

Odor nuisances can be minimized by locating feedlots downwind from nearby towns and residences. Buffer zones of trees between the feedlot and adjacent homes will help decrease odor problems. In the drier climates, such as the southwest, dust problems occur. Natural and artificial windbreaks may help this problem. Many of the gross water and air pollution problems from feedlots can be minimized by judicious selection of feedlot sites.

A critical feedlot management item is to divert all water outside the feedlot, thereby preventing contact with the feedlot wastes. Diversion of extraneous water provides for a drier environment for the animals, possibly reduced odor, and a marked decrease in the volume of feedlot runoff to be managed. Even with proper site selection and lot management some runoff will result and will require suitable management.

The most practical approach for feedlot runoff control appears to be a combination of protective dikes and levees to prevent the entry of rainfall from outside the feedlot area, plus collecting dikes, levees, and holding ponds to collect the rainwater falling directly on the lot (Fig. 15.6). After collection and retention, some acceptable means of disposing of the liquid and solid material, such as fertilization and irrigation on available pastureland and cropland, should

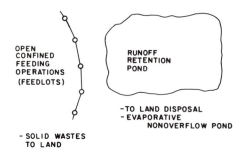

OPEN
CONFINED
FEEDING
OPERATIONS
(FEEDLOTS)

RUNOFF
RETENTION
POND

-TO LAND DISPOSAL
- EVAPORATIVE
        NONOVERFLOW POND

- SOLID WASTES
  TO LAND

Fig. 15.6. Runoff retention ponds and liquid and solid disposal for beef feedlots. [From (5).]

be employed. Overflows from the retention ponds may be allowed only during times of rainfall that exceed the design criteria.

Retention ponds will minimize the pollution caused by runoff from animal production units housed in the open. In dry areas, they may be the only treatment needed. Where retention ponds are not satisfactory to protect the receiving waters, a combination of lagoons, oxidation ponds, and aerated systems may be possible. With proper operation, runoff retention ponds are a very feasible waste management technique for open and partially confined feedlots.

Disposal of the collected feedlot runoff by irrigation is the most common disposal method used. The disposal should be done at such a rate and time so as to reclaim the majority of the nutrients in the production of a crop.

Grass terraces offer another opportunity for runoff control and can be incorporated with runoff retention ponds. Runoff flowing over grass or agricultural land prior to retention ponds can be expected to be reduced in volume by soil percolation and reduced in pollutional characteristics by solids removal on the land and by oxidation of some of the organic matter in the stubble or grass.

Runoff control systems using a settling channel, porous dams, and a detention pond for holding the liquid portion of the runoff have been used to remove settleable solids from the runoff. This approach is feasible for separating discrete particulate material from beef feedlot runoff. It may be less successful with wastes that contain fine particulate matter such as dairy or poultry manure slurries. Such fine matter may be difficult to remove from the settling areas and may plug the porous dams.

Disposal of animal wastes on land is the method of choice. This approach should not be considered a panacea, and the disposal of accumulated feedlot wastes on land should be done with care. Incorporation of the solid waste after field spreading will reduce subsequent runoff and odors. Wastes should not be placed on snow or frozen ground if runoff and thawing may cause pollution. Waste-spreading areas should not be adjacent to streams or other surface waters.

The potential for impairment of groundwater quality around animal production operations should receive close scrutiny, since once the groundwater becomes contaminated, it is difficult to return it to its original condition or to use it for potable purposes. Most states have provisions in their statutes to protect the groundwater and maintain its quality.

Pollution of groundwater can also result from the direct movement of runoff into inadequately cased or poorly protected wells located in the immediate vicinity of a feedlot. Feedlot drainage should be away from all wells, and wells should not be drilled in geological formations likely to intercept feedlot runoff.

Enclosed Confined Operations

In contrast to feedlots, most other animal production operations utilize enclosed confined facilities. These facilities represent sources of wastes which are less subject to the probabilistic nature of hydrologic conditions. With few exceptions, the ultimate disposal of the untreated or treated wastes will be on the land. Greater operational flexibility is possible when discharge to the land is practiced, since application of the wastes can be matched to crop needs, the soil assimilative capacity, and the work and seasonal schedule of the producer.

Although processes are possible, only a few are in general use. For liquid wastes these are the aerobic oxidation pond, the aerated lagoon, the oxidation ditch, the rotating biological contactor, and the anaerobic lagoon. For concentrated waste slurries it is a combination of anaerobic stabilization, storage, and/or intermittent cleaning with disposal on the land. For drier wastes it is land disposal, composting, and drying for use as an agricultural or home garden soil conditioner.

A waste management system for enclosed confined facilities will deal with wastes having a consistency of "as-defecated," liquid, or "dry" material. The most feasible approach will depend upon the nature of the waste, the ultimate disposal method, and labor requirements and availability.

The water content of the manure determines the handling system alternatives that are available and suitable. Semisolid manure generally contains greater than 15% solids and may include bedding materials. Handling is by front-end loader or equivalent equipment. Manure slurries generally contain 10–20% solids, require the addition of water to the manure, cannot be efficiently handled with a front-end loader, and require special solids handling pumps. Liquid manure generally contains less than 10% solids and can be pumped with most conventional liquid handling pumps.

Daily hauling and storage of manures can be considered as parts of a waste management system. Daily hauling requires out-of-doors work in all kinds of weather, has a greater potential for water pollution with nutrient and organic matter losses during runoff, and can have an overall yearly manure value loss of 40–60%. Availability of land for spreading may be a problem during the crop growing season.

Storage allows spreading of manure during optimum field conditions, tends to minimize loss of organic matter and nutrients, and increases capital investment. A storage tank permits flexibility during periods when the manure cannot be disposed of on the land, such as when the land is wet or frozen or when crops are being grown. In northern climates, a storage capacity of 3–5 months is suggested. Such storage results in anaerobic conditions that cause partial decomposition and odor. The odor nuisance is caused more by the movement of the stored slurry, i.e., pumping and disposal, than by its storage.

Another development has been the use of manure stackers i.e., barn cleaner elevators up to 60 ft in length, running on a semicircular track, which are used to stack the dairy manure in a concrete storage area during the winter months. This system does not require a large capital input and is flexible. The investment in this system is intermediate between daily hauling and liquid systems. The advantages include storage during inclement weather and when fields are unavailable. A tractor with a manure loader is required. The stacking area should be fenced off, and drainage and runoff from the area should be kept from nearby streams.

The primary wastes associated with concentrated poultry operations are manure, litter, and other wastes such as dead birds. The most common methods of disposing of the other wastes are landfill, incineration, and rendering. Almost all of the manure from broiler production is dropped on litter, which serves to absorb moisture from the manure and to assist in drying. Many materials are used as litter, such as sawdust, wood shavings, peanut hulls, chopped straw, and bark and chips. Litter is one of the most significant wastes from commercial turkey operations, since the first 8 weeks or so of the turkey's life are spent on litter. The commercial egg production in the United States uses no litter. Consequently the wastes from egg-laying operations and the remainder of a turkey production operation consist almost entirely of manure.

Various methods have been developed to handle, treat, and dispose of the wastes from commercial poultry operations. In most egg-laying facilities the wastes are permitted to accumulate, in either a slurry or drier form, in pits under the birds for periods of 1 or more years before being cleaned. The odors from these buildings and cleaning operations have been a source of nuisance complaints when the facilities are near residential areas. The major method of treatment and disposal of poultry wastes has been land spreading. Lagooning, composting, and drying have been used with various degrees of success. Manure with litter generally is disposed of on the land.

Swine wastes are characterized by their fluid nature, a noticeable odor even when fresh, and a rapid decomposition rate. The pollution potential of swine waste is associated with surface runoff, odors, and land disposal. Daily removal of swine manure is not common in open and partially confined lots. The manure is removed as convenient for the operator and disposed of on nearby land. In total confinement systems either a solid floor, a slotted floor, or a gutter arrangement is used to accommodate manure removal.

In a solid-floor operation, the wastes are scraped and removed to a storage unit or lagoon. The slotted-floor system eliminates manure handling in the production areas. The manure, spilled food, and water fall through a slotted floor into a storage unit. Manure stored under the slotted floor will become anaerobic unless the unit is aerated, such as with an in-house oxidation ditch. The gutter system collects the wastes by scraping or flushing across the floor and into a gutter. The wastes then are moved hydraulically to a lagoon, storage tanks, or an aerated system. All swine waste handling systems require regular removal and/or careful storage of manure to avoid odors, flies, and surface water and groundwater pollution. The ultimate disposal of the stored or aerated manure should be on the land. Diversion of runoff from all surface areas containing manure is desirable.

Odors have caused problems when the contents of anaerobic storage units have been agitated. The released gases have caused the death of some hogs and the unconsciousness of a few humans. Air movement in these facilities should be directed away from the animal area to minimize these problems.

Land disposal of stored wastes should be done by incorporating the wastes in the soil as soon as possible. Anaerobic, stored waste should not be disposed of in close proximity to residences, motels, or other places frequented by humans, especially during seasons of outdoor recreation. All liquid or slurry wastes should be transported in liquid-tight containers maintained in a clean and neat condition. The advantage of the manure storage and land disposal system is its simplicity and low cost. It is also one of the traditional methods of waste handling and disposal and as such has been accepted by the agricultural community.

Surface-spreading of manures is the most common method of manure disposal (Chapter 11). It can be used with manures of all consistencies but has potential odor, runoff, and insect problems. Crop damage may result from heavy applications. Other disposal methods are plow-furrow-cover and injection. Plow-furrow-cover refers to the process where manure is applied directly into the plow furrow. The method may not be suitable for use during the growing season and winter. It requires a special spreader and plow combination. Good control of odors and insects is achieved.

Injection refers to direct injection of the liquid or slurry manure into the ground either on sod or between the rows of crops. Disposal of manure during the growing season can occur. With careful operation, control of flies and odors can be obtained. A tractor, tank spreader, and injector attachment is required.

Milking center wastes are liquid and contain little manure. They can be incorporated into the system used to handle the dairy cattle manure or handled separately. When manure slurry or liquid systems are used, inclusion of milking center wastes in these systems is a recommended choice. Oxidation ponds, spray irrigation, and other direct land application approaches can be used if the milking center wastes are handled separately. Septic tanks are not recommended, because the soil leaching beds can plug rapidly. Discharge of the treated wastes must meet the effluent requirements of the specific state.

Certain animal wastes, such as duck production wastewaters, are a liquid rather than a slurry and can be treated by conventional liquid waste treatment methods. As with many industrial wastes, the conventional methods have to be adapted to the characteristics of the specific waste. Duck production wastewater has the characteristics of strong domestic sewage and has been successfully treated by a combination of aerated lagoons and settling ponds, followed by chlorination.

Separation at the source is both a feasible approach and a waste management philosophy. The treatment of any waste is a direct function of both the strength and volume of the wastes. Since animal wastes are generated as semisolids, it is logical to separate these wastes from wash or clean-up waters at the source and to treat the semisolid wastes and the liquids in separate units. The liquids can be treated in liquid systems, whereas the solids can be disposed of directly on the land or by systems that handle semisolids, such as drying, composting, and incineration.

Drying and composting are feasible with animal wastes. Poultry waste, because of its lower initial moisture content, requires less energy to dry than do wastes containing a higher water content. In-house drying of poultry wastes by greater air circulation or the addition of heat or by both offers the possibility of least-cost drying systems for these wastes. The success of drying and composting requires a market or ultimate disposal for the product.

Odor Control

Odors associated from animal production result primarily from manure storage and management. Other possible sources of odor are feed storage units (silage) and related animal products (milking parlor wastewaters, dead animals, and spoiled feed). The odors from manure storage and poor management result from anaerobic conditions. Approaches which inhibit anaerobic action are effective in controlling, minimizing, or reducing odor production. Odor control approaches which have been tried with some success are (a) improved manure management to achieve aerobic (rather than anaerobic) decomposition, (b) chemical or biological treatment of manure to prevent or overcome odorous gases, (c) subsoil injection of liquid manure, and (d) judicious site selection with respect to surrounding land use patterns and climatic factors.

A number of commercial, proprietary chemicals have been reported to be effective in preventing and/or suppressing odors. Not all commercial chemicals are equally effective for odor control, and care should be taken when making decisions to use these compounds. Odor control chemicals have limited use at farms and animal production operations. Because of their expense and lack of proven performance and effective means to measure performance, short-term application of such chemicals is used primarily for offensive weather conditions and areas.

Many of the obvious cases of odor nuisances would have been minimized if the production facilities had been in locations less susceptible to air movement that can carry the odors to residential areas. When waste management systems that may produce odors are selected, proper site location will help minimize odor problems. Knowledge of wind movement, dispersal of the odors in the atmosphere, and the general meteorology of an area greatly can assist in planning for odor control at a site. Consideration of air drainage basins in developing land-use plans can be helpful in minimizing odor nuisances.

Fresh animal manures are rarely offensive. Drying of the manures to conditions that prevent microbial action has been shown to be an effective odor control method. Natural air movement without added heat, increased air movement using fans or blowers, and the combination of heat and air movement have effectively controlled odors in poultry houses. Commercial dryers also have been used with animal wastes. Storage of the dried material under conditions that do not produce odors is essential. Poultry wastes which have been dried to a moisture content of approximately 30% or lower are less odorous than those not dried. The odor emitted and its intensity depends to a considerable extent on the temperature at which the material is stored and the temperature and humidity conditions at the time it is spread.

The key to a successful aeration system for odor control is to supply oxygen at a rate equal to or greater than the oxygen demand. If the oxygen supplied is considerably less than the oxygen demand, odors may still result. The oxygen transfer relationships of possible aeration equipment are affected by the characteristics, solids content, and temperature of the wastes (Chapter 7).

The greatest benefit of aeration as a method of odor control occurs when it is used from the time storage is first utilized. Sufficient aeration under these conditions will prevent odor production. Aeration of stored manure after it has been permitted to accumulate under anaerobic conditions will strip the existing odorous compounds from the wastes, with possible severe odor problems.

When odorous animal wastes result, proper application to the land can minimize odor problems. Spreading thinly creates fewer problems, because the wastes dry much more rapidly than when spread in a thick layer. With proper management and suitable soil type, plowing thinly spread waste into the soil has alleviated odor problems. Soil injection of wastes or the use of the plow-furrow-cover method will minimize odor release during disposal.

Although chemicals, modification of storage and treatment processes, and proper application to land can be feasible approaches, effective long-range solutions to odor problems occur by consideration of their source and generation and the total animal production and manure management system. Many odors result from poor decisions about farm, feed, and animal management as well as from poor manure management.

Food Processing Wastes

The type of waste management suitable for a food processing operation will depend upon the characteristics of the wastes, the location of the operation, the degree of waste treatment required, and available points of discharge for the treated liquid wastes and collected solid wastes. As with almost all industrial waste situations, a unique waste management solution may exist for each operation.

One of the first approaches is to investigate the opportunities for in-field and in-plant modifications to reduce waste quality. Techniques to reduce the pollutional load from fruit and vegetable processing operations include harvesting equipment that leaves more of the stems, leaves, and culls in the field; field washing and/or processing of the crop, with the residue returned to the cropland; air and liquid separators to remove extraneous material from the crop; modification of peeling and pitting operations to use less water and to waste less crop; recovery and reuse of the process water throughout the plant; and separation of waste process streams at the source for potential by-product utilization and separate waste treatment and disposal of wastes with different characteristics.

Two major waste streams contribute to wastewater flows in a food processing plant: water used in food product processing and water used for cooling. Water conservation can be practiced by the use of low-volume, high-pressure sprays for plant cleanup, elimination of excess overflow from water supply tanks, use of automatic shut-off valves on water lines, use of mechanical conveyors to replace water flumes, and investigations into the minimum water requirements of a specific operation. The reuse of cooling water as product wash water is an example of a separation and reuse possibility.

Other modifications to food processing operations for waste product separation become apparent when materials and waste load balances are conducted at a plant and when by-product utilization opportunities are available. The development of dry caustic methods of potato and fruit peeling has reduced both the volume and strength of wastes from these operations. Separation of blood for use as animal feed, fat and grease for rendering, and paunch manure for land disposal can be practiced at a meat slaughtering and meat-packing installation to reduce the pollutional characteristics and treatment costs of the resultant flow. Other examples of potential by-product recovery include protein recovery from shellfish wastes, potato starch separation, and evaporation recovery of protein-containing wastes, fungal and microbial solids production in biological waste treatment, and whey solids separation. A waste survey of a food processing facility, including a water balance and mass balances of product, BOD, solids, and nutrients, can reveal other opportunities for waste management by in-plant modifications.

Use of waste-monitoring equipment such as temperature recorders, conductivity measurements, flow measurement devices, and turbidity measurements can be used to make the materials balances at a processing plant and to identify potential improvements in processing operations. Such waste production indices can be used to measure the performance of both management and production workers. By using these techniques, waste handling costs can be reduced, water conservation and reuse can be enhanced, and savings of raw materials may result.

After in-plant waste management opportunities have been explored, decisions can be made on the appropriate methods of treatment and disposal of the resultant wastes. The discharge of untreated or screened food processing wastes directly to a municipal waste treatment plant may be a desirable method if satisfactory waste surcharge arrangements can be made. The seasonal nature of many food processing operations, such as canneries, and the strength of these wastes can cause serious problems at municipal plants. The strength of typical food processing wastes is greater than that of municipal wastes. In addition, the food processing plants often are adjacent to small communities whose treatment plants may not have the capacity to absorb the food processing wastes without difficulty. The result is large slug loads on the municipal treatment plant, caused by seasonal, diurnal, and clean-up conditions. It is difficult to operate or to design an efficient municipal waste treatment plant under such conditions.

For the preceding reasons, either pretreatment of the food processing waste before discharge to the municipal system or complete treatment of the waste by the processing plant is common. In either case, a waste equalization unit is valuable to reduce the variability of waste volume and strength during operation. The objective of waste equalization is to dampen variations in flow and concentration and to attempt to achieve a nearly constant flow and contaminant waste loading to subsequent treatment units. Equalization can improve the performance of existing treatment facilities and can reduce the size of subsequent treatment units.

Biological methods of many types can be used with food processing wastes since they are organic in nature. Certain wastes may need nutrient supplementation for optimum biological treatment. Feasible processes include aerated lagoons, oxidation ponds, anaerobic ponds, trickling filters, activated sludge modifications, rotating biological contactors, chemical precipitation, and land treatment. The methods can be used separately or in combination. The advantages and disadvantages of each process and examples of their use with food processing wastes have been noted earlier. Reduction of the solids content by screening, sedimentation, or flotation may be advisable with certain wastes before biological treatment.

While all of the preceding methods have been used successfully with food processing wastes, the most common methods of treatment and disposal have been oxidation ponds, aerated lagoons, and land treatment. These are used because of the availability of land near most processing plants and the minimum maintenance and control that are needed. For cannery wastes, the oxidation pond may also act as a surge equilization unit for the wastes which are seasonally produced. Where odors occur due to temporary heavy loadings, and aeration equipment does not appear justified, the addition of sodium nitrate in quantities usually sufficient to satisfy 20–30% of the BOD can be added for odor control.

Solids from food processing operations can be used for animal feed and compost and can be disposed of by landfill. Landfilling is the most common method for solids disposal.

Summary

The agricultural producer has both immediate and long-term waste management problems. The immediate problems have arisen because of changes in production, the national emphasis on enhancement of environmental quality, and lack of knowledge of how to deal adequately with the waste produced from these operations. Current solutions to many of the problems will change as better methods become available. In the interim, there are a number of good practice or management guidelines that can minimize some of the environmental problems caused by agriculture.

The long-term approach for agriculture production must be based upon not only optimal production of the product, such as egg production, animal weight gain, or maximum fruit or vegetable production, but also on the management of the entire production scheme such that it is consistent with the maintenance of acceptable environmental quality, not only to the animals and to the producers, but to society as a whole. No longer can wastes be discharged indiscriminately. Such discharge invariably intrudes upon someone else's living space.

The agricultural waste problem usually is viewed as a treatment and disposal problem. While this is the most obvious aspect of the problem, waste management systems should consider all aspects of an agricultural production operation rather than concentrate only on the treatment and disposal aspects. Sufficient knowledge is available to minimize the gross adverse environmental effects that occurred in the past through ignorance.

References

1. Anonymous, "Agricultural Code of Practice for Ontario." Ministry of the Environment and Ministry of Agriculture and Food, Ottawa, Canada, 1973.
2. Water Pollution Control Federation, "Regulation of Sewer Use," MOP-3. Washington, D.C., 1975.
3. Maystre, Y., and Geyer, J. C., Charges for treating industrial wastewater in municipal plants. *J. Water Pollut. Contr. Fed.* **42,** 1277–1291 (1970).
4. Effluent Guidelines and Standards, Part 412, Feedlots Point Source Category, *Fed. Regist.* **39,** 5704–5710 (1974).
5. Loehr, R. C., Alternatives for the treatment and disposal of animal wastes. *J. Water Pollut. Contr. Fed.* **43,** 669–678 (1971).

Appendix: Characteristics of Agricultural Wastes

TABLE A.1

Waste Characteristics from Vegetables, Fruit, and Wine Processing[a,b]

Product	COD (mg/liter)	BOD (mg/liter)	BOD/COD (mean)	pH
Beets	1,800–13,200	1,200–6,400	0.51	5.6–11.9
Beans, green	100–2,200	40–1,360	0.55	6.3–8.6
Beans, wax	200–600	60–320	0.58	6.5–8.2
Carrots	1,750–2,900	800–1,900	0.52	7.4–10.6
Corn	3,400–10,100	1,600–4,700	0.50	4.8–7.6
Peas	700–2,200	300–1,350	0.61	4.9–9.0
Sauerkraut	500–65,000	300–4,000	0.66	3.6–6.8
Tomatoes	650–2,300	450–1,600	0.72	5.6–10.8
Apples	400–37,000	240–19,000	0.55	4.1–7.7
Cherries	1,200–3,800	660–1,900	0.53	5.0–7.9
Grape juice	550–3,250	330–1,700	0.59	6.5–8.2
Wine	30–12,000	30–7,600	0.60	3.1–9.2

[a]From D. F. Splittstoesser and D. L. Downing, Analysis of effluents from fruit and vegetable processing factories. *N.Y., Agric. Exp. Stn., Geneva, Res. Circ.* **17** (1969).
[b]Minimum amount of wash and cooling waters.

TABLE A.2

Fruit Waste[a]

Product	Flow (gal/ton raw product)	BOD (lb/ton raw product)	COD (lb/ton raw product)	P/BOD	N/BOD
Pears	5800	78	105	0.0012	0.006
Peaches	7500	66	98	0.0027	0.011
Purple plums	7000	64	95	—	—
Apples	8200	53	71	0.0010	0.0023

[a]From L. A. Esvelt, Aerobic treatment of liquid fruit processing waste. *In* "Proceedings of the First National Symposium on Food Processing Wastes," pp. 119–143. Pacific Northwest Water Lab., Federal Water Quality Administration, U.S. Dept. of the Interior, Washington, D.C., 1970.

TABLE A.3

Waste Characteristics from Potato Processing[a]

Characteristic	Waste production per ton of potatoes processed
Process water (gal)	4200
BOD (lb)	50–90
COD (lb)	210
Suspended solids (lb)	60–110
Total phosphate as PO_4 (lb)	0–6
Total nitrogen as N (lb)	3.5

[a]From K. A. Dostal, "Aerated Lagoon Treatment of Food Processing Wastes," Rep. 12060. Water Quality Office, U.S. Environmental Protection Agency, Washington, D.C., 1968.

TABLE A.4

Raw Waste Loads[a] for the Fruit, Vegetable, and Specialties Processing Industry[b]

Category	Flow (gal/ton)	BOD (lb/ton)	Total suspended solids (lb/ton)
Fruit			
Apple processing	690	4.1	0.6
Apple products except juice	1,290	12.8	1.6
Apricots	5,660	30.9	8.5
Caneberries	1,400	5.7	1.2
Cherries			
Sweet	1,860	19.3	1.2
Sour	2,880	34.3	2.1
Brined	4,780	43.5	2.9
Cranberries	2,950	19.9	2.8
Citrus, all products	2,420	6.4	2.6
Dried fruit	3,180	24.8	3.7
Grape juice			
Canning	1,730	21.4	2.5
Pressing	370	3.8	0.8
Olives	9,160	87.4	15.0
Peaches			
Canned	3,130	28.1	4.6
Frozen	1,300	23.4	3.7
Pears	2,840	42.3	6.5

(continued)

TABLE A.4 (*Continued*)

Category	Flow (gal/ton)	BOD (lb/ton)	Total suspended solids (lb/ton)
Pickles			
Fresh packed	2,050	19.0	3.8
Process packed	2,300	36.7	6.5
Salting stations	250	15.9	0.8
Pineapples	3,130	20.6	5.5
Plums	1,190	8.2	0.7
Raisins	670	12.1	3.3
Strawberries	3,150	10.6	2.7
Tomatoes			
Peeled	2,150	8.2	12.3
Products	1,130	2.6	5.3
Vegetables			
Asparagus	16,520	4.2	6.9
Beets	1,210	39.4	7.9
Broccoli	10,950	19.6	11.2
Brussel sprouts	8,720	6.9	21.6
Carrots	2,910	39.0	23.9
Cauliflower	21,470	10.5	5.1
Corn			
Canned	1,070	28.8	13.4
Frozen	3,190	40.4	11.2
Dehydrated onion and garlic	4,770	13.0	11.8
Dehydrated vegetables	5,300	15.8	11.3
Dry beans	4,310	30.7	8.8
Lima beans	6,510	27.8	20.7
Mushrooms	5,380	17.4	9.6
Onions, canned	5,520	45.1	18.7
Peas			
Canned	4,720	44.2	10.8
Frozen	3,480	36.6	9.8
Pimentos	6,910	54.5	5.8
Sauerkraut			
Canning	840	7.0	1.2
Cutting	103	2.5	0.4
Snap beans			
Canned	3,690	6.3	4.0
Frozen	3,820	12.1	6.0
Spinach			
Canned	9,040	16.4	13.0
Frozen	7,020	9.6	4.0
Squash	1,340	33.6	4.6
Sweet potatoes	990	60.2	22.9

(*continued*)

TABLE A.4 (*Continued*)

Category	Flow (gal/ton)	BOD (lb/ton)	Total suspended solids (lb/ton)
White potatoes	1,990	54.6	74.8
Potatoes			
Frozen	2,710	45.8	38.8
Dehydrated	2,100	22.1	14.7
Specialty foods			
Baby food	1,770	9.1	3.2
Chips			
Corn	2,880	70.4	59.8
Potato	5,630	74.0	84.4
Tortilla	4,880	59.4	72.1
Jams and jellies	630	11.7	1.9
Mayonnaise and dressings	550	10.9	5.13
Soups	7,340	29.7	19.5

[a]The raw waste load is in terms of the quantity of wastewater parameter per ton of raw material processed for fruits and vegetables and per ton of final product for the specialty foods. The values noted are the average raw waste loads for the available data in the categories noted. Except where noted, the raw waste loads are those generated from canning processing.

[b]From the U.S. Environmental Protection Agency, "Development Document for Interim Final and Proposed Effluent Limitations Guidelines and New Source Performance Standards for the Fruits, Vegetables, and Specialties Segment of the Canned and Preserved Fruits and Vegetables Point Source Category," EPA 440/1-75/046, Group I, Phase II. Washington, D.C., 1975; and the U.S. Environmental Protection Agency, "Development Document for Effluent Limitations Guidelines and New Source Performance Standards for the Apple, Citrus, and Potato Processing Segment of the Canned and Preserved Fruits and Vegetable Point Source Category," EPA-440/1-74-027a. Washington, D.C., 1974.

TABLE A.5

Characteristics of Meat-Packing Wastes[a]

Characteristic	Average	Range
BOD (mg/liter)	1240	600–2720
COD (mg/liter)	2940	960–8290
Organic nitrogen (mg/liter)	85	22–240
Grease (mg/liter)	1010	250–3000
Suspended solids (mg/liter)	1850	300–4200
Volatile suspended solids (%)	92	80–97

[a]From M. C. Dart, The treatment of meat trade effluents. *Effluent Water Treat. J.* **7,** 29–33 (1967).

TABLE A.6

Raw Waste Characteristics[a] for Slaughterhouses and Packinghouses[b]

Type of plant[c]	Flow	BOD$_5$	Suspended solids	Grease	Kjeldahl nitrogen as N	Chlorides as Cl	Total phosphorus as P
Simple slaughterhouses	5330 (1330–14,640)	6.0 (1.5–14.3)	5.6 (0.6–12.9)	2.1 (0.2–7.0)	0.7 (0.2–1.4)	2.6 (0.01–5.4)	0.05 (0.14–0.86)
Complex slaughterhouses	7380 (3630–12,500)	10.9 (5.4–18.8)	9.6 (2.8–20.5)	5.9 (0.7–16.8)	0.8 (0.1–2.1)	2.8 (0.8–7.9)	0.33 (0.05–1.2)
Low processing packinghouses	7840 (2020–17,000)	8.1 (2.3–18.4)	5.9 (0.6–13.9)	3.0 (0.8–7.7)	0.5 (0.04–1.3)	3.6 (0.5–4.9)	0.13 (0.03–0.43)
High processing packinghouses	12,510 (5440–20,260)	16.1 (6.2–30.5)	10.5 (1.7–22.5)	9.0 (2.8–27.0)	1.3 (0.6–2.7)	15.6 (0.8–36.7)	0.38 (0.2–0.63)

[a]The raw waste loads are in terms of liters per 1000 kg of liveweight killed (LWK) for flow and kilograms per 1000 kg LWK for all other parameters. Data in parentheses represent the range of values reported; the single number is the average value for that parameter and type of plant.

[b]From the U.S. Environmental Protection Agency, "Development Document for Effluent Limitations Guidelines and New Source Performance Standards for the Red Meat Processing Segment of the Meat Product and Rendering Processing Point Source Category." EPA-440/1-74-012a. Washington. D.C., 1974.

[c]Simple slaughterhouse, very limited by-product processing; complex slaughterhouse, extensive by-product processing such as rendering, blood processing, hide or hair processing, or paunch and viscera handling; low processing packinghouse, processes no more than the total animals killed at that plant; high processing packinghouse, processes both animals slaughtered at the site and carcasses from outside sources.

TABLE A.7

Summary of Raw Waste Characteristics[a] for Rendering Industry[b]

Parameter	Average	Range
Flow	3260	470–20,000
BOD_5	2.15	0.1–5.8
Suspended solids	1.13	0.3–5.2
Grease	0.72	0–4.2
COD	8.04	1.6–37.0
Total volatile solids	3.34	0.04–13.1
Total dissolved solids	3.47	0.01–11.7
Total Kjeldahl nitrogen as N	0.48	0.12–1.20
Ammonia as N	0.30	0.08–0.74
Nitrate as N	0.008	0–0.06
Nitrite as N	0.003	0–0.04
Chloride as Cl	0.80	0.08–2.56
Total phosphorus as P	0.04	0.003–0.28

[a]The raw waste loads are in terms of liters per 1000 kg raw material (RM) processed for flow and in terms of kilograms per thousand kilograms RM for the other parameters. All raw waste data represent the effluent following any in-plant materials recovery such as catch basins, skimmers, etc.

[b]From the U.S. Environmental Protection Agency, "Development Document for Effluent Limitations Guidelines and New Source Performance Standards for the Renderer Segment of the Meat Products and Rendering Processing Point Source Category," EPA-440/1-74/031d, Group I, Phase II. Washington, D.C., 1975.

TABLE A.8

Poultry Processing Plant Waste[a,b]

Item	Range (mg/liter)
BOD	150–2400
COD	2–3200
Suspended solids	100–1500
Dissolved solids	200–2000
Volatile solids	250–2700
Total solids	350–3200
Total alkalinity	40–350
Total nitrogen	15–300
BOD (lb/1000 birds processed)	25
Suspended solids (lb/1000 birds processed)	13

[a]From R. Porges and E. J. Struzeski, Characteristics and treatment of poultry processing wastes. *Proc. Purdue Ind. Waste Conf.* **17**, 583–601 (1962).

[b]Combined wastes from 13 processing plants.

TABLE A.9
Summary of Waste Characteristics[a] for Chicken, Turkey, Fowl, and Duck Processors[b,c]

Parameter	Chicken Average	Chicken Range	Turkey Average	Turkey Range	Fowl Average	Fowl Range	Duck Average	Duck Range
Flow	34.4	15.9–87.0	118	36.3–270	48.9	11.0–159	74.9	71.5–78.3
BOD$_5$	9.9	3.3–19.9	4.9	1.0–9.1	15.2	11.8–23.1	7.1	6.6–7.5
Suspended solids	6.9	0.13–22.1	3.2	0.6–10.9	10.1	6.1–14.9	4.4	3.5–5.2
Grease	4.2	0.12–14.0	0.9	0.3–1.8	2.3	0.7–3.3	1.9	0.7–3.1
COD	19.7	2.0–56.8	7.4	3.1–10.9	41.4	24.3–58.5	14.1	13.6–14.6
Total volatile solids	13.3	3.5–47.2	8.4	2.2–19.2	18.4	13.1–23.7	7.1	6.7–7.5
Total dissolved solids	11.7	3.5–45.8	13.5	1.5–38.5	24.9	9.1–40.6	8.3	4.0–12.6
Total Kjeldahl nitrogen as N	1.8	0.15–12.2	0.9	0.4–1.9	0.3	—	1.4	0.8–2.0
Ammonia as N	0.23	0.005–0.73	0.15	0.06–0.37	0.1	—	0.8	0.06–1.5
Nitrate as N	0.008	0–0.14	0.04	0.005–0.09	0.004	—	0.03	0.02–0.04
Nitrite as N	0.007	0–0.037	0.001	0.0001–0.002	0.0005	—	0.01	0.001–0.02
Chlorides as Cl	1.97	0.006–9.16	2.5	0.4–5.4	4.0	—	1.4	0.8–2.1
Total phosphorus as P	0.40	0.05–2.46	0.1	0.03–0.18	0.3	—	0.08	0.07–0.10

[a]The raw waste characteristics are in terms of liters per bird processed for flow and in terms of kilograms per 1000 kg per liveweight killed (LWK) for the other parameters.

[b]Chicken processing includes processing of broilers and chickens not classified as mature chickens or fowl: turkey processing includes processing of hens or tom turkeys of varying age and size; fowl processing includes processing mature chickens larger in size and older than broilers such as geese, capons, and roosters.

[c]From the U.S. Environmental Protection Agency, "Development Document for Proposed Effluent Limitations Guidelines and New Source Performance Standards for the Poultry Segment of the Meat Product and Rendering Process Point Source Category." EPA 440/1-75/031b. Group I, Phase II. Washington, D.C., 1975.

TABLE A.10

Milk Plant Wastewater Coefficients[a]

Type of product	Waste volume (lb/lb milk processed)		BOD (lb/1000 lb milk processed)	
	Average	Range	Average	Range
Milk	3.25	0.1–5.4	4.2	0.2–7.8
Cheese	3.14	1.6–5.7	2.0	1.0–3.5
Ice cream	2.8	0.8–5.6	5.7	1.9–20.4
Condensed milk	2.1	1.0–3.3	7.6	0.2–13.3
Butter	0.8	—	0.85	—
Powdered milk	3.7	1.5–5.9	2.2	0.02–4.6
Cottage cheese	6.0	0.8–12.4	34.0	1.3–71.2[b]
Cottage cheese and milk	1.84	0.05–7.2	3.47	0.7–8.6[b]
Cottage cheese, ice cream, and milk	2.52	1.4–3.9	6.37	2.3–12.9
Mixed products	2.34	0.8–4.6	3.09	0.9–6.9
Overall	2.43	0.1–12.4	5.85	0.2–71.2

[a]From W. J. Harper and J. L. Blaisdell, State of the art of dairy food plant wastes and waste treatment. *In* "Proceedings of the Second National Symposium on Food Processing Wastes," pp. 509–545. Pacific Northwest Water Lab., U.S. Environmental Protection Agency, Corvallis, Oregon, 1971.

[b]Whey included, whey excluded from all other operations manufacturing cottage cheese.

TABLE A.11

Milk Plant Processing Waste Data from Five Plants[a]

	Range of averages (mg/liter)	Total range of data (mg/liter)
BOD	940–4790	400–9440
COD	1240–7800	360–15,300
Ammonia	7–36	1–76
Organic nitrogen	36–150	9–250
Alkalinity	81–505	0–1080
pH	4.8–6.8	4.2–9.5
Total solids	2280–6490	1210–11,990
Suspended solids	360–1040	270–1980
Volatile suspended solids	300–1000	200–1840

[a]From G. W. Lawton, L. E. Engelhart, G. A. Rohlich, and N. Porges, Effectiveness of spray irrigation as a method for disposal of dairy plant wastes. *Wisc. Eng. Exp. Stn., Res. Rep.* **15** (1960).

TABLE A.12

General Characteristics of Fish Processing Wastewater[a]

Type of fish processing	Flow		BOD			COD		Total solids		Suspended solids		Grease	
	gal/min	gal/ton of fish	mg/liter	lb/ton of fish	lb/ton of fish product	mg/liter	lb/ton of fish	mg/liter	lb/ton of fish	mg/liter	lb/ton of fish	mg/liter	lb/ton of fish
General	—	465–9100	2700–3440	8–120	21–24	—	—	4200–21,800	—	2200–3020	—	—	—
Bottom fish	105–450	—	90–1726	—	74	—	—	—	—	300	—	—	—
Herring, menhaden, and anchovies	680–1440	240–1020	620–1000	—	—	—	—	—	—	—	—	—	—
Salmon	—	—	173–3900	—	3.2–80	5920	—	88–7400	—	40–4780	—	—	—
Sardine	40–1000	—	100–2000	—	—	170–1700	—	—	—	100–2100	—	60–1340	—
Tuna	—	—	890	48	—	2270	129	17,900	950	1090	58	287	15

[a] From M. R. Soderquist, K. J. Williamson, G. I. Blanton, D. C. Phillips, D. K. Law, and D. L. Crawford, "Current Practice in Seafoods Processing Waste Treatment," Proj. 12060ECF. Water Quality Office, U.S. Environment Protection Office, Washington, D.C., 1970.

TABLE A.13

Catfish Processing Waste Characteristics[a]

Parameter	Average value
Flow	2.0 gal/fish
BOD$_5$	8.0 lb/1000 fish
COD	10.8 lb/1000 fish
Total suspended solids	5.1 lb/1000 fish
Total volatile suspended solids	4.5 lb/1000 fish
Grease	1.7 lb/1000 fish

[a]L. A. Mulkey, and T. N. Sargent, Waste loads and water management in catfish processing. *J. Water Pollut. Contr. Fed.* **46,** 2193–2201 (1974).

TABLE A.14

Duck Wastewater Characteristics (from Two Studies)[a]

Parameter	lb/1000 ducks/day		mg/liter	
	Average (range)	Range	Average (range)	Range
BOD	—	16–30	—	80–220
COD	82 (18–260)	—	810 (140–7520)	—
Total solids	96 (26–320)	—	1010 (330–6340)	—
Suspended solids	54 (10–237)	52–123	340 (17–4630)	—
Total nitrogen	—	4.7–6.4	—	—
Phosphates (as PO$_4$)				
Total	—	4.7–10.5	—	—
Soluble	5.5 (1.0–13.4)[b]	2.5–4.6	42.5 (31–130)[b]	—
Water use (gal/duck/day)	15 (5–27)	10–34	—	—

[a]From W. Sanderson, Studies of the character and treatment of wastes from duck farms. *Proc. Purdue Ind. Waste Conf.* **8,** 170–176 (1953); and R. C. Loehr and K. J. Johanson, "Removal of Phosphates from Duck Wastewaters—1970 Laboratory Study," Rep. 71-2. Agric. Waste Manage. Program, Cornell University, Ithaca, New York, 1971.

[b]Orthophosphate.

TABLE A.15

Hog Manure Nutrient Characteristics[a]

Element	Feces[b] (mg/liter)	Urine[b] (mg/liter)	Feces (mg/gm solids)	Urine (mg/gm solids)
Ca	25,100	340	71.0	8.3
Mg	8,020	88	22.5	2.2
Zn	510	2.3	1.4	0.05
Cu	108	0.16	0.3	0.004
Fe	456	1.1	1.3	0.03
Mn	176	0.3	0.5	0.008
Na	2,630	1,300	7.4	31.0
K	10,200	2,300	29.0	56.0
P	16,700	178	47.0	4.4
S	1,040	1,100	2.9	27.1
N	34,600	5,000	97.7	122.6

[a]From P. O. Ngoddy, J. P. Harper, R. K. Collins, G. D. Wells, and F. A. Heidor, "Closed System Waste Management for Livestock," Final Rep., Proj. 13040 DKP. U.S. Environmental Protection Agency, Washington, D.C., 1971.

[b]Feces at 65% moisture, urine at 96% moisture.

TABLE A.16

Dairy Cattle Manure from Stanchion Barns, Fresh Manure Including Bedding and Urine[a,b]

Parameter	Average	Range
Total solids (%)	13.5	10.6–16.6
Volatile solids (% of TS)	35.5	81–90
COD (mg/liter)	136,500	97,000–166,000
Total nitrogen (mg/liter as N)	4,500	3,100–6,500
Ammonia (mg/liter as N)	1,750	360–3,300
Total phosphorus (mg/liter as P)	1,250	880–1,730
Potassium (mg/liter as K)	3,250	2,050–4,900
pH	6.9	5.4–8.0

[a]From C. O. Cramer, J. C. Converse, G. H. Tenpas, and P. A. Schlough, "The Design of Solid Manure Storages for Dairy Herds," Paper 71-910. Am. Soc. Agric. Eng., St. Joseph, Michigan, 1971.

[b]Data collected from November through August during two trials in separate years: about 30% less bedding was used during trial 2; average values of COD, total nitrogen, ammonia nitrogen, and percentage of volatile solids were statistically different between the two trials.

TABLE A.17

Characteristics of Milking Center Wastes (24 Dairy Farms)[a]

Characteristic	Average	Range
Flow (gal/cow/day)	4.0	1.8–16.8
BOD (lb/cow/day)	0.127	0.01–0.24
Total dry solids (gm/liter)	5.0	0.8–10.4
Phosphorus (mg/liter), soluble as P	57.6	6–183
Ammonia nitrogen, soluble (mg/liter)	132.	5–625
Nitrate plus nitrate nitrogen, soluble (mg/liter)	1.6	0.3–6.5

[a]From R. R. Zall, Characteristics of milking center waste effluents from New York State dairy farms. *J. Milk Food Technol.* **35,** 53–55 (1972).

TABLE A.18

Characteristics of Waste Material Removed from Outdoor, Unpaved Beef Cattle Feedlots[a]

Parameter (tons/acre except where noted)	Animal stocking rate			
	100 ft²/animal		200 ft²/animal	
	Mean	Range	Mean	Range
Moisture (% dry solids)	99	50–186	84	32–170
Total solids (dry solids)	379	209–815	263	66–635
Volatile solids (dry solids)	129	76–175	64	28–134
Total nitrogen	3.6	0.6–7.6	1.8	0.05–5.7
NH₄-N	0.61	0.35–0.70	0.2	0.01–0.46
Total Phosphorus	0.44	0.18–0.69	0.2	0.02–0.55

Parameter (ppm dry weight)	Mean	Range
Na	1050	550–1660
K	4620	530–10,200
Ca	2560	300–4770
Mg	1700	315–3145
Zn	45	25–62
Cu	6	1.5–11.3
Fe	4090	195–9910
Mn	510	24–1630

[a]From T. M. McCalla, J. R. Ellis, C. B. Gibertson, and W. R. Woods, Chemical studies of runoff, soil profile and groundwater from beef feedlots at Mead, Nebraska. *In* "Proceedings of the Agricultural Waste Management Conference," pp. 211–223. Cornell University, Ithaca, New York, 1972.

TABLE A.19

Composition of Animal Wastes[a,b]

Animal	Weight (lb/head)	Feed (dry) (lb/head/day)	Hydraulic load (gal/day)[c]	5-day BOD (lb/day)[d]	Total solids (lb/day)	Susp. solids (lb/day)	N (% by weight)		Phosphate (% by weight)[b]		Potassium oxide (% by weight)[b]		Carbohydrates (% by weight)[b]	
							Solid	Liquid	Solid	Liquid	Solid	Liquid	Solid	Liquid
Horse	950–1400	30–45	20.0	6.50	15.0	12.5	0.50	1.2	0.30	Trace	0.30	1.60	27.1	1.9
Cattle	900–1250	25–40	20.0	6.50	15.0	12.5	0.30	0.9	0.21	0.03	0.18	0.93	16.7	1.3
Calf	450–600	15–20	15.0	3.50	10.0	8.0	0.32	0.95	0.20	0.01	0.15	0.80	15.9	0.9
Pig	100–300	8–12	5.0	0.70	1.7	1.1	0.6	0.3	0.50	0.15	0.12	0.50	15.5	0.3
Small pig	20–90	3–8	3.0	0.40	1.1	0.6	0.3	0.4	0.13	0.13	0.10	0.44	12.2	0.7
Rabbit	4.0–4.5	0.25–0.33	0.4	0.08	1.2	0.9	0.4	0.2	0.10	Trace	0.07	0.50	30.0	0.3
Rat	1.5–2.0 oz	0.04–0.05	0.2	0.04	0.08	0.02	0.2	0.3	0.05	Trace	—	0.10	16.0	0.2
Mouse	0.4–1.2 oz	0.01–0.015	0.1	0.01	0.02	0.01	0.4	0.3	0.05	—	—	0.10	15.0	0.2
Sheep	100–150	4.0–6.0	6.0	1.20	2.50	2.0	0.65	1.7	0.51	0.02	0.03	0.25	30.7	0.9
Monkey	4.0–5.0	0.25–0.40	2.5	0.10	0.20	0.10	0.9	1.2	0.30	0.02	0.05	0.30	17.0	0.1
Dog	50–70	0.25–0.40	1.0	0.08	0.30	0.18	0.75	1.1	0.30	Trace	0.02	0.58	17.0	0.1
Cat	4.0–4.5	0.17–0.37	0.2	0.06	0.15	0.07	0.70	1.2	0.20	Trace	0.02	0.50	17.5	0.1
Turkey	20–25	0.75–0.80	0.4	0.12	0.35	0.20	1.2	0.5	1.00	—	—	0.60	31.0	0.1
Chicken	2.0–4.0	0.25–0.30	0.2	0.07	0.12	0.05	1.4	0.5	0.90	—	—	0.50	30.0	0.1
Duck	3.5–5.0	0.25–0.30	0.4	0.10	0.15	0.07	1.5	0.8	0.89	—	—	0.40	12.0	0.1
Goose	7.5–9.5	0.50–0.75	1.0	0.20	0.35	0.10	1.8	0.8	1.20	—	—	0.70	14.0	0.1

[a]From R. H. L. Howe, Research and practice in animal wastes treatment. *Water Wastes Eng.* **6**, A14–A18 (1969).
[b]All observed animals were kept in cages or stalls. COD for all wastes was 1.30–1.50 times higher than 5-day BOD. Data obtained on confined animals at a scientific laboratory.
[c]Including spilled water.
[d]Including spilled food.

TABLE A.20

Average Characteristics of Mink Wastes as Related to Different Feeds[a]

	Adults		
Parameter	Male	Female	Kits
Typical feed[b]			
Volume (liter/day)	0.17	0.11	0.21
Total solids (gm/day)	35.7	22.3	26.2
Volatile solids (% total solids)	77.3	77.2	75.1
COD (% total solids)	93.5	93.4	96.0
TKN (% total solids)	12.7	12.8	12.2
Commercial pelleted feed[c]			
Volume (liter/day)	0.19	0.13	0.21
Total solids (gm/day)	24.1	15.5	21.0
Volatile solids (% total solids)	83.5	82.2	82.0
COD (% total solids)	118.4	118.4	152.2
TKN (% total solids)	20.8	20.8	17.4
BOD (% total solids)	49.5	49.6	60.3

[a]From J. H. Martin, T. E. Pilbeam, R. C. Loehr, and H. F. Travis. The characteristics and management of mink wastes. *Trans. Am. Soc. Agric. Eng.* **20**, 515–517 (1977).

[b]Dry matter, 32.2%; protein, 32.2%; fat, 26.0%; ash, 7.6%.

[c]Dry matter, 96.2%; protein, 40.5%; fat, 24.3%; ash, 8.0%.

TABLE A.21

Average Analysis of Piggery Wastes[a]

Origin of sample	Total solids (% WB)[b]	Nitrogen		COD (mg/liter)	COD/BOD
		(% WB)[b]	(% DB)[b]		
Hog (45 kg)	16	0.72	4.5	179,500	3.7
Hog feedlot	31.5	1.22	3.86	330,400	3.0
Fattening hog					
Water on demand	10.7	0.86	8.0	—	—
Water ratio 4/1	6.6	0.63	9.5	64,000	3.0
Fattening hog	23	1.07	4.35	61,400	—
Fattening hog	16	0.97	6.1	93,000	—

[a]Adapted from D. E. Brown, T. D. Boardman, and S. W. Reddington, A microbial recycle process for piggery waste treatment, *Agric. Wastes* **2**, 185–197 (1980).

[b]WB = wet solids basis; DB = dry solids basis.

TABLE A.22

**Average Characteristics of Wastes Collected from
Livestock Feedlots**[a]

Parameter[b]	Beef cattle feedlot	Hog feedlot
Total solids[c]	468	315
Volatile solids	688	773
Ash	312	176
COD	795	1050
BOD	165	345
Carbohydrates—total neutral	272	298
Lignin	66	16
TKN	29	38
Amino acids		
Total	101	151
Free	8	10
Lipids	72	168
Potassium	18	13
Sodium	10	6
Calcium	17	23
Copper[d]	39	218
Chloride	15	3
Sulfur	4	3
Boron[d]	28	18

[a]Adapted from G. R. Hrubant, R. A. Rhodes, and J. H. Sloneker, "Specific Composition of Representative Feedlot Wastes: A Chemical and Microbial Profile," SEA-NC-59. Northern Regional Research Center, U.S. Department of Agriculture, Peoria, Illinois, 1978.

[b]Milligram per gram dry weight of whole waste except as noted.

[c]Milligram per gram wet weight.

[d]Microgram per gram dry weight.

TABLE A.23

Average Characteristics of Wastewaters from Milking Centers, Milkhouses, and Cowsheds[a]

Wastewater source[b,c]	Volume (liter/cow/day)	Total phosphorus (mg/liter)	TKN (mg/liter)	BOD (mg/liter)	COD (mg/liter)	Total solids (mg/liter)	Suspended solids (mg/liter)
Center (80)	36	—	—	1,030	2,880	—	1,900
Mixed (N.A.)	15	274	248	3,760	—	5,000	—
Cowshed (N.A.)	50	35	205	1,500	6,600	7,170	—
Center (60)	18	135	280	2,290	4,060	3,830	1,160
Center (125)	10	80	560	2,270	6,100	6,460	3,870
Center (63)	32	44	60	370	880	1,200	290
Pipeline (45)	16	70	130	1,790	5,290	3,420	—
Pipeline (85)	10	140	55	460	1,470	2,110	—
Cowshed (330)	97	21	190	725	—	—	—
Pipeline (45)	18	220	5	—	—	—	90
Pipeline (44)	22	255	10	—	—	—	150
Portable collection (40)	9	510	20	—	—	—	1,240

[a]Adapted from D. F. Sherman, "Selection of Treatment Alternatives for Milkhouse and Milking Center Wastewaters," Master's Degree Project Report, Department of Agricultural Engineering, Cornell University, Ithaca, New York, 1981.

[b]Center = milking center; pipeline = wastewater from cleaning of milk pipelines.

[c]Number in parentheses indicates the number of cows milked at each site; N.A. indicate that no data were available.

TABLE A.24

Summary of Veal Calf Manure Characteristics (Kilograms per 1000 kg of Liveweight per Day)[a]

Parameter	Animal size (kg)			Mean value[b]
	45	90	135	
Raw waste	77	60	49	65
Total solids	1.5	1.9	2.1	1.8
Volatile solids	0.81	0.95	0.99	0.89
BOD_5	0.37	0.39	0.46	0.41
COD	1.4	1.7	1.9	1.6
Kjeldahl nitrogen	0.29	0.21	0.25	0.25
Phosphorus	0.02	0.04	0.04	0.03
Potassium	0.20	0.24	0.33	0.24

[a]Adapted from D. J. Hills, W. A. Ward, C. P. Wingett, and R. H. Bennett, Veal calf characteristics. *Agric. Wastes* **7**, 61–64 (1983).

[b]The mean value usually can be used for design purposes, since the animal confinement period covers all three animal weight periods.

Index

457